Henry L. Manning

Reactive Power Control
In Electric Systems

Large electric arc furnace used in steel production. Such loads on the electricity supply system often require reactive power control equipment of the type described in Chapter 9.

Reproduced by kind permission of United States Steel, American Bridge Division.

REACTIVE POWER CONTROL IN ELECTRIC SYSTEMS

T. J. E. MILLER

General Electric Company
Corporate Research and Development Center
Schenectady, New York

With a Foreword by

CHARLES CONCORDIA

1807 175 YEARS OF PUBLISHING 1982

A Wiley-Interscience Publication

JOHN WILEY & SONS

New York Chichester Brisbane Toronto Singapore

Library of Congress Cataloging in Publication Data

Main entry under title:

Reactive power control in electric systems

"A Wiley-Interscience publication."
Bibliography: p.
Includes index.
1. Reactive power (Electrical engineering)

I. Miller, T. J. E. (Timothy John Eastham), 1947-
TK3226.R38 1982 621.319 82-10838
ISBN 0-471-86933-3

Printed in the United States of America

10 9 8 7 6 5 4 3 2 1

PLANTS OF CHARLES P. STEINMETZ
MEMORIAL PARK, SCHENECTADY

Drawing by JANET MILLER

CONTRIBUTORS

Larry E. Bock
Capacitor Products Department, Hudson Falls, N.Y.

Philip G. Brown
Electric Utility Systems Engineering Department, Schenectady, N.Y.

Ronald L. Hauth
Electric Utility Systems Engineering Department, Schenectady, N.Y.

Ronald W. Lye
Canadian General Electric Company, Peterborough, Ontario

T. J. E. Miller
Corporate Research and Development Center, Schenectady, N.Y.

Arthur H. Moore
Industrial Power Systems Engineering Operation, Schenectady, N.Y.

A. Robert Oltrogge
Industrial Sales Department, Pittsburgh, Pa.

Raymond L. Rofini
High-Voltage DC Projects Operation, Philadelphia, Pa.

Kim A. Wirgau
Electric Utility Systems Engineering Department, Schenectady, N.Y.

FOREWORD

Reactive power has been recognized as a significant factor in the design and operation of alternating current electric power systems for a long time. In a very general and greatly oversimplified way, it has been observed that, since the impedances of the network components are predominantly reactive, the transmission of active power requires a difference in *angular phase* between the voltages at the sending and receiving points (which is feasible within rather wide limits), whereas the transmission of reactive power requires a difference in *magnitude* of these same voltages (which is feasible only within very narrow limits).

But why should we want to transmit reactive power anyway? Is it not just a troublesome concept, invented by the theoreticians, that is best disregarded? The answer is that reactive power is consumed not only by most of the network elements, but also by most of the consumer loads, so it must be supplied somewhere. If we can't transmit it very easily, then it ought to be generated where it is needed.

Of course, the same might be said about active power, but the constraints on its transmission are much less severe and the penalties on inappropriate generator siting (and sizing) much more severe. Still, the differences are only quantitative.

It is important to recognize that, when we speak of transmission of electricity, we must speak of *electrical* distances. For example, the reactance of a transformer may be as great as that of 50 miles of transmission line. Thus, when we consider that the average transmission distance in the United States is only of the order of 100 miles, it is evident that we do not really avoid transmission entirely unless the generation of reactive power is at the same voltage level as is the consumption to be supplied. This partially explains the superficially strange fact that we can often observe in the same network, shunt compensation in the form of capacitors in the distribution system and shunt inductors in the transmission system.

There is a fundamental and important interrelation between active and reactive power transmission. We have said that the transmission of active power requires a phase displacement of voltages. But the magnitudes of these voltages are equally important. Not only are they necessary for power transmission, but also they must be high enough to support the loads and low enough to avoid equipment breakdown. Thus, we have to control, and, if necessary, to support or constrain, the voltages at all the key points of the network. This control may be accomplished in large part by the supply or consumption of reactive power at these points.

Although these aspects of reactive power have long been recognized, they have recently acquired increased importance for at least two reasons: first, the increasing pressures to utilize transmission capacity as much as possible; and second, the development of newer static types of controllable reactive-power compensators. Many years ago, when the growing extent of electric power networks began to justify it, synchronous condensers were used for voltage support and power transfer capability improvement. At the same time, shunt capacitors began to be installed in distribution circuits to improve voltage profiles and reduce line loading and losses by power-factor improvement. The rapid development and relative economy of these shunt capacitors led to their almost displacing synchronous condensers in the transmission system also. It was found that practically as much could be gained by switched capacitors as by synchronous condensers, and at a much lower cost. Now there are signs that the tide may have turned again, and controlled reactive-power supplies are beginning to return in the form of static devices. However, from an economic point of view, it is still the job of the system engineer to determine how much can be accomplished by fixed capacitors (and inductors), how much needs to be switched, and finally how much needs to be rapidly and continuously controlled, as, for example, during disturbances. Of course, there also remains the question of how much, if any, reactive power should be obtained from the synchronous generators themselves.

We have so far discussed reactive power as being supplied to, or from, the network. However, at the beginning of this Foreword it was pointed out that some of the consumption of reactive power is in the network series elements themselves, for example, in transmission lines and transformer leakage reactances. Thus, a direct way of increasing power transfer capacity in transmission systems, and of reducing voltage drop in distribution systems, is to compensate part of the series inductive reactance by series capacitors. Ordinarily, this is looked at from the standpoint of reducing the net inductive reactance, rather than in terms of reactive-power supply. Operating problems have been encountered, and applications to distribution systems have become rare, but the series capacitor remains the best way to increase transmission capacity in many cases. It has also been used to balance line loadings among network branches. As a corollary to this, in a meshed network the application of series capacitors must be coordinated in order to preserve, or obtain, a proper distribution of line loadings, while a shunt capacitor (or other means of voltage support) can often be applied with benefit at a single point.

The present book includes both theoretical and practical accounts of reactive power control and compensation, as well as a detailed chapter on harmonics. There is no doubt that the subject of reactive power control is increasing in importance, and the book is therefore particularly appropriate at this time. The authors are all practising power-system engineers who have had a total of many decades of experience in the technologies related to reactive power.

CHARLES CONCORDIA

Venice, Florida
October 1982

PREFACE

About thirty percent of all primary energy resources worldwide are used to generate electrical energy, and almost all of this is transmitted and distributed by alternating current at 50 or 60 Hz. It is now more important than ever to design and operate power systems with not only the highest practicable efficiency but also the highest degree of security and reliability. These requirements are motivating a wide range of advances in the technology of ac power transmission and the purpose of this book is to describe some of the more important theoretical and practical developments.

Because of the fundamental importance of reactive power control, and because of the wide range of subjects treated, as well as the method of treatment, this book should appeal to a broad cross section of electrical, electronics, and control engineers. Practising engineers in the utility industry and in industrial plants will find both the theory and the description of reactive power control equipment invaluable in solving problems in power-factor correction, voltage control and stabilization, phase balancing and the handling of harmonics. In universities the book should form an ideal basis for a postgraduate or even an undergraduate course in power systems, and several sections of it have already been used for this purpose at the University of Wisconsin and in the General Electric Power Systems Engineering Course.

Reactive power control, which is the theme of the book, has grown in importance for a number of reasons which are briefly as follows. First, the requirement for more efficient operation of power systems has increased with the price of fuels. For a given distribution of power, the losses in the system can be reduced by minimizing the total flow of reactive power. This principle is applied throughout the system, from the simple power-factor correction capacitor used with a single inductive load, to the sophisticated algorithms described in Chapter 11 which may be used in large interconnected networks controlled by computers. Second, the extension of transmission networks has been curtailed in general by high interest rates, and in particular cases by the difficulty of acquiring right-of-way. In many cases the power transmitted through older circuits has been increased, requiring the application of reactive power control measures to restore stability margins. Third, the exploitation of hydropower resources has proceeded spectacularly to the point where remote, hostile generation sites have been developed, such as those around Hudson Bay and in mountainous regions of Africa and South America. In spite of the parallel development of dc transmission technology, ac transmission has been preferred in many of these schemes. The problems of stability and voltage control are identifiable as problems in reactive power control, and a wide range of different solutions has been developed, ranging from the use of fixed shunt reactors and capacitors, to series capacitors, synchronous condensers, and the modern static compensator. Fourth, the requirement for a high quality of supply has increased because of the increasing use of electronic equipment (especially computers and color television receivers), and because of the growth in continuous-process industries.

Voltage or frequency depressions are particularly undesirable with such loads, and interruptions of supply can be very harmful and expensive. Reactive power control is an essential tool in maintaining the quality of supply, especially in preventing voltage disturbances, which are the commonest type of disturbance. Certain types of industrial load, including electric furnaces, rolling mills, mine hoists, and dragline excavators, impose on the supply large and rapid variations in their demand for power and reactive power, and it is often necessary to compensate for them with voltage stabilizing equipment in the form of static reactive power compensators. Fifth, the development and application of dc transmission schemes has created a requirement for reactive power control on the ac side of the converters, to stabilize the voltage and to assist the commutation of the converter.

All these aspects of ac power engineering are discussed, from both a theoretical and a practical point of view. Chapters 1 through 3 deal with the theory of ac power transmission, starting with the simplest case of power-factor correction and moving on to the detailed principles on which the extremely rapid-response static compensator is designed and applied. In Chapter 2 the principles of transmitting power at high voltages and over longer distances are treated, and Chapter 3 deals with the important aspects of the dynamics of ac power systems and the effect of reactive power control. The unified approach to the "compensation" problem is particularly emphasized in Chapter 2, where the three fundamental techniques of compensation by sectioning, surge-impedance compensation, and line-length compensation are defined and compared. Chapter 1 is also unified in its approach to the compensation or reactive power control of loads; the compensating network is successively described in terms of its power-factor correction attributes, its voltage-stabilizing attributes, and finally its properties as a set of sequence networks capable of voltage stabilization, power-factor correction, and phase balancing, both in terms of phasors and instantaneous voltages and currents.

Chapter 4 introduces and describes in detail the principles of modern static reactive power compensators, including the thyristor-controlled reactor, the thyristor-switched capacitor, and the saturated reactor. Particular attention is given to control aspects, and a detailed treatment of switching phenomena in the thyristor-switched capacitor is included.

The modern static compensator receives further detailed treatment in Chapters 5 and 6. Chapter 5 describes the high-power ac thyristor controller and associated systems, while Chapter 6 gives a complete description of a modern compensator installation including details of the control system and performance testing.

In Chapter 7 the series capacitor is described. The solution of the subsynchronous resonance (SSR) problem, together with the introduction of virtually instantaneous reinsertion using metal-oxide varistors, have helped restore the series capacitor to its place as an economic and very effective means for increasing the power transmission capability and stability of long lines. Both the varistor and the means for controlling SSR are described in this chapter.

Chapter 8 on synchronous condensers has been included because of the continuing importance of this class of compensation equipment. As a rotating machine the synchronous condenser has a natural and important place in the theory of reactive power control, and several of the most recent installations have been very large and technically advanced. Rapid response excitation systems and new control strategies have steadily enhanced the performance of the condenser.

In Chapter 9 there is a detailed treatment of reactive power control in connection with electric arc furnaces, which present one of the most challenging load compensation problems, requiring large compensator ratings and extremely rapid response to minimize "flicker." Chapter 9 will be of interest to the general reader for its exploration of the limits of the speed of response of different methods of reactive power control. It also shows clearly the advantages of compensation on the steel-producing process, which exemplifies the principle that the performance of the load is often significantly improved by voltage and reactive power control, even when these are required for other reasons (such as the reduction of flicker).

The subject of reactive power control is closely connected with the subject of harmonics, because reactive power compensation and control is often required in connection with loads which are also sources of harmonics. A separate reason for the importance of harmonics in a book on reactive power control is that reactive compensation almost always influences the resonant frequencies of the power system, at least locally, and it is important that capacitors, reactors, and compensators be deployed in such a way as to avoid problems with harmonic resonances. Chapter 10 deals with these matters, and includes a treatment of filters with practical examples.

The final chapter, Chapter 11, deals with the relatively new subject of reactive power coordination, and describes a number of systematic approaches to the coordinated control of reactive power in a large interconnected network. Minimization of system losses is one of several possible optimal conditions which can be determined and maintained by computer analysis and control. This promising new subject is given the last word in leaving the reader to the future.

Many people have contributed to the writing and production of this book, and the editor would like to record his warmest thanks for all contributions great and small. Special thanks are due to Dr. Eike Richter of the General Electric Research and Development Center for his generosity and sustained support of the project, and also to Dr. F. J. Ellert, D. N. Ewart, D. Swann, Dr. P. Chadwick, R. J. Moran, D. Lamont, and Dr. E. P. Cornell.

Special thanks are also due to Mrs. Barbara West for her tremendous assistance in the preparation (and often the repair) of the manuscript. Mrs. Christine Quaresimo, Miss Kathy Kinch, and Holly Powers also helped substantially with the manuscript, and Dean Klimek with the figures. Composition was by the Word Processing Unit of General Electric's Corporate Research and Development Center, and thanks are due to A. E. Starbird and his staff. Acknowledgment is also due to the staff of John Wiley and Sons for their expert guidance throughout the writing and production of the book.

Dr. John A. Mallick made several helpful suggestions in connection with Chapters 1 and 2. We are also especially grateful to Dr. Laszlo Gyugyi of Westinghouse Electric for permission to use some of his ideas in Section 9 of Chapter 1. Also to S. A. Miske, Jr. for his comments on the whole work and on Chapters 2 and 7 in particular, and to him and R. J. Piwko and Dr. F. Nozari (all of the GE Electric Utility Systems Engineering Department) for contributions to Chapter 3. Acknowledgment is also due to the U.S. Department of Energy and to Dr. H. Boenig of Los Alamos Laboratory in connection with studies which formed the basis of some sections in Chapter 3. Philadelphia Electric Co. is acknowledged for the photograph of the synchronous condenser station in Chapter 8. Others who helped at various stages include D. J. Young, S. R. Folger, D. Demarest, R. A. Hughes, H. A. Happ, L. Kirchmayer, W. H. Steinberg, Dr. W. Berninger and Dr. James Lommel. United States Steel is acknowledged for the Frontispiece arc furnace photograph, and the Institute of Electrical and Electronics Engineers for the use of several figures and material throughout the book. The Institution of Electrical Engineers is acknowledged for the use of material and figures in Section 3 of Chapter 4 which are taken from IEE Conference Publication CP205, "Thyristor and Variable Static Equipment for AC and DC Transmission."

No book is the sole work of its named authors, and acknowledgment is made to the many unnamed contributors to the technology and application of reactive power control. This book does not attempt to set out hard-and-fast rules for the application of any particular type of equipment, and in particular the authors accept no responsibility for any adverse consequences arising out of the interpretation of material in the book. The chapters are all written from the individual points of view of the authors, and do not represent the position of any manufacturing company or other institution.

T. J. E. MILLER

Schenectady, New York
October 1982

CONTENTS

2. THE THEORY OF STEADY-STATE REACTIVE POWER CONTROL IN ELECTRIC TRANSMISSION SYSTEMS 49

Reactive Power Control
In Electric Systems

Chapter 1

THE THEORY OF LOAD COMPENSATION

T. J. E. MILLER

PRINCIPAL SYMBOLS

Note: Lower case symbols for voltage, current, etc., denote instantaneous values.

Boldface symbols denote complex numbers (i.e., impedances, admittances, and phasor voltages and currents). The asterisk denotes complex conjugation. *Plain italic type* denotes the magnitude of a phasor voltage or current.

Symbols

B	Susceptance, S
E	Source voltage, V
G	Conductance, S
h	Complex operator $e^{j2\pi/3}$
I	Current, A
j	$\sqrt{-1}$
K	Slope of voltage/current characteristic, Ohm
P	Power, W
Q	Reactive power, VAr (inductive positive)
R	Resistance, Ohm
S	Apparent power, VA
V	Voltage, V
X	Reactance, Ohm

Y	Admittance, S
Z	Impedance, Ohm

Greek Symbols

ϕ	Power-factor angle, ° or radian
Δ	A small change in...
ω	Radian frequency, radian/sec

Subscripts

a,b,c	The three phases of the power system
k	Knee-point
ℓ	Load
mp	Maximum permitted
R	Real or resistive component
s	Supply system
sc	Short circuit
X	Imaginary or reactive component

Greek subscript

γ	Compensator

1.1. INTRODUCTION: THE REQUIREMENT FOR COMPENSATION

In an ideal ac power system, the voltage and frequency at every supply point would be constant and free from harmonics, and the power factor would be unity. In particular these parameters would be independent of the size and characteristics of consumers' loads. In an ideal system, each load could be designed for optimum performance at the given supply voltage, rather than for merely adequate performance over an unpredictable range of voltage. Moreover, there could be no interference between different loads as a result of variations in the current taken by each one.

We can form a notion of the quality of supply in terms of how nearly constant are the voltage and frequency at the supply point, and how near to unity is the power factor. In three-phase systems, the degree to which the phase currents and voltages are balanced must also be included in the notion of quality of supply. A definition of "quality of supply" in numerical terms involves the specification of such quantities as the maximum fluctuation in rms supply voltage averaged over a stated period of time. Specifications of this kind can be made more precise through the use of statistical concepts, and these are especially helpful in problems where voltage fluctuations can take place very rapidly (for example, at the supply to arc furnaces).

In this chapter we identify some of the characteristics of power systems and their loads which can deteriorate the quality of supply, concentrating on those which can be corrected by compensation, that is, by the supply or absorption of an appropriately variable quantity of reactive power. A study of how the quality of supply can be degraded by such loads will lead to the definition of the "ideal" compensator. Section 1.4 discusses the types of load which require compensation and outlines the effects of modern trends in the electrical characteristics of industrial plant design. In later sections the theory of compensation is developed for steady-state and slowly varying conditions.

1.2. OBJECTIVES IN LOAD COMPENSATION

Load compensation is the management of reactive power to improve the quality of supply in ac power systems. The term *load compensation* is used where the reactive power management is effected for a single load (or group of loads), the compensating equipment usually being installed on the consumer's own premises near to the load. The techniques used, and indeed some of the objectives, differ considerably from those met in the compensation of bulk supply networks (transmission compensation).

In load compensation there are three main objectives:

1. Power-factor correction.
2. Improvement of voltage regulation.
3. Load balancing.

We shall take the view that power-factor correction and load balancing are desirable even when the supply voltage is very "stiff" (i.e., virtually constant and independent of the load).

Power-factor correction usually means the practice of generating reactive power as close as possible to the load which requires it, rather than supplying it from a remote power station. Most industrial loads have lagging power-factors; that is, they absorb reactive power. The load current therefore tends to be larger than is required to supply the real power alone. Only the real power is ultimately useful in energy conversion and the excess load current represents a waste to the consumer, who has to pay not only for the excess cable capacity to carry it but also for the excess Joule loss produced in the supply cables. The supply utilities also have good reasons for not transmitting unnecessary reactive power from generators to loads: their generators and distribution networks cannot be used at full efficiency, and the control of voltage in the supply system can become more difficult. Supply tariffs to industrial customers almost always penalize low power-factor loads, and have done so for many years; the result has been the extensive development of power-factor correction systems for industrial plants.

Voltage regulation becomes an important and sometimes critical issue in the presence of loads which vary their demand for reactive power. All loads vary their demand for reactive power, although they differ widely in their range and rate of variation. In all cases, the variation in demand for reactive power causes variation (or *regulation*) in the voltage at the supply point, which can interfere with the efficient operation of all plants connected to that point, giving rise to the possibility of interference between loads belonging to different consumers. To protect against this, the supply utility is usually bound by statute to maintain supply voltages within defined limits. These limits may vary from typically $\pm 5\%$ averaged over a period of a few minutes or hours, to the much more stringent constraints imposed where large, rapidly varying loads could produce voltage dips hazardous to the operation of protective equipment, or flicker annoying to the eye.† Compensating devices have a vital role to play in maintaining supply voltages within the intended limits.

The most obvious way to improve voltage regulation would be to "strengthen" the power system by increasing the size and number of generating units and by making the network more densely interconnected. This approach would in general be uneconomic and would introduce problems associated with high fault levels and switchgear ratings. It is much more practical and economic to size the power system according to the maximum demand for *real* power, and to manage the reactive power by means of compensators and other equipment which can be deployed more flexibly than generating units and which make no contribution to fault levels.

The third main concern in load compensation is *load balancing*. Most ac power systems are three-phase, and are designed for balanced operation. Unbalanced operation gives rise to components of current in the wrong phase-sequence (i.e., negative- and zero-sequence components). Such components can have undesirable effects, including additional losses in motors and generating units, oscillating torque in ac machines, increased ripple in rectifiers, malfunction of several types of equipment, saturation of transformers, and excessive neutral currents. Certain types of equipment (including several types of compensators) depend on balanced operation for suppression of triplen harmonics. Under unbalanced conditions, these would appear in the power system.

† The supply authority is usually also bound by statute to maintain frequency within defined limits. Frequency variations are not considered in this chapter.

The harmonic content in the voltage supply waveform is an important parameter in the quality of supply, but it is a problem specialized by the fact that the spectrum of fluctuations is entirely above the fundamental power frequency. Harmonics are usually eliminated by filters, whose design principles differ from those of compensators as developed in this chapter.[2] Nevertheless, harmonic problems often arise together with compensation problems and frequent reference will be made to harmonics and their filtration. Moreover, many types of compensator inherently generate harmonics which must be either suppressed internally or filtered externally.

1.3. THE IDEAL COMPENSATOR

Having outlined the main objectives in load compensation, it is now possible to form a concept of the ideal compensator. This is a device which can be connected at a supply point (i.e., in parallel with the load) and which will perform the following three main functions: (1) correct the power factor to unity, (2) eliminate (or reduce to an acceptable level) the voltage regulation, and (3) balance the load currents or phase voltages. The ideal compensator will not be expected to eliminate harmonic distortion existing in the load current or the supply voltage (this function being assigned to an appropriate harmonic filter); but the ideal compensator would not itself generate any extra harmonics. A further property of the ideal compensator is the ability to respond instantaneously in performing its three main functions. Strictly, the concept of instantaneous response requires the definition of instantaneous power-factor and instantaneous phase-unbalance. The ideal compensator would also consume zero average power; that is, it would be lossless.

The three main functions of the ideal compensator are interdependent. In particular, the power-factor correction and phase-balancing themselves tend to improve voltage regulation. Indeed, in some instances, especially those where load fluctuations are slow or infrequent, a compensator designed for power-factor correction or phase-balancing is not required to perform any specific voltage regulating function.

The ideal compensator can be specified more precisely by stating that it must

1. Provide a controllable and variable amount of reactive power precisely according to the requirements of the load, and without delay.
2. Present a constant-voltage characteristic at its terminals.
3. Be capable of operating independently in the three phases.

The burden of responsibility for providing compensation divides between supplier and consumer according to several factors including the size and nature of the load and any future extensions projected for it; the national standards in force; local practice; and the degree to which other consumers may be affected. It is often the case that the consumer assumes responsibility for power-factor correction and for balancing the load currents; indeed, most supply tariffs require consumers to do this. Voltage regulation, on the other hand, is more usually the concern of the supplier.

1.4. PRACTICAL CONSIDERATIONS

1.4.1. Loads Requiring Compensation

Whether a given load should have power-factor correction in the steady state is an economic question whose answer depends on several factors including the supply tariff, the size of the load, and its uncompensated power-factor. It is typical that for sizeable industrial loads, power-factor correction is economically advantageous if the uncompensated power-factor is less than 0.8.

Loads which cause fluctuations in the supply voltage may have to be compensated not only for power factor but for voltage regulation also. The degree of voltage variation is assessed at the "gauge point" or "point of common coupling" (PCC), which is usually the point in the network where the customer's and the supplier's areas of responsibility meet; for example, the high-voltage side of the distribution transformer supplying a particular factory.

Typical of loads requiring compensation are arc furnaces, induction furnaces, arc welders, induction welders, steel rolling mills, mine winders, very large motors (particularly those which start and stop frequently), opencast excavators, wood chip mills, and high-energy physics experiments (e.g., synchrotrons) which require pulsed high-power supplies. These loads can be classified into those which are inherently nonlinear in operation and those which cause disturbance by being switched on and off. The nonlinear loads usually generate harmonics as well as fundamental-frequency voltage variations. Arc furnace compensators, for example, virtually always include harmonic filters, usually for harmonic orders 3, 5, 7 and often for orders 2, 4, 11, and 13 as well.[1]

Both power-factor and voltage regulation can be improved if some of the drives in a plant are synchronous motors instead of induction motors, because the synchronous motor can be controlled to supply (or absorb) an adjustable amount of reactive power. By virtue of its rotating mass it also stores kinetic energy which tends to support the supply system against suddenly applied loads. In many cases, voltage dips caused by motor starts are avoided by having "soft starts"; that is, the motor is started through an adjustable transformer or by an electronic drive with facilities for a gradual start.

The modern trend with large dc drives which are used in an "on-off" mode is to use thyristor controls, which themselves exacerbate the compensation problem because they generate harmonics, require reactive power for commutation, and have no rotational inertia. Figure 1 shows the typical form of variation in the reactive power requirement of a steel rolling mill. Not only are the reactive power variations large (about 50 MVAr), but they are also sudden, so the compensator must be able to respond rapidly.

A first idea of the compensation requirement can be formed by characterizing the load according to the following headings:

1. Type of drive (dc or ac; thyristor-fed or transformer-fed).
2. Duty-cycle in terms of the real and reactive power requirements (e.g., Figure 1).
3. Rate of change of real and reactive power (or the time taken for the real or reactive power to swing from maximum to minimum).
4. Generation of harmonics.
5. Concurrence of maximum real and reactive power requirements in multiple-load plants.

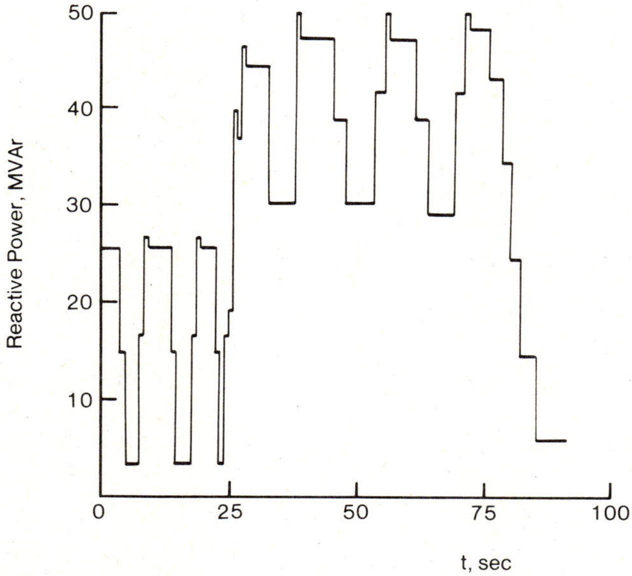

FIGURE 1. Typical variation in the reactive power requirements of a steel rolling mill.

1.4.2. Acceptance Standards for the Quality of Supply

The first objectionable effect of supply voltage variations in a distribution system is the disturbance to the lighting level produced by tungsten filament lamps. The degree to which variations are objectionable depends not only on the magnitude of the light variation but also on its frequency or rate of change, because of the sensitivity characteristics of the human eye. Very slow variations of up to 3% may be tolerable, while the rapid variations caused by arc furnaces or welding plant can coincide with the maximum visual sensitivity (between 1 and 25 Hz) and must be limited in magnitude to 0.25% or less.

Several other types of loads are sensitive to supply voltage variations, especially computers, certain types of relays employed in control and protective schemes, induction motors, and lamps of the discharge and fluorescent types.

Very often the variation in supply voltage is detrimental to the performance of the load which is causing the variation. Compensation may therefore be applied to improve this performance as well as to benefit other consumers.

Table 1 is representative of the type of standards which might be prescribed for the performance of a system with a disturbing load. In the case of the welding plant, the permitted voltage variation is inversely related to the sensitivity of the human eye to light fluctuations as a function of frequency.

TABLE 1
Typical Voltage Fluctuation Standards[a]

Type of Load	Limits Permitted in Voltage Fluctuation
Large motor starts	1–3% depending on frequency
Mine hoists, excavators, steel rolling mills, large thyristor-fed dc drives	1–3% at distribution voltages ½–1½% at transmission voltages
Welding plant	¼–2% depending on frequency
Induction furnaces	Up to 1% depending on time between steps of "soft start"
Arc furnaces	See Chapter 9

[a] See Reference 3.

1.4.3. Specification of a Load Compensator

The parameters and factors which need to be considered when specifying a load compensator are summarized in the following list. The list is not intended to be complete, but only to give an idea of the kind of practical considerations which are important.

1. Maximum continuous reactive power requirement, both absorbing and generating.
2. Overload rating and duration (if any).
3. Rated voltage and limits of voltage between which the reactive power ratings must not be exceeded.
4. Frequency and its variation.
5. Accuracy of voltage regulation required.
6. Response time of the compensator for a specified disturbance.
7. Special control requirements.
8. Protection arrangements for the compensator and coordination with other protection systems, including reactive power limits if necessary.
9. Maximum harmonic distortion with compensator in service.
10. Energization procedure and precautions.
11. Maintenance; spare parts; provision for future expansion or rearrangement of plant.
12. Environmental factors: noise level; indoor/outdoor installation; temperature, humidity, pollution, wind and seismic factors; leakage from transformers, capacitors, cooling systems.
13. Performance with unbalanced supply voltages and/or with unbalanced load.
14. Cabling requirements and layout; access, enclosure, grounding.
15. Reliability and redundancy of components.

In the case of arc furnace compensation, the "improvement ratio" or "flicker reduction ratio" may be specified as one principal measure of the compensator's performance.

1.5. FUNDAMENTAL THEORY OF COMPENSATION: POWER-FACTOR CORRECTION AND VOLTAGE REGULATION IN SINGLE-PHASE SYSTEMS

The first purpose of a theory of compensation must be to explain the relationships between the supply system, the load, and the compensator. In

Section 1.5.1 we begin with the principle of power-factor correction, which, in its simplest form, can be studied without reference to the supply system. In later sections, we examine voltage regulation and phase unbalance, and build up a quantitative concept of the ideal compensator.

The supply system, the load, and the compensator can be characterized, or modeled, in various ways. Thus the supply system can be modeled as a Thévenin equivalent circuit with an open-circuit voltage and either a series impedance, its current, or its power and reactive power (or power-factor) requirements. The compensator can be modeled as a variable impedance; or as a variable source (or sink) of reactive current; or as a variable source (or sink) of reactive power. The choice of model used for each element can be varied according to requirements, and in the following sections the models will be combined in different ways as appropriate to give the greatest physical insight, as well as to develop equations of practical utility. The different models for each element are, of course, essentially equivalent, and can be transformed into one another.

The theory is developed first for stationary or nearly stationary conditions, implying that loads and system characteristics are understood to be either constant or changing slowly enough so that phasors can be used. This simplifies the analysis considerably. In most practical instances, the phasor or quasi-stationary equations are adequate for determining the rating and external characteristics of the compensator. For loads whose power and reactive power vary rapidly (such as electric arc furnaces), the phasor equations are not entirely valid; special analysis methods have been developed for these.

1.5.1. Power Factor and Its Correction

Figure 2*a* shows a single-phase load of admittance $\mathbf{Y}_\ell = G_\ell + \mathrm{j}B_\ell$ supplied from a voltage $\mathbf{V}$. The load current is $\mathbf{I}_\ell$ and

$$\mathbf{I}_\ell = \mathbf{V}(G_\ell + \mathrm{j}B_\ell) = VG_\ell + \mathrm{j}VB_\ell = \mathrm{I}_\mathrm{R} + \mathrm{j}I_\mathrm{X}\,. \tag{1}$$

Both $\mathbf{V}$ and $\mathbf{I}_\ell$ are phasors, and Equation 1 is represented in the phasor diagram (Figure 2*b*) in which $\mathbf{V}$ is the reference phasor. The load current has a "resistive" component, I_R, in phase with $\mathbf{V}$, and a "reactive" component, $I_\mathrm{X} = VB_\ell$, which is in phase quadrature with $\mathbf{V}$; in the example shown, I_X is negative; $\mathbf{I}_\ell$ is lagging, and the load is inductive (this is the commonest case). The angle between $\mathbf{V}$ and $\mathbf{I}_\ell$ is ϕ. The apparent power supplied to the loads is †

$$\begin{aligned}\mathbf{S}_\ell &= \mathbf{V}\mathbf{I}_\ell^* \\ &= V^2G_\ell - \mathrm{j}V^2B_\ell \\ &= P_\ell + \mathrm{j}Q_\ell.\end{aligned} \tag{2}$$

† Note that S_ℓ, P_ℓ, and Q_ℓ are not phasor quantities.

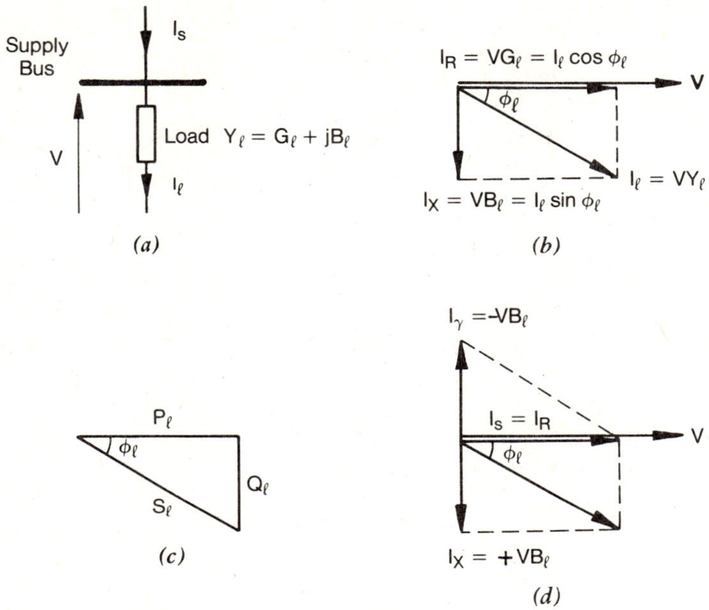

FIGURE 2. (*a*) through (*d*). Power-factor correction.

The apparent power thus has a real component P_ℓ (i.e., the power which is usefully converted into heat, mechanical work, light, or other forms of energy); and a reactive component Q_ℓ (the reactive power, which cannot be converted into useful forms of energy but whose existence is nevertheless an inherent requirement of the load). For example, in an induction motor, Q_ℓ represents the magnetizing reactive power. The relationship between S_ℓ, P_ℓ, and Q_ℓ is shown in Figure 2*c*. For lagging (inductive) loads B_ℓ is negative and Q_ℓ is positive, by convention.

The current $\mathbf{I}_s = \mathbf{I}_\ell$ supplied by the power system is larger than is necessary to supply the real power alone, by the factor

$$I_\ell / I_R = 1/\cos \phi_\ell \; . \tag{3}$$

Here, $\cos \phi_\ell$ is the power factor, so called because

$$\cos \phi_\ell = P_\ell / S_\ell , \tag{4}$$

that is, $\cos \phi_\ell$ is that fraction of the apparent power which can be usefully converted into other forms of energy.

The Joule losses in the supply cables are increased by the factor $1/\cos^2 \phi_\ell$. Cable ratings must be increased accordingly, and the losses must be paid for by the consumer.

The principle of power-factor correction is to compensate for the reactive power; that is, to provide it locally by connecting in parallel with the load a compensator having a purely reactive admittance $-jB_\ell$. The current supplied by the power system to the combined installation of load and compensator becomes

$$\begin{aligned} \mathbf{I}_s &= \mathbf{I}_\ell + \mathbf{I}_\gamma \\ &= \mathbf{V}(G_\ell + jB_\ell) - \mathbf{V}(jB_\ell) = VG_\ell = I_R, \end{aligned} \tag{5}$$

which is in phase with **V**, making the *overall* power-factor unity. Figure 2*d* shows the phasor relationships. The supply current $\mathbf{I}_s$ now has the smallest value capable of supplying full power P_ℓ at the voltage **V**, and all the reactive power required by the load is supplied locally by the compensator: the load is thus totally compensated. Relieved of the reactive requirements of the load, the supply now has excess capacity which is available for supplying other loads.

The compensator current is given by †

$$\mathbf{I}_\gamma = \mathbf{V}\mathbf{Y}_\gamma = -j\mathbf{V}B_\ell. \tag{6}$$

The apparent power exchanged with the supply system is

$$\begin{aligned} \mathbf{S}_\gamma &= P_\gamma + jQ_\gamma \\ &= \mathbf{V}\mathbf{I}_\gamma^* \\ &= jV^2B_\ell. \end{aligned} \tag{7}$$

Thus $P_\gamma = 0$ and $Q_\gamma = V^2B_\ell = -Q_\ell$. The compensator requires no mechanical power input. Most loads are inductive, requiring capacitive compensation (B_γ positive, Q_γ negative).

From Figure 2*c*, we can see that for total compensation of reactive power, the reactive power rating of the compensator is related to the rated *power* P_ℓ of the load by

$$Q_\ell = P_\ell \tan \phi_\ell, \tag{8}$$

and to the rated *apparent power* S_ℓ of the load by

$$Q_\gamma = S_\ell \sin \phi_\ell = S_\ell \sqrt{1 - \cos^2 \phi_\ell}. \tag{9}$$

Table 2 shows the compensator rating per unit of S_ℓ for various power-factors. The rated current of the compensator is given by Q_γ/V, which equals the reactive current of the load at rated voltage.

The load may be partially compensated (i.e., $|Q_\gamma| < |Q_\ell|$ or $|B_\gamma| < |B_\ell|$), the degree of compensation being decided by an economic trade-off between the capital cost of the compensator (which depends on its rating) and the capitalized cost of obtaining the reactive power from the supply system over a period of time.

† The subscript γ is used to denote compensator quantities. "*c*" would be confused with phase "*c*" in later sections.

TABLE 2
Reactive Power Required for Complete Compensation at Various Power Factors

Load Power-Factor ($\cos \phi_\ell$)	Compensator Rating Q_γ (per unit of rated load apparent power)
1.0	0
0.95	0.312
0.90	0.436
0.80	0.600
0.60	0.800
0.40	0.917
0	1.0

As developed so far, the compensator is a fixed admittance (or susceptance) incapable of following variations in the reactive power requirement of the load. In practice a compensator such as a bank of capacitors (or inductors) can be divided into parallel sections, each switched separately, so that discrete changes in the compensating reactive power may be made, according to the requirements of the load. More sophisticated compensators (e.g., synchronous condensers or static compensators) are capable of stepless variation of their reactive power. (See Chapter 4.)

The foregoing analysis has taken no account of the effect of supply voltage variations on the effectiveness of the compensator in maintaining an overall power-factor of unity. In general the reactive power of a fixed-reactance compensator will not vary in sympathy with that of the load as the supply voltage varies, and a compensation "error" will arise. In the next section the effects of voltage variations are examined, and we find out what extra features the ideal compensator must have to perform satisfactorily when both the load and the supply system parameters can vary. We also see how the improvement of power-factor by itself tends to improve voltage regulation.

1.5.2. Voltage Regulation

We begin this section by determining the voltage regulation that would be obtained without a compensator, identifying the most important parameters of the load and the supply system. Then we introduce the concept of an ideal compensator that maintains constant supply-point voltage by maintaining the supply-system reactive power approximately constant. The characteristics of the compensator are developed both graphically and mathematically.

Voltage regulation can be defined as the proportional (or per-unit) change in supply voltage magnitude associated with a defined change in load current (e.g., from no load to full load). It is caused by the voltage drop in the supply impedance carrying the load current. If the supply system is represented by the single-phase Thévenin equivalent circuit shown in Figure 3*a*, then the voltage regulation is given by $(|\mathbf{E}|-|\mathbf{V}|)/|\mathbf{V}| = (|\mathbf{E}| - V)/V$, $\mathbf{V}$ being the reference phasor.

In the absence of a compensator, the supply voltage change caused by the load current $\mathbf{I}_\ell$ is shown in Figure 3*b* as $\Delta\mathbf{V}$, and

$$\Delta\mathbf{V} = \mathbf{E} - \mathbf{V} = \mathbf{Z}_s\mathbf{I}_\ell. \tag{10}$$

Now $\mathbf{Z}_s = R_s + jX_s$, while from Equation 2,

$$\mathbf{I}_\ell = \frac{P_\ell - jQ_\ell}{V}, \tag{11}$$

so that

$$\begin{aligned}\Delta\mathbf{V} &= (R_s + jX_s)\left(\frac{P_\ell - jQ_\ell}{V}\right)\\ &= \frac{(R_sP_\ell + X_sQ_\ell)}{V} + j\,\frac{X_sP_\ell - R_sQ_\ell}{V} \qquad (12)\\ &= \Delta V_R + j\Delta V_X.\end{aligned}$$

The voltage change has a component ΔV_R in phase with $\mathbf{V}$ and a component ΔV_X in quadrature with $\mathbf{V}$; these are illustrated in Figure 3*b*. It is clear that both the magnitude and the phase of $\mathbf{V}$, relative to the supply voltage $\mathbf{E}$, are functions of the magnitude and phase of the load current; or, in other words, the voltage change depends on both the real and reactive power of the load.

By adding a compensator in parallel with the load, it is possible to make $|\mathbf{E}| = |\mathbf{V}|$; that is, to make the voltage regulation zero, or to maintain the supply voltage magnitude constant at the value E in the presence of the load. This is shown in Figure 3*c* for a purely reactive compensator. The reactive power Q_ℓ in Equation 12 is replaced by the sum $Q_s = Q_\gamma + Q_\ell$, and Q_γ is adjusted in such a way as to rotate the phasor $\Delta\mathbf{V}$ until $|\mathbf{E}| = |\mathbf{V}|$. From Equations 10 and 12,

$$|\mathbf{E}|^2 = \left[V + \frac{R_sP_\ell + X_sQ_s}{V}\right]^2 + \left[\frac{X_sP_\ell - R_sQ_s}{V}\right]^2. \tag{13}$$

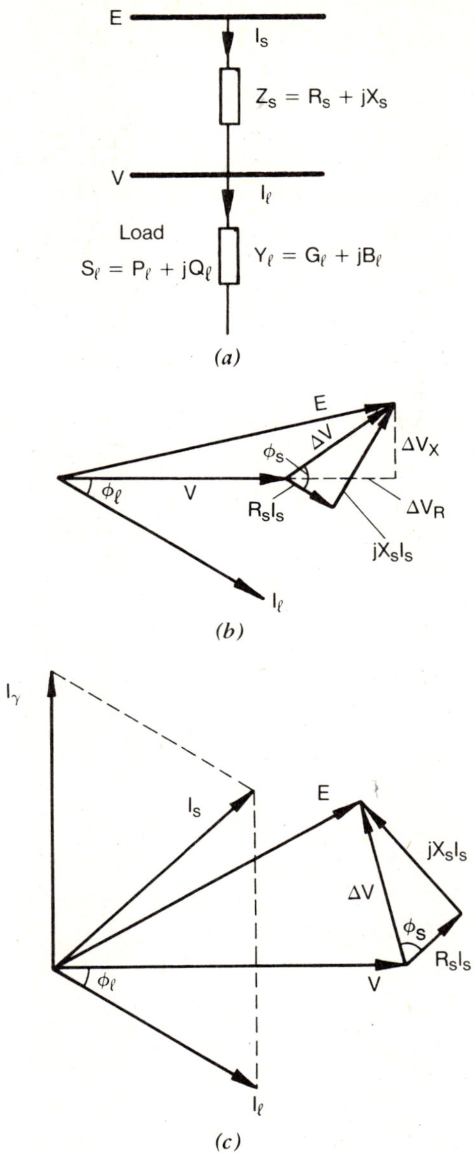

FIGURE 3. (*a*) Equivalent circuit of load and supply system. (*b*) Phasor diagram for Figure 3*a* (uncompensated). (*c*) Phasor diagram for Figure 3*a* (compensated for constant voltage).

The required value of Q_γ is found by solving this equation for Q_S with $|\mathbf{E}| = V$; then $Q_\gamma = Q_S - Q_\ell$. The algebraic solution for Q_γ is given in Section B of the Appendix. In an actual compensator, the value would be determined automatically by a control loop. What is important here is that there is always a solution for Q_S whatever the value of P_ℓ. This leads to the following important conclusion:

A purely reactive compensator can eliminate supply-voltage variations caused by changes in both the real and the reactive power of the load.

Provided that the reactive power of the compensator can be controlled smoothly, over a sufficient range (both lagging and leading, in general) and at an adequate rate, the compensator can perform as an ideal voltage regulator. It should be realized that only the *magnitude* of the voltage is being controlled; its *phase* varies continuously with the load current.

It is instructive to consider this principle from a different point of view. We have already seen in Section 1.5.1 how the compensator can reduce to zero the reactive power supplied by the system. That is, instead of acting as a voltage regulator, the compensator acts as a power-factor corrector. If the compensator is designed to do this, we can replace Q_ℓ in Equation 12 by $Q_s = Q_\ell + Q_\gamma$, which is zero. The voltage change phasor is then

$$\Delta \mathbf{V} = \frac{R_s P_\ell + jX_s P_\ell}{V} = (R_s + jX_s)\,\frac{P_\ell}{V}, \tag{14}$$

which is independent of Q_ℓ and *not under the control of the compensator.* Therefore

The purely reactive compensator cannot maintain both constant voltage and unity power-factor at the same time.

The only exception to this rule is when $P_\ell = 0$, but this is generally not of practical interest. It is important to note that the principle refers to *instantaneous* power-factor: it is quite possible for a purely reactive compensator to maintain both constant voltage and unity *average* power-factor (see Section 1.6).

Approximate Formula for the Voltage Regulation. The expressions for ΔV_R and ΔV_X in Equation 12 are sometimes given in a useful alternative form, as follows. If the system is short-circuited at the load busbar, the "short-circuit apparent power" will be

$$\mathbf{S}_{sc} = P_{sc} + jQ_{sc} = \mathbf{E}\mathbf{I}^*_{sc} = \frac{E^2}{\mathbf{Z}^*_{sc}}, \tag{15}$$

where $\mathbf{Z}_{sc} = R_s + jX_s$ and $\mathbf{I}_{sc}$ is the short-circuit current. Since $|\mathbf{Z}^*_{sc}| = |\mathbf{Z}_{sc}|$, we have

$$R_s = |\mathbf{Z}_{sc}| \cos \phi_{sc} = \frac{E^2}{S_{sc}} \cos \phi_{sc} \tag{16}$$

and

$$X_s = |\mathbf{Z}_{sc}| \sin \phi_{sc} = \frac{E^2}{S_{sc}} \sin \phi_{sc} , \tag{17}$$

with

$$\tan \phi_{sc} = \frac{X_s}{R_s} , \tag{18}$$

that is, the X:R ratio of the supply system. Substituting in Equation 12 for R_s and X_s, normalizing ΔV_R and ΔV_X to V, and assuming that $E/V \simeq 1$, we have

$$\frac{\Delta V_R}{V} \simeq \frac{1}{S_{sc}} [P_\ell \cos \phi_{sc} + Q_\ell \sin \phi_{sc}] \tag{19}$$

and

$$\frac{\Delta V_X}{V} \simeq \frac{1}{S_{sc}} [P_\ell \sin \phi_{sc} - Q_\ell \cos \phi_{sc}] . \tag{20}$$

Very often ΔV_X is ignored on the grounds that it tends to produce only a *phase* change in the supply point voltage (relative to $\mathbf{E}$), the bulk of the change in magnitude being represented by ΔV_R. Equation 19 is frequently quoted in the literature. Although approximate, the formulas are useful in that they are expressed in terms of quantities that are in common parlance: fault level or short-circuit level S_{sc}, X:R ratio (i.e., $\tan \phi_{sc}$), and the power and reactive power of the load, P and Q . For accurate results, the expressions in Equations 19 and 20 should be multiplied by E^2/V^2.

So far the equations have been written as though ΔV were associated with a full-scale change from 0 to P_ℓ or from 0 to Q_ℓ in the load. Equations 12, 19, and 20 are also valid for small changes in P_ℓ and Q_ℓ; thus, for example,

$$\frac{\Delta V_R}{V} = \frac{1}{S_{sc}} [\Delta P_\ell \cos \phi_{sc} + \Delta Q_\ell \sin \phi_{sc}] \tag{21}$$

for small changes.

If the supply resistance R_s is much less than the reactance X_s it may be permissible to neglect the voltage changes caused by swings in the real power ΔP_ℓ, so that the voltage regulation is governed by the equation

$$\frac{\Delta V}{V} \simeq \frac{\Delta V_R}{V} = \frac{\Delta Q_\ell}{S_{sc}} \sin \phi_{sc} \simeq \frac{\Delta Q_\ell}{S_{sc}} . \tag{22}$$

That is, the per-unit voltage change or swing is equal to the ratio of the reactive power swing to the short-circuit level of the supply system. This relationship can be represented graphically, as in Figure 4, which shows the *supply system voltage characteristic* (or system load line) as approximately linear. An alternate representation is

$$V \simeq \frac{E}{1 + Q_\ell / S_{sc}} \simeq E \left(1 - \frac{Q_\ell}{S_{sc}}\right) \tag{23}$$

if $Q_\ell << S_{sc}$.

Although the characteristic is only approximate, it is very useful in visualizing the action of the compensator, as will appear below.

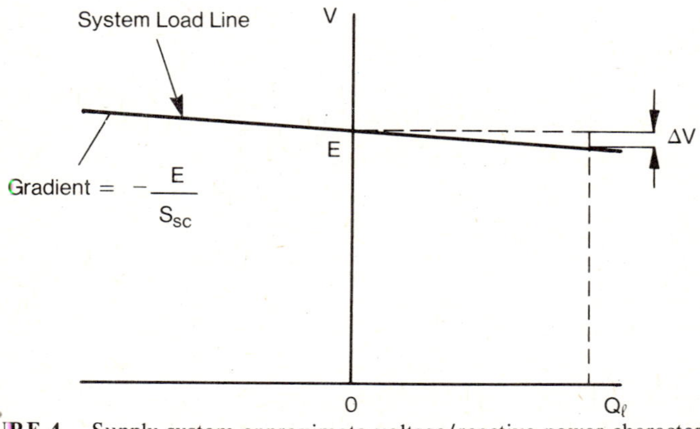

FIGURE 4. Supply system approximate voltage/reactive power characteristic.

1.6. APPROXIMATE REACTIVE POWER CHARACTERISTICS

1.6.1. Voltage Regulation with a Varying Inductive Load

In this section, we shall deduce the properties of an ideal compensator intended for voltage regulation improvement with a variable inductive load.

The load will be assumed to be three-phase, balanced, and to vary sufficiently slowly so that the per-phase phasor or quasi-stationary equations may be used. The load variations are assumed to be small so that $\Delta V << V$, and it is also assumed that $R_s << X_s$ so that the approximate Equations 22 and 23 are applicable. Figure 5*a* shows the arrangement of the system, compensator, and load; the system characteristic is drawn in

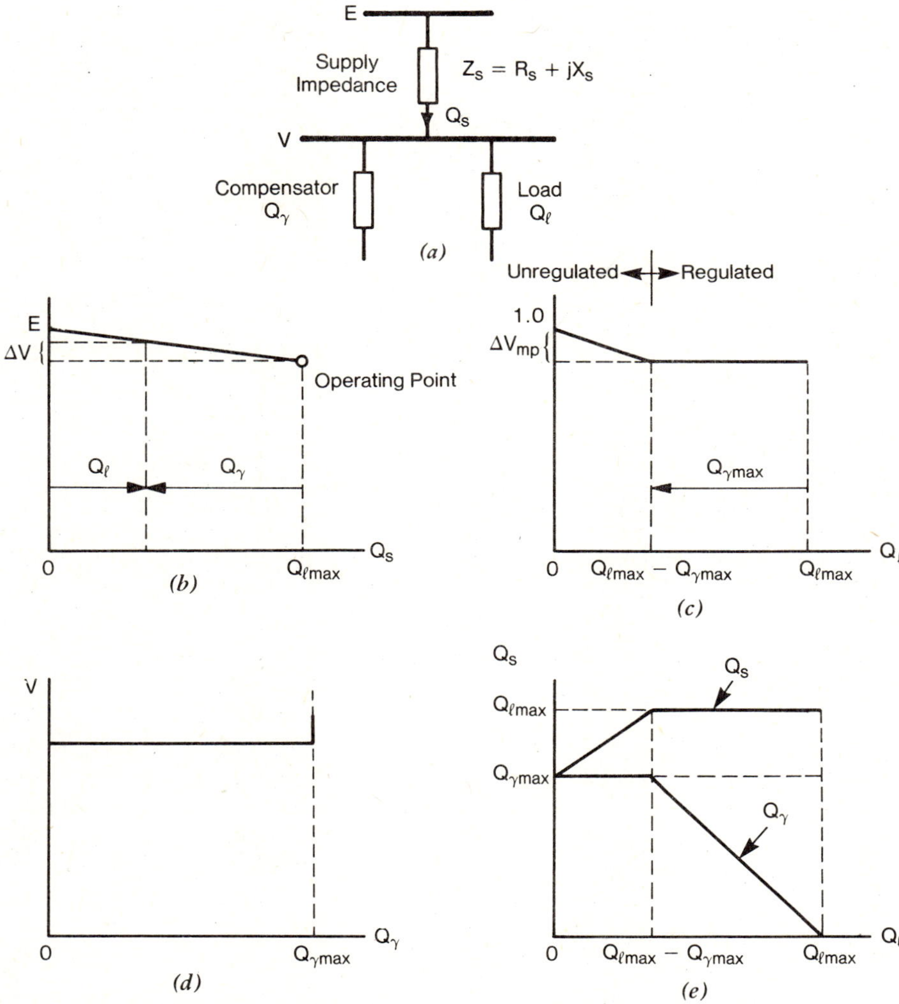

FIGURE 5. (*a*) Single-phase equivalent circuit of compensated load. (*b*) Approximate voltage/reactive power characteristic of completely compensated system. (*c*) Approximate voltage/reactive power characteristic of partially compensated system. (*d*) Ideal compensator voltage/reactive power characteristic (approximate). (*e*) Reactive power balance diagram (variation of Q_s and Q_γ with Q_ℓ).

Figure 5*b*. This characteristic is drooping; that is, an increase in the reactive power Q_s supplied by the system decreases the voltage at the supply point. Replacing Q_ℓ in Equation 23 by $Q_s(= Q_\ell + Q_\gamma)$,

$$V \simeq E\left(1 - \frac{Q_s}{S_{sc}}\right) \tag{24}$$

or

$$\frac{\Delta V}{V} \simeq \frac{Q_s}{S_{sc}} . \tag{25}$$

The reactive power Q_s supplied by the system is given by

$$Q_s = Q_\ell + Q_\gamma , \tag{26}$$

and it is clear that if the compensator reactive power Q_γ could be varied in such a way as to keep Q_s constant, the supply voltage could be constant. In particular, if

$$Q_s = Q_{\ell\max} = \text{constant}, \tag{27}$$

then V is constant with the value $E(1 - Q_{\ell\max}/S_{sc})$, as shown in Figure 5*b*. When the load reactive power Q_ℓ increases, the compensator reactive power absorption decreases, their sum remaining constant. When $Q_\ell = 0$, the compensator is fully on and absorbs $Q_{\ell\max}$; when $Q_\ell = Q_{\ell\max}$ the compensator is fully off and absorbs no reactive power. Note that we have a purely *inductive* compensator holding constant supply voltage with an inductive load.

The compensation shown in Figure 5*b* is said to be *complete,* because constant voltage is maintained throughout the reactive power range of the load.

The voltage regulation $\Delta V/V$ can be kept zero only if the reactive power rating of the compensator equals or exceeds $Q_{\ell\max}$. If the compensator reactive power is limited to $Q_{\gamma\max}$ (less than $Q_{\ell\max}$), then when $Q_\ell = 0$ the compensator will absorb $Q_{\gamma\max}$ and the voltage regulation will be

$$\frac{\Delta V}{V} = \frac{Q_{\ell\max} - Q_{\gamma\max}}{S_{sc}} . \tag{28}$$

This situation is illustrated in Figure 5*c*; the compensation is said to be *partial.* This equation illustrates the "leverage" which the compensator has on the system voltage, in that the maximum value of $\Delta V/V$ which can be caused by a change in the compensator's reactive power from zero to maximum is given by $-Q_{\gamma max}/S_{sc}$. For example, on a 10-kV busbar with a short-circuit level of 250 MVA, the smallest compensator capable of forcing a 1% change in voltage is rated 0.01 × 250 = 2.5 MVA. The minimum compensator rating can be chosen so that $-Q_{\gamma max}/S_{sc}$ just corresponds to the maximum permitted voltage swing ΔV_{mp}: thus

$$Q_{\gamma max} = Q_{\ell\, max} - S_{sc}\frac{\Delta V_{mp}}{V}. \tag{29}$$

It is now instructive to split Figure 5*b* into two separate diagrams as shown in Figures 5*c* and 5*d*; this can be done with the aid of the reactive power balance diagram shown in Figure 5*e*. Figure 5*c* shows the variation of the supply-point voltage with Q_ℓ: it represents the compensated system characteristic and should be compared with the uncompensated characteristic of Figure 5*b*. The compensator is rated at $Q_{\gamma max} < Q_{\ell\, max}$ and is controlled ideally in such a way as to maintain Q_s constant as in Equation 27, provided that its rating is not exceeded; that is, the compensator acts as an ideal voltage regulator.

Figure 5*c* shows that the compensator reactive-power rating need be no larger than the *variation* in load reactive power in order to maintain constant supply voltage as the load varies. This affords a useful economy in compensator rating where the load reactive power varies between maximum and some fractional value, say 0.5 pu. Provided that the compensator is rated according to Equation 29, then whatever the load reactive power, the supply voltage variation does not exceed ΔV_{mp}.

The two segments of Figure 5*c* can be identified as an unregulated range for $0 < Q_\ell < (Q_{\ell\, max} - Q_{\gamma max})$; and a regulated range for $(Q_{\ell\, max} - Q_{\gamma max}) < Q_\ell < Q_{\ell\, max}$. Throughout the unregulated range, the compensator absorbs $Q_{\gamma max}$ and limits the voltage rise to the maximum permitted level ΔV_{mp}. In the regulated range, the compensator maintains Q_s = constant and $\Delta V = 0$.

The control characteristic of the compensator is shown in Figure 5*d*. Since there is no voltage change for $0 < Q_\gamma < Q_{\gamma max}$ the characteristic is flat in the regulated range; if Q_ℓ falls below the regulated range, the compensator merely absorbs a constant $Q_{\gamma max}$ irrespective of the voltage.

1.6.2. Power Factor Improvement

The average power-factor of the inductively compensated inductive load is substantially *worse* than that of the load itself. If, for example, the average reactive power of the load $\bar{Q}_\ell$ were one-half its maximum, then the average reactive power supplied by the system to the compensated load would be $2\bar{Q}_\ell$, that is, twice as much.

To achieve ideal voltage regulation as well as unity average power-factor, it is clear that a capacitive compensator is required. Instead of keeping $Q_s =$ constant $= Q_{\ell \max}$ as in Equation 27, the compensator should keep

$$Q_s = \text{constant} = 0. \tag{30}$$

Neglecting the effect of variations in load power, a procedure similar to that of Section 1.6.1 will bring out the voltage/reactive power characteristic of the ideal compensator which achieves this. Figures 6*a* through 6*d* illustrate the procedures; Figure 6*c* shows the ideal compensator characteristic. The minimum *capacitive* rating of the compensator is given by

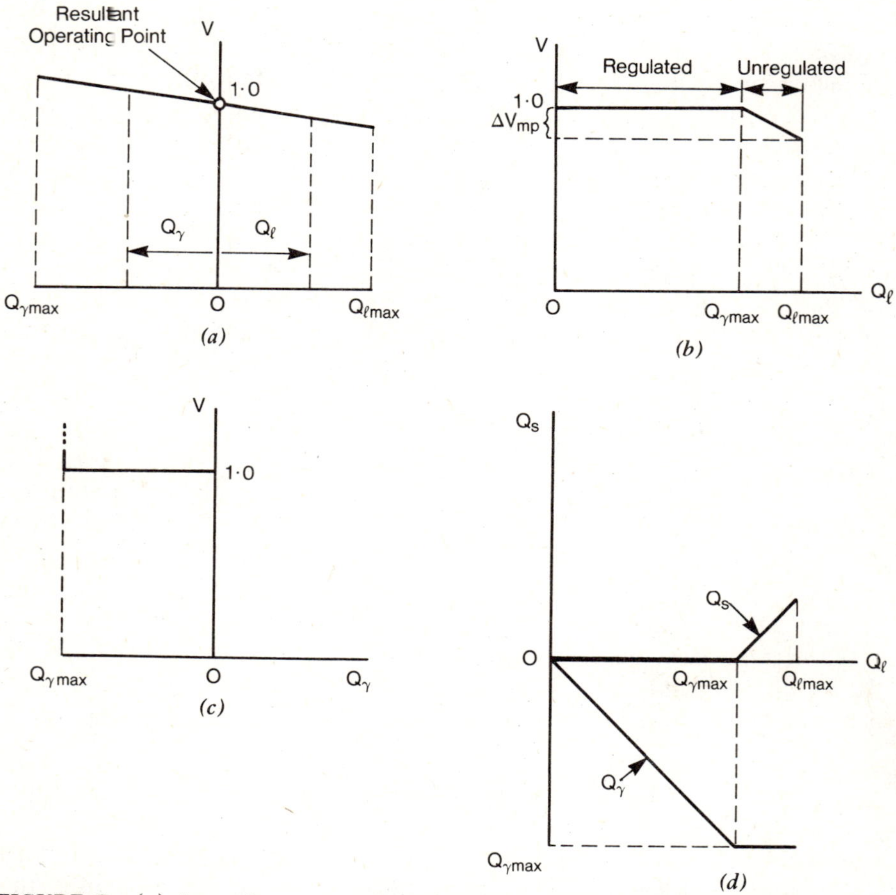

FIGURE 6 (*a*) Approximate voltage/reactive power characteristic of uncompensated system. (*b*) Approximate voltage/reactive power characteristic of compensated system. (*c*) Ideal compensator voltage/reactive power characteristic (approximate). (*d*) Reactive power balance diagram (variation of Q_s and Q_γ with Q_ℓ).

Equation 29, and outside its regulating range the compensator is assumed to generate a constant reactive power $Q_{\gamma max}$. Unity voltage is now defined so as to correspond to the fully compensated condition defined by Equation 30, and the mean operating point is at $V = 1.0$ per unit with $Q_s = 0$.

Instead of absorbing just enough reactive power to make up the total $Q_\ell + Q_\gamma$ to $Q_{\ell max}$, the compensator now generates whatever the load absorbs; the compensator is purely capacitive. If the compensator is designed as an ideal voltage regulator, then Q_s is not quite constantly zero because of load power variations. Generally this effect will be small. (See the worked example in Section 1.7.)

1.6.3. Reactive Power Bias

If the load reactive power can vary from leading to lagging, then what is required is a compensator whose regulated $V - Q$ characteristic extends into both quadrants as in Figure 7*a*. The inductive compensator characteristic of Figure 5*c* can be biased in this way by means of a fixed shunt capacitor as shown in Figure 7*b*. In the same way, the capacitive compensator can be biased into the lagging quadrant by a fixed shunt inductor, as shown in Figure 7*c*. If the shunt capacitor of Figure 7*b* is sufficiently large, then the inductive compensator can be biased so that its characteristic is wholly in the leading quadrant. When combined with a shunt capacitor, the inductive compensator becomes capable of keeping both constant voltage and unity average power-factor of an inductive load.

The distinction between an inductive and a capacitive compensator may now seem a little artificial, but it is important from a practical point of view because all real compensators except the synchronous condenser work by controlling the currents in either a capacitor bank or in an arrangement of inductors. The saturated reactor compensator, for example, is usually biased at least part way into the leading quadrant by means of shunt capacitors. A fixed shunt reactance is cheaper than a variable compensator having the same reactive power rating, and it is sometimes economic to size the compensator to match only the *variations* in load reactive power, while biasing it with a fixed shunt reactance to achieve the desired average power factor.

In Figures 5, 6, and 7 the voltage/reactive power characteristics of both the compensator and the supply system are not truly straight but are quadratic. They are shown approximately as straight lines under the assumption that V does not deviate appreciably from 1.0 pu. More exact calculation requires the exact forms of Equations 12 or 13. Alternatively, the working can be done in terms of the currents instead of the reactive powers.

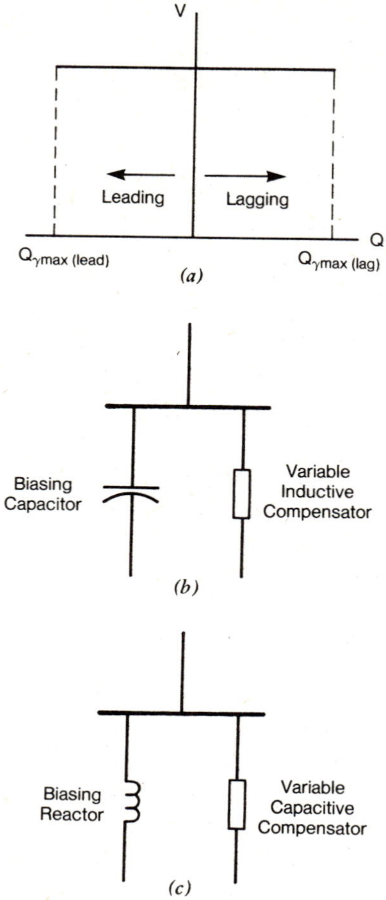

FIGURE 7. (*a*) Approximate voltage/reactive power characteristic of ideal compensator capable of leading and lagging reactive power. (*b*) Capacitive bias of variable inductive compensator. (*c*) Inductive bias of variable capacitive compensator.

1.7. AN EXAMPLE

Consider a supply system at 10 kV line-neutral voltage with a short-circuit level of 250 MVA and an X_s:R_s ratio of 5, supplying a wye-connected inductive load whose mean power is 25 MW and whose reactive power varies from zero to 50 MVAr; all quantities are expressed *per phase*.

The Thévenin impedance of the supply system is $Z_s = E_s^2/S_{sc} = (10\text{ kV})^2/250\text{ MVA} = 0.4$ Ohm/phase. With $\tan\phi_{sc} = 5$, we have $\phi_{sc} = 78.69°$, so that $X_s = Z_s \sin\phi_{sc} = 0.3922$ Ohm and $R_s = 0.0784$ Ohm. By phasor construction or "load flow" calculation,† we have, with $E = 1.0$ pu $= 10$ kV,

$$V = 6.782 \text{ kV}(l\text{-}n)$$

and

$$\Delta\mathbf{V} = 3.1806 + \mathrm{j}0.8678 \text{ kV} .$$

Now the line current is given by $\mathbf{I} = (P_\ell - \mathrm{j}Q_\ell)/\mathbf{V} = 3.686 - \mathrm{j}7.372 = 8.242 \angle -63.44°$ kA at full load, with a power-factor of 0.447 lagging. The phasor diagram is drawn in Figure 8*a*. The voltage magnitude is depressed by $10 - 6.782 = 3.218$ kV. The accurate forms of Equations 19 and 20 have been used for this calculation.

1.7.1. Compensation for Constant Voltage

Following the method of Section B in the Appendix, we have, with $V = 10$,

$$\begin{aligned}
a &= R_s^2 + X_s^2 = 0.160 \\
b &= 2V^2X_s = 2 \times 10^2 \times 0.3922 = 78.44 \\
c &= (V^2 + R_sP_\ell)^2 + X_s^2P_\ell{}^2 - E^2V^2 \\
&= (10^2 + 0.0784 \times 25)^2 + (0.3922 \times 25)^2 - 10^2 \times 10^2 \\
&= 491.98 .
\end{aligned}$$

Hence,

$$Q_s = \frac{-78.44 \pm \sqrt{78.44^2 - 4 \times 0.160 \times 491.98}}{2 \times 0.16}$$

$$= -6.35 \text{ or } -484 \text{ MVAr} .$$

Only the first solution gives $E = 10.0$ kV in Equation 13, so that

$$\Delta V_R = \frac{R_sP_\ell + X_sQ_s}{V} = \frac{0.0784 \times 25 + 0.3922 \times (-6.35)}{10} = -0.0532 \text{ kV}$$

$$\Delta V_X = \frac{X_sP_\ell - R_sQ_s}{V} = \frac{0.3922 \times 25 - 0.0784 \times (-6.35)}{10} = 1.030 \text{ kV} .$$

† See Section B in Appendix at the end of this chapter.

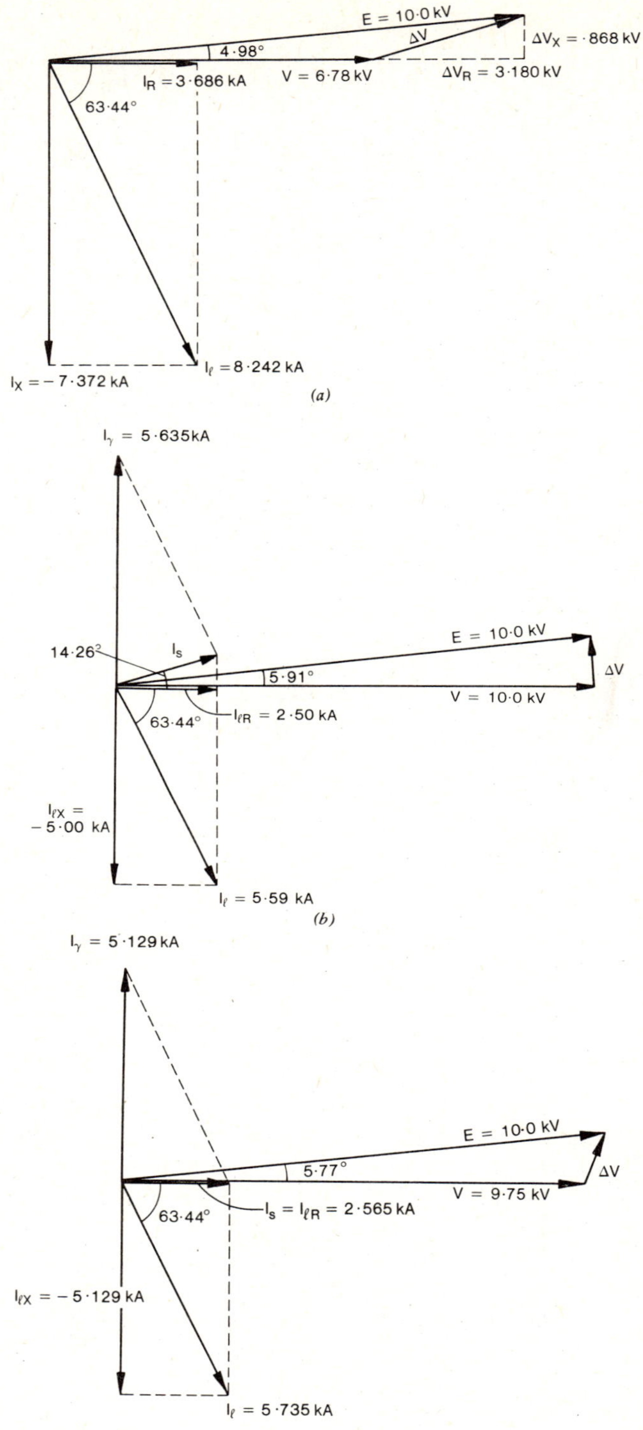

(a)

(b)

(c)

FIGURE 8. (*a*) Uncompensated load (25 + j50 MVA). (*b*) Load compensated for constant voltage. (*c*) Load compensated for unity power-factor.

The total current in the supply lines is $(P_\ell - jQ_s)/V = 2.50 + j0.635$ kA, and the compensator current is given by $-jQ_\gamma/V = +j5.635$ kA with $Q_\gamma = -56.35$ MVAr. The phasor diagram is shown in Figure 8*b*. It has a number of interesting features. The voltage magnitude is held exactly to 1.0 pu and it is now not permissible to calculate ΔV as ΔV_R, neglecting ΔV_X. Instead, both components must be calculated accurately.

The compensator reactive power is not equal to the load reactive power, but exceeds it by 6.35 MVAr as a result of the compensation of the voltage regulation caused by the real load power P_ℓ. Consequently the system power-factor is not unity but $25/\sqrt{25^2 + 6.35^2} = 0.969$ leading. In the diagram, the supply line current $\mathbf{I}_s$ can be seen leading the voltage by $\cos^{-1}(0.969) = 14.3°$.

1.7.2. Compensation for Unity Power Factor

With $Q_\gamma = Q_\ell$, the phasor diagram is as shown in Figure 8*c*, with $\mathbf{I}_\gamma = j\,5.129$ kA $= -\mathbf{I}_\ell$ and $Q_s = 0$. The voltage is $V = 9.748$ kV† with $\Delta V_R = 0.201$ kV and $\Delta V_X = 1.006$ kV; the voltage depression is therefore $9.748 - 10.0 = -0.252$ kV, or approximately 2.5%. The power-factor correction thus improves the voltage regulation enormously compared with the uncompensated case. In many situations this degree of improvement is adequate and the compensator can be designed to provide the reactive power requirement of the load rather than as an ideal voltage regulator.

1.8. LOAD COMPENSATOR AS A VOLTAGE REGULATOR

The control characteristics shown in Figures 5*d* and 6*c* can be characterized by three numbers:

1. The knee-point voltage V_k.
2. The maximum or rated reactive power $Q_{\gamma max}$.
3. The gain K_γ.

The gain may be defined as the change in reactive power Q_γ divided by the change in voltage V: thus

$$K_\gamma = \frac{dQ_\gamma}{dV}, \qquad (31)$$

† Calculated by the method of Section C in Appendix.

and if the control characteristic is linear, then for $Q_\gamma < Q_{\gamma\max}$ it is represented by the equation

$$V = V_k + Q_\gamma / K_\gamma \,. \tag{32}$$

In Figures 5*d* and 6*c*, the gain K_γ is infinite; the compensator absorbs or generates exactly the right amount of reactive power to maintain the supply point voltage constant as the load varies.

We now determine the voltage regulating properties of a compensator having a finite voltage-regulating gain K_γ, operating on a supply system with a finite short-circuit level S_{sc}. As in Section 1.6.1, the central question is how the supply-point voltage magnitude varies with the load (in particular, with the load reactive power).

As developed so far, the ideal compensator in its voltage-regulating mode has had infinite gain K_γ. Very high values of K_γ are rare in practice. They tend to weaken the stability of the compensator's operating point, and in certain types of compensator it is inherently expensive to design for a high value. So the performance with finite K_γ is important.

We continue to work on a per-phase basis, assuming balanced conditions. The X_s:R_s ratio of the supply system is assumed to be high, and load power fluctuations are neglected. The system is as shown in Figure 5*c*. Reactive power balance is expressed by

$$Q_\ell + Q_\gamma = Q_s \,. \tag{33}$$

The system voltage characteristic or load line is given by Equation 24; approximately,

$$V \simeq E\left(1 - \frac{Q_s}{S_{sc}}\right)$$

(see Figure 9*a*). The gradient of the load line represents the intrinsic sensitivity of the supply voltage to variations in the reactive power Q_s; thus

$$\frac{dV}{dQ_s} = -\frac{E}{S_{sc}} \,. \tag{34}$$

A high short-circuit level S_{sc} reduces the voltage sensitivity, making the load line flat; the system is then termed "stiff."

In the uncompensated case, $Q_\gamma = 0$ and $Q_s = Q_\ell$, so that the voltage sensitivity to the load reactive power Q_ℓ is the same as its intrinsic sensi-

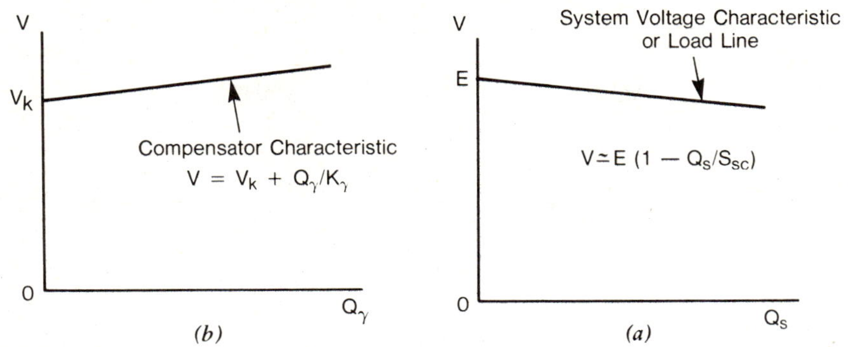

FIGURE 9. (*a*) Approximate voltage/reactive power characteristic of supply system. (*b*) Ideal compensator voltage/reactive power characteristic.

tivity $-E/S_{sc}$. In the presence of a compensator, from Equations 24 and 33,

$$V \simeq E\left[1 - \frac{Q_\ell + Q_\gamma}{S_{sc}}\right], \tag{35}$$

and since Q_γ is a function of V, the sensitivity will be modified.

The compensator reactive power Q_γ is determined by the voltage difference $V - V_k$ according to Equation 32 (see Figure 9*b*), so that

$$Q_\gamma = K_\gamma\,(V - V_k)\ . \tag{36}$$

We have already seen that a high compensator gain K_γ implies a flat V/Q characteristic; that is a "stiff," constant voltage characteristic. In per unit (pu) terms, a gain of, say, 40 pu means that the compensator reactive power changes from zero to 1 pu for a change in $V - V_k$ (or V) equal to 1/40 or 0.025 pu. In the following, it will be convenient to use a per-unit system in which $Q_{\gamma max}$ is the base reactive power and E the base voltage.

The influence of the compensator is determined by substituting for Q_γ from Equation 36 in Equation 35 and rearranging, so that

$$V = E\left[\frac{1 + K_\gamma V_k/S_{sc}}{1 + K_\gamma E/S_{sc}} - \frac{Q_\ell/S_{sc}}{1 + K_\gamma E/S_{sc}}\right]. \tag{37}$$

This equation shows how the supply point voltage V varies with load reactive power Q_ℓ in the presence of the compensator, provided of course that $Q_\gamma < Q_{\gamma max}$. Although approximate, it directly shows the influence of all the major parameters: the load reactive power itself, the compensator characteristic V_k and K_γ, and the system characteristic E and S_{sc}.

If the load is uncompensated, we have $K_\gamma = Q_\gamma = 0$ and Equation 37 reduces to Equation 24. The compensator has two effects immediately apparent from Equation 39: it *alters the no-load supply-point voltage* and it *modifies the sensitivity of the supply-point voltage to the load reactive power*. If the compensator gain K_γ *is positive, then the voltage sensitivity is reduced*:

$$\frac{dV}{dQ_\ell} = -\frac{E/S_{sc}}{1 + K_\gamma E/S_{sc}} . \tag{38}$$

For instance, suppose $E = 1.00$ pu and $S_{sc} = 25$ pu (based on $Q_{\gamma\max}$), then the uncompensated voltage sensitivity to load reactive power variations is (from Equation 34) -0.04 pu. A compensator with $K_\gamma = 100$ pu would reduce this sensitivity to

$$\frac{-0.04}{1 - 100 \times 0.04} = \frac{-0.04}{5} = -0.008 \text{ pu} . \tag{39}$$

The no-load supply-point voltage is expressed by the first term on the right-hand side of Equation 37. It can be made equal to the uncompensated no-load voltage E by making $V_k = E$.

It is useful to express the gradient $-E/S_{sc}$ in a form analogous to that of K_γ: if we write

$$\frac{1}{K_s} = -\frac{E}{S_{sc}} , \tag{40}$$

then K_s represents the system "gain" in that it equals the rate at which reactive power must be absorbed from the system in order to depress the system voltage by unit amount. K_s is then analogous to K_γ for the compensator, and the leverage which the compensator has in determining the overall sensitivity of the supply-point voltage to the load reactive power is clearly a function of the ratio K_γ/K_s, provided that $Q_\gamma < Q_{\gamma\max}$.

The compensator reactive power corresponding to a given value of Q_ℓ can be determined from Equations 36 and 37 as

$$Q_\gamma = \frac{K_\gamma}{1 + K_\gamma E/S_{sc}} \left[E \left(1 - \frac{Q_\ell}{S_{sc}} \right) - V_k \right] , \tag{41}$$

and if $E = V_k$

$$Q_\gamma = -\frac{K_\gamma E/S_{sc}}{1 + K_\gamma E/S_{sc}} Q_\ell . \tag{42}$$

Figure 10 shows the relationship between these characteristics, using the example with $S_{sc} = 250$ MVA and assuming $Q_{\ell\max} = 10$ MVAr $= -Q_{\gamma\max}$ that is, the compensator is capacitive. Because the compensator

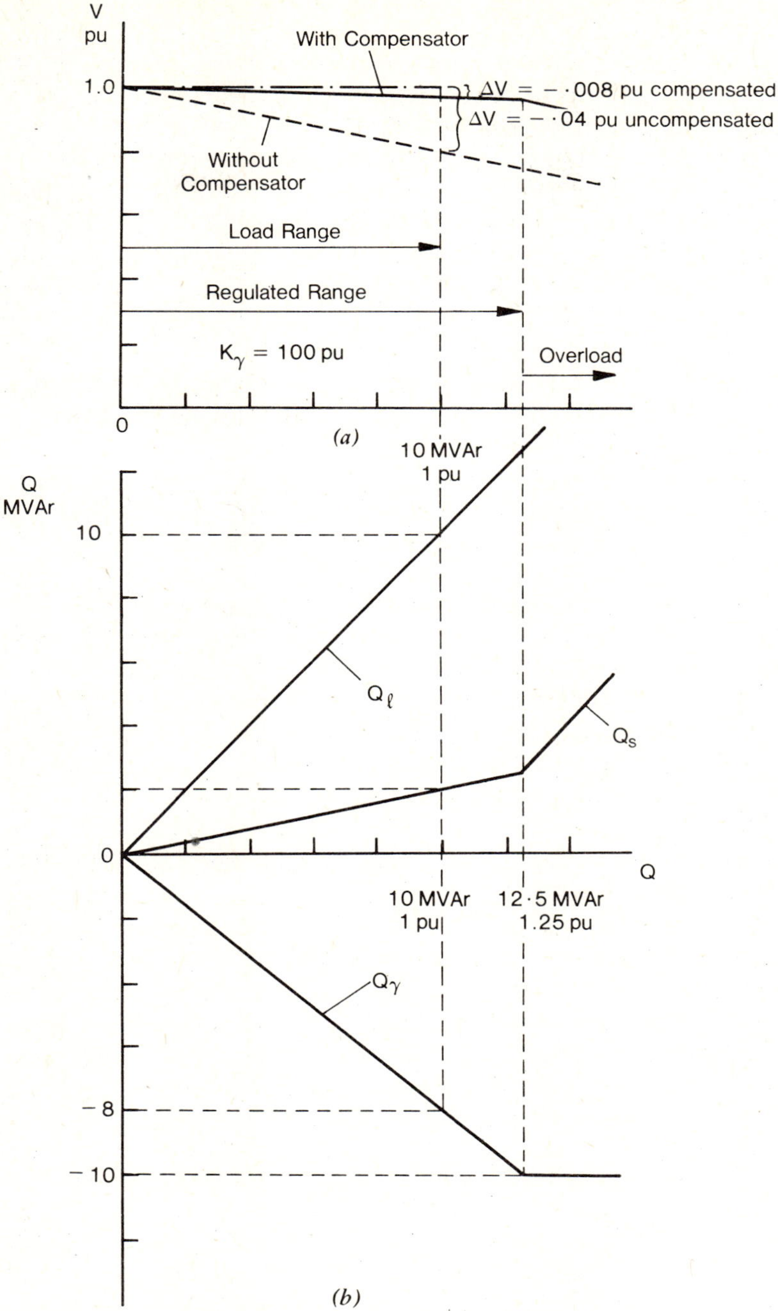

FIGURE 10. (*a*) Construction of compensated system voltage/reactive power characteristic. (*b*) Reactive power balance diagram.

gain is finite, the reactive power compensation is imperfect and Q_s varies from 0 to 2 MVAr (0.20 pu) as Q_ℓ varies from 0 to 10 MVAr. The compensator reactive power correspondingly varies from zero to only 8 MVAr over this range while the voltage regulation is held to 0.008 pu (Figure 10*a*). When the *load* reactive power equals the compensator's rated reactive power, the compensator still has 2 MVAr (20%) in hand so that the regulated range extends in this example into the overload range up to a load reactive power of 12.5 MVAr (1.25 pu).

1.9. PHASE BALANCING AND POWER-FACTOR CORRECTION OF UNSYMMETRICAL LOADS

Our discussion of load compensation has so far been on a per-phase or single-phase basis, and we come now to the third fundamental objective in load compensation: the balancing of unbalanced (unsymmetrical) three-phase loads.

In developing the concept of the ideal compensator as used for power-factor correction or for voltage regulation, we have modeled the compensator either as a controllable source of reactive power or as a reactive device with a constant-voltage control characteristic. Although the models are ultimately equivalent, the one may be more convenient or more illuminating in a given application than the other. In considering unbalanced loads it is helpful to begin by modeling both the load and the compensator in terms of their admittances and impedances. In taking this point of view, as indeed throughout this section, we shall follow the excellent paper by Gyugyi, Otto and Putman,[4] to which the reader should refer for greater detail. The analysis will be made sufficiently general to include power-factor correction at the same time, because this helps to preserve continuity with the earlier analysis. More importantly, the simultaneous treatment of phase balancing and power-factor correction in terms of load and compensator admittances leads to a fundamental view of load compensation which is different from the aspects developed so far and which gives further insight into the nature of the problem.

1.9.1. The Ideal Compensating Admittance Network

Supply voltages will be assumed balanced. The load is represented by the delta-connected network of Figure 11 in which the admittances $\mathbf{Y}_\ell^{ab}$, $\mathbf{Y}_\ell^{bc}$, and $\mathbf{Y}_\ell^{ca}$ are complex and unequal. Any ungrounded wye-connected load can be represented by Figure 11*a* by means of the wye-delta transformation (Section A in Appendix). Changes in the load are assumed to be

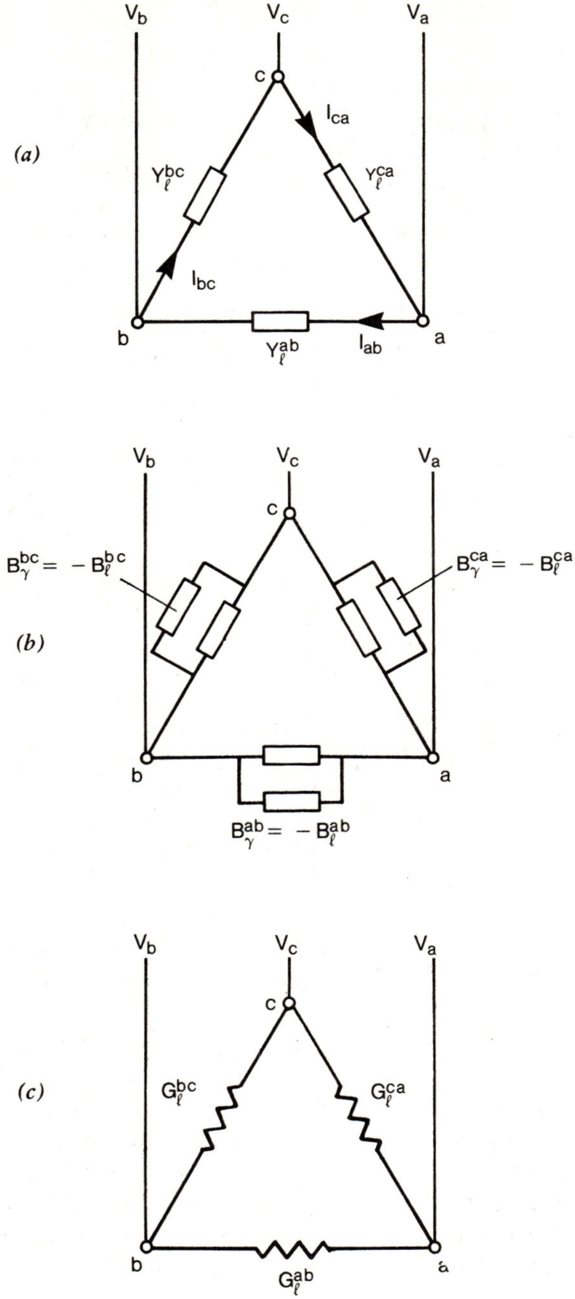

FIGURE 11. (*a*) General unbalanced three-phase load. (*b*) Connection of power-factor correcting susceptances in individual phases. (*c*) Resultant load; unbalanced but with unity power-factor.

sufficiently slow, or "quasi-stationary," so that phasor analysis is permitted, and the load is assumed to be linear.

The ideal load compensator (if one exists) is conceived as any passive three-phase admittance network which, when combined in parallel with the load, will present a real and symmetrical load to the supply.

Beginning with the power-factor correction concept of Equation 5, each load admittance can first be made purely real by connecting in parallel a compensating susceptance equal to the negative of the load susceptance in that branch of the delta. Thus if

$$\mathbf{Y}_\ell^{ab} = G_\ell^{ab} + jB_\ell^{ab} , \tag{43}$$

the compensating susceptance is

$$B_\gamma^{ab} = - B_\ell^{ab} . \tag{44}$$

Similarly the compensating susceptances $B_\gamma^{bc} = - B_\ell^{bc}$ and $B_\gamma^{ca} = - B_\ell^{ca}$ are connected in parallel with $\mathbf{Y}_\ell^{bc}$ and $\mathbf{Y}_\ell^{ca}$ respectively, as shown in Figure 11*b*. The resulting load admittances are shown in Figure 11*c*. They are real, giving an overall power-factor of unity; but they remain unbalanced.

As a first step towards balancing this real, unbalanced load, consider the single-phase load G_ℓ^{ab} (Figure 12*a*). The three-phase positive-sequence line currents can be balanced by connecting between phases *b* and *c* the capacitive susceptance

$$B_\gamma^{bc} = \frac{G_\ell^{ab}}{\sqrt{3}} \tag{45}$$

together with the inductive susceptance

$$B_\gamma^{ca} = \frac{- G_\ell^{ab}}{\sqrt{3}} \tag{46}$$

between phases *c* and *a*. This is illustrated in Figure 12*b*. The construction of the line currents $\mathbf{I}_a$, $\mathbf{I}_b$, and $\mathbf{I}_c$ for positive-sequence voltages V_{ab}, V_{bc} and V_{ca} is shown in Figure 13*a*. The line currents are not only balanced, but are also in phase with their respective phase voltages, so that each phase of a wye-connected supply system would supply one-third of the total power and no reactive power. For positive-sequence voltages, then, the equivalent circuit is three wye-connected resistors, each having the conductance G_ℓ^{ab}, as shown in Figure 12*c*. The total power is $3V^2G_\ell^{ab}$, where V is the rms value of the line-neutral supply voltage, assumed balanced. Both the overall power-factor and the power-factor in each phase of the supply are unity. Although the currents in the three branches of the delta are unbalanced, there is a reactive power equilib-

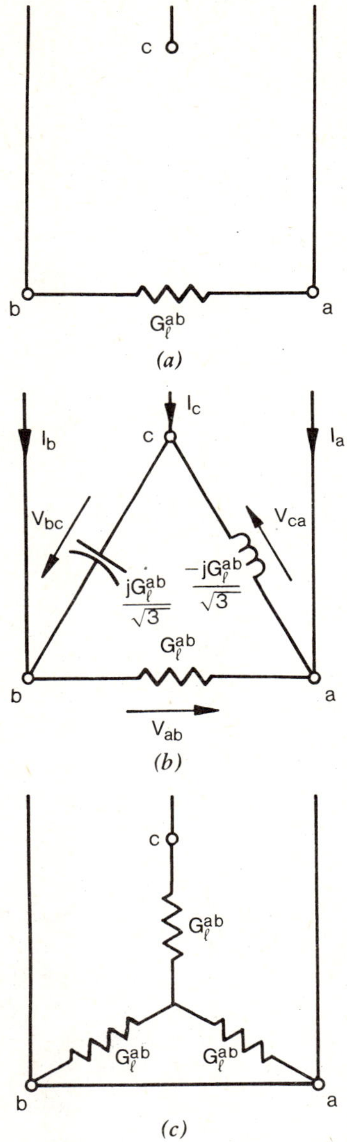

FIGURE 12. (*a*) Single-phase, unity-power-factor load before positive-sequence balancing. (*b*) Positive-sequence balancing of single-phase, unity-power-factor load. (*c*) Positive-sequence equivalent circuit of compensated single-phase load.

rium within the delta, in which the reactive power generated by the capacitor between lines *b* and *c* equals that absorbed by the inductor between lines *c* and *a,* so that none is generated or absorbed in the supply system.

To emphasize the fact that the balance depends on the phase sequence,

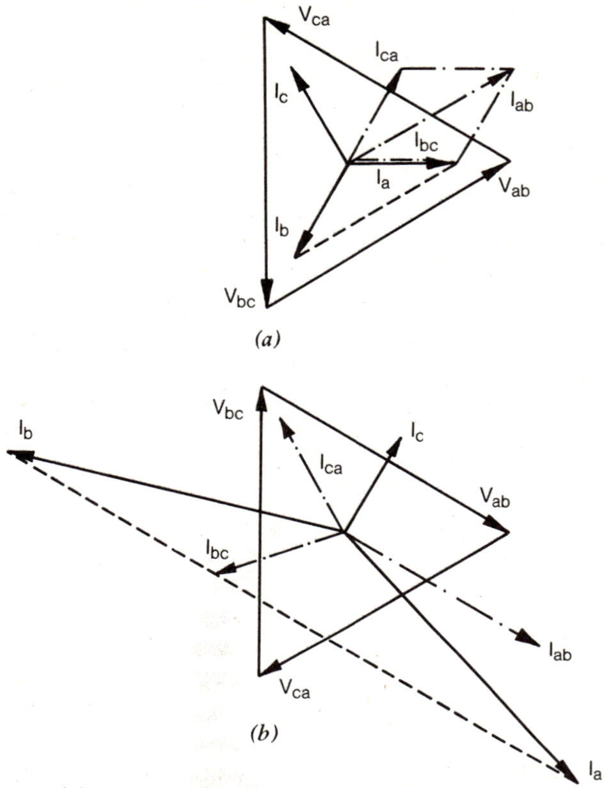

FIGURE 13. (*a*) Positive-sequence phasor diagram corresponding to Figure 12*c*. (*b*) Negative-sequence phasor diagram corresponding to Figure 12*c*.

Figure 13b shows the line currents obtained with pure *negative*-sequence supply voltages. Both the line currents and the currents in the three branches of the delta are unbalanced, although the total power remains the same ($=3V^2G_\ell^{ab}$) and no *net* reactive power is supplied or absorbed by the supply system. The power-factor in all three phases of the supply is, however, different from unity.

The real admittances in the remaining phases *bc* and *ca* can be balanced in turn by the same procedure. Thus G_ℓ^{bc} is balanced by the compensating susceptances $B_\gamma^{ca} = G_\ell^{bc}/\sqrt{3}$ and $B_\gamma^{ab} = -G_\ell^{ca}/\sqrt{3}$ between lines *a* and *b* and lines *b* and *c*, respectively. Together with the power-factor-correcting susceptances given by Equation 44 *et seq.*, each branch of the delta now has three parallel compensating susceptances which can be added together to give the three-phase, delta-connected *ideal compensating network*, attributed to C.P. Steinmetz:

$$
\begin{aligned}
B_{\gamma}^{ab} &= -B_{\ell}^{ab} + (G_{\ell}^{ca} - G_{\ell}^{bc})/\sqrt{3} \\
B_{\gamma}^{bc} &= -B_{\ell}^{bc} + (G_{\ell}^{ab} - G_{\ell}^{ca})/\sqrt{3} \\
B_{\gamma}^{ca} &= -B_{\ell}^{ca} + (G_{\ell}^{bc} - G_{\ell}^{ab})/\sqrt{3} \,.
\end{aligned} \tag{47}
$$

This is illustrated in Figure 14*a*. The resulting compensated load admittances are purely real and balanced, as shown in the equivalent circuit of Figure 14*b*. This equivalent circuit is valid only for positive-sequence voltages.

If the load conductances are balanced (implying that the load requires the same power in all three phases), then $G_{\ell}^{ca} - G_{\ell}^{bc} = 0$, and so on, and the compensating network does no more than cancel the reactive power in each branch of the load.

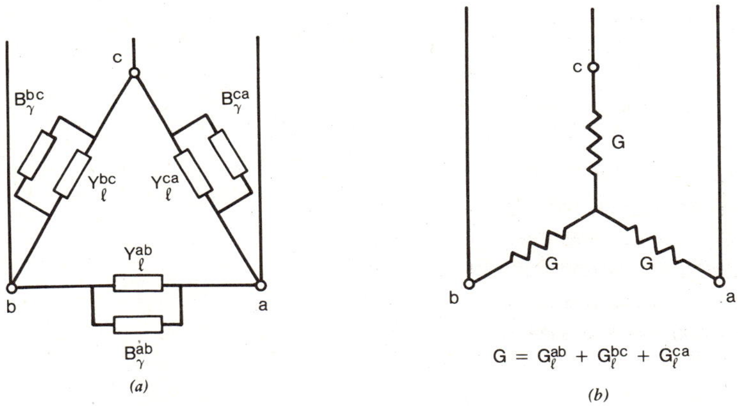

FIGURE 14. (*a*) General ideal three-phase compensating network. (*b*) Equivalent circuit for positive-sequence voltages.

We can summarize this approach to load compensation in the following important principles:

1. *Any unbalanced linear ungrounded three-phase load can be transformed into a balanced, real three-phase load without changing the real power exchange between source and load, by connecting an ideal compensating network in parallel with it.*
2. *The ideal compensating network can be purely reactive.*

If the load admittances vary, then so must the susceptances of the compensating network if the compensation is to remain perfect.

1.9.2. Load Compensation in Terms of Symmetrical Components

The principles of load compensation summarized in Equation 47 and as stated in the foregoing summary are a useful theoretical statement of what can be done with a reactive compensator. In the design of a compensator, however, it would not be convenient to make Equation 47 the basis of its control system because, for one thing, the desired compensator susceptances are prescribed in terms of the load admittances which are not so readily measured as are the separate line currents and voltages. What is needed instead is a formula for the desired compensating susceptances in terms of these currents and voltages. In this section, we show how one such formula can be derived.

The formula will directly show how to generate electrical signals representing to the compensator the demand for capacitive or inductive susceptance in each phase. The compensator's control system must adjust the effective susceptances of the compensator to satisfy this demand. Depending on the type of compensator, the functional elements of the control system may be realized explicitly in electronic circuits or implicitly in the equations describing the saturation of iron in the saturable-reactor type of compensator. Such considerations are the subjects of later chapters.

The analysis of the unbalanced load has so far been developed implicitly in terms of the actual line currents and voltages, that is, in "phase coordinates." Since these are the currents and voltages most readily measured, the formula for the desired compensating susceptances will also be developed in terms of them. An analytical difficulty with this approach is that there is no concise way of specifying mathematically that the currents in the compensated system must be balanced. This difficulty is removed by first transforming the currents and voltages into their symmetrical components. Later, the inverse transformation will be applied so that the desired compensating susceptances can be expressed in terms of actual line currents and voltages.

The use of symmetrical components is also useful in determining the performance of different types of compensator with unbalanced loads. Compensators differ in their negative-sequence characteristics.

The unbalanced load of Figure 11*a* is supplied by a balanced three-phase set of voltages with positive phase sequence. The rms line-neutral voltages are

$$\mathbf{V}_a = \mathbf{V};\ \mathbf{V}_b = \mathbf{h}^2\mathbf{V};\ \mathbf{V}_c = \mathbf{h}\mathbf{V}\ , \tag{48}$$

where

$$\mathbf{h} = e^{j2\pi/3} = -\frac{1}{2} + j\frac{\sqrt{3}}{2}\ . \tag{49}$$

The line-line voltages are

$$\begin{aligned} \mathbf{V}_{ab} &= \mathbf{V}_a - \mathbf{V}_b = (1 - \mathbf{h}^2)\,\mathbf{V} \\ \mathbf{V}_{bc} &= \mathbf{V}_b - \mathbf{V}_c = (\mathbf{h}^2 - \mathbf{h})\,\mathbf{V} \\ \mathbf{V}_{ca} &= \mathbf{V}_c - \mathbf{V}_a = (\mathbf{h} - 1)\,\mathbf{V}\,. \end{aligned} \tag{50}$$

The load currents in the three branches of the delta are

$$\begin{aligned} \mathbf{I}_{ab} &= \mathbf{Y}_\ell^{ab}\,\mathbf{V}_{ab} = \mathbf{Y}_\ell^{ab}\,(1 - \mathbf{h}^2)\,\mathbf{V} \\ \mathbf{I}_{bc} &= \mathbf{Y}_\ell^{bc}\,\mathbf{V}_{bc} = \mathbf{Y}_\ell^{bc}\,(\mathbf{h}^2 - \mathbf{h})\,\mathbf{V} \\ \mathbf{I}_{ca} &= \mathbf{Y}_\ell^{ca}\,\mathbf{V}_{ca} = \mathbf{Y}_\ell^{ca}\,(\mathbf{h} - 1)\,\mathbf{V}\,, \end{aligned} \tag{51}$$

and the line currents are

$$\begin{aligned} \mathbf{I}_a &= \mathbf{I}_{ab} - \mathbf{I}_{ca} = [\mathbf{Y}_\ell^{ab}\,(1 - \mathbf{h}^2) - \mathbf{Y}_\ell^{ca}\,(\mathbf{h} - 1)]\,\mathbf{V} \\ \mathbf{I}_b &= \mathbf{I}_{bc} - \mathbf{I}_{ab} = [\mathbf{Y}_\ell^{bc}\,(\mathbf{h}^2 - \mathbf{h}) - \mathbf{Y}_\ell^{ab}\,(1 - \mathbf{h}^2)]\,\mathbf{V} \\ \mathbf{I}_c &= \mathbf{I}_{ca} - \mathbf{I}_{bc} = [\mathbf{Y}_\ell^{ca}\,(\mathbf{h} - 1) - \mathbf{Y}_\ell^{bc}\,(\mathbf{h}^2 - \mathbf{h})]\,\mathbf{V}\,. \end{aligned} \tag{52}$$

The symmetrical components of the line currents are given by†

$$\begin{aligned} \mathbf{I}_0 &= (\mathbf{I}_a + \mathbf{I}_b + \mathbf{I}_c)/\sqrt{3} \\ \mathbf{I}_1 &= (\mathbf{I}_a + \mathbf{h}\mathbf{I}_b + \mathbf{h}^2\mathbf{I}_c)/\sqrt{3} \\ \mathbf{I}_2 &= (\mathbf{I}_a + \mathbf{h}^2\mathbf{I}_b + \mathbf{h}\mathbf{I}_c)/\sqrt{3}\,, \end{aligned} \tag{53}$$

where $\mathbf{I}_0$, $\mathbf{I}_1$, and $\mathbf{I}_2$ are the reference phasors of the zero-, positive- and negative-sequence sets, respectively. With $\mathbf{I}_a$, $\mathbf{I}_b$, and $\mathbf{I}_c$ given by Equation 52, we get

$$\begin{aligned} \mathbf{I}_0 &= 0 \\ \mathbf{I}_1 &= (\mathbf{Y}_\ell^{ab} + \mathbf{Y}_\ell^{bc} + \mathbf{Y}_\ell^{ca})\,\mathbf{V}\sqrt{3} \\ \mathbf{I}_2 &= -(\mathbf{h}^2\mathbf{Y}_\ell^{ab} + \mathbf{Y}_\ell^{bc} + \mathbf{h}\mathbf{Y}_\ell^{ca})\,\mathbf{V}\sqrt{3}\,. \end{aligned} \tag{54}$$

The third line of Equation 54 shows that with a balanced load there is no negative-sequence current, since with $\mathbf{Y}_\ell^{ab} = \mathbf{Y}_\ell^{bc} = \mathbf{Y}_\ell^{ca}$, $(\mathbf{h}^2 + 1 + \mathbf{h})$ becomes a factor of $\mathbf{I}_2$, and this is, of course, zero.

The symmetrical components of the line currents to a delta-connected reactive compensator are given similarly by

$$\begin{aligned} \mathbf{I}_{0(\gamma)} &= 0 \\ \mathbf{I}_{1(\gamma)} &= \mathrm{j}(B_\gamma^{ab} + B_\gamma^{bc} + B_\gamma^{ca})\,\mathbf{V}\sqrt{3} \\ \mathbf{I}_{2(\gamma)} &= -\mathrm{j}(\mathbf{h}^2 B_\gamma^{ab} + B_\gamma^{bc} + \mathbf{h}B_\gamma^{ca})\,\mathbf{V}\sqrt{3}\,. \end{aligned} \tag{55}$$

† The $1/\sqrt{3}$ factor is included to make the symmetrical component transformation unitary, that is, $C^{-1} = C'^*$. This guarantees power invariance and makes inversion simple.

The compensated load will be balanced if its negative-sequence current is zero, requiring that

$$\mathbf{I}_{2(\ell)} + \mathbf{I}_{2(\gamma)} = 0 \,, \tag{56}$$

where the subscript ℓ has been added to emphasize that $\mathbf{I}_{2(\ell)}$ is the load current. This equation applies to both the real and imaginary parts of $\mathbf{I}_{2(\ell)}$ and $\mathbf{I}_{2(\gamma)}$. The overall power-factor of the compensated load will be unity if the imaginary part of the positive-sequence line current is zero. This requires that

$$\operatorname{Im}\,[\mathbf{I}_{1(\ell)} + \mathbf{I}_{1(\gamma)}] = 0 \,. \tag{57}$$

The extreme conciseness of Equations 56 and 57 should be noted. If the compensator currents $\mathbf{I}_{1(\gamma)}$ and $\mathbf{I}_{2(\gamma)}$ are substituted from Equation 55 into Equations 56 and 57, and these equations are solved for $B_\gamma^{\rm ab}$, $B_\gamma^{\rm bc}$ and $B_\gamma^{\rm ca}$, then the following formula results for the susceptances of the ideal compensator:

$$\begin{aligned}
B_\gamma^{\rm ab} &= -\frac{1}{3\sqrt{3}\,V}\,[\operatorname{Im}\mathbf{I}_{1(\ell)} + \operatorname{Im}\mathbf{I}_{2(\ell)} - \sqrt{3}\operatorname{Re}\mathbf{I}_{2(\ell)}] \\
B_\gamma^{\rm bc} &= -\frac{1}{3\sqrt{3}\,V}\,[\operatorname{Im}\mathbf{I}_{1(\ell)} - 2\operatorname{Im}\mathbf{I}_{2(\ell)}] \\
B_\gamma^{\rm ca} &= -\frac{1}{3\sqrt{3}\,V}\,[(\operatorname{Im}\mathbf{I}_{1(\ell)} + \operatorname{Im}\mathbf{I}_{2(\ell)} + \sqrt{3}\operatorname{Re}\mathbf{I}_{2(\ell)}] \,.
\end{aligned} \tag{58}$$

The right-hand side must now be transformed back into phase coordinates by means of the inverse of the symmetrical components transformation (Equation 53). If there is no zero-sequence current the result is

$$\begin{aligned}
B_\gamma^{\rm ab} &= -\frac{1}{3V}\,[\operatorname{Im}\mathbf{I}_{\rm a(\ell)} + \operatorname{Im}\mathbf{h}\mathbf{I}_{\rm b(\ell)} - \operatorname{Im}\mathbf{h}^2\mathbf{I}_{\rm c(\ell)}] \\
B_\gamma^{\rm bc} &= -\frac{1}{3V}\,[\operatorname{Im}\mathbf{h}\mathbf{I}_{\rm b(\ell)} + \operatorname{Im}\mathbf{h}^2\mathbf{I}_{\rm c(\ell)} - \operatorname{Im}\mathbf{I}_{\rm a(\ell)}] \\
B_\gamma^{\rm ca} &= -\frac{1}{3V}\,[\operatorname{Im}\mathbf{h}^2\mathbf{I}_{\rm c(\ell)} + \operatorname{Im}\mathbf{I}_{\rm a(\ell)} - \operatorname{Im}\mathbf{h}\mathbf{I}_{\rm b(\ell)}]
\end{aligned} \tag{59}$$

This equation expresses the desired compensating susceptances in terms of the phasor line currents $\mathbf{I}_{\rm a(\ell)}$, $\mathbf{I}_{\rm b(\ell)}$, and $\mathbf{I}_{\rm c(\ell)}$ of the load.

Desired Compensating Susceptances Expressed in Terms of Instantaneous Currents and Voltages. The desired compensating susceptances can be expressed in terms of the instantaneous values of voltage and load current. This can be done in two ways: one is by a sampling process, whereas the other is by an averaging process. Taking the sampling approach first, the term Im $\mathbf{I}_{a(\ell)}$ is related to the instantaneous line current $i_{a(\ell)}$ by the equation

$$\begin{aligned} i_a &= \sqrt{2}\ \mathrm{Re}\ [(I_{aR} + \mathrm{j}I_{aX})e^{\mathrm{j}\omega t}] \\ &= \sqrt{2}(I_{aR}\cos\omega t - I_{aX}\sin\omega t)\ . \end{aligned} \tag{60}$$

Now Im $\mathbf{I}_{a(\ell)}$, that is, I_{aX}, is equal to i_a at the instant when $\sin\omega t = -1$ and $\cos\omega t = 0$. To define this instant, a reference phasor is necessary, and it is convenient to choose the line-neutral voltage $\mathbf{V}_a$ since

$$\begin{aligned} v_a &= \sqrt{2}\ \mathrm{Re}\ [V\ e^{\mathrm{j}\omega t}] \\ &= \sqrt{2}\ V\cos\omega t. \end{aligned} \tag{61}$$

The required instant is therefore defined by

$$v_a = 0 \text{ and } \frac{\mathrm{d}v_a}{\mathrm{d}t} > 0. \tag{62}$$

The condition that the derivative be positive is necessary to distinguish the required instant from the one when $\cos\omega t = 0$ and $\sin\omega t = +1$, which occurs one half-cycle later. We can now write

$$\mathrm{Im}\ \mathbf{I}_a = I_{aX} = \frac{i_a}{\sqrt{2}}\Bigg|_{\substack{v_a = 0 \\ \mathrm{d}v_a/\mathrm{d}t > 0.}} \tag{63}$$

Im $\mathbf{I}_a$ can therefore be measured by sampling i_a at the instant when $v_a = 0$ and $\mathrm{d}v_a/\mathrm{d}t$ is positive, that is, at a positive-going zero-crossing of the voltage v_a.

All terms on the right-hand side of Equation 59 can be similarly expressed in terms of instantaneous values of the line currents and the line-neutral voltages and their derivatives, giving the following result:

$$\begin{aligned} B_\gamma^{ab} &= -\frac{1}{3\sqrt{2}V}\left[i_a\Big|_{\substack{v_a=0 \\ \mathrm{d}v_a/\mathrm{d}t>0}} + i_b\Big|_{\substack{v_b=0 \\ \mathrm{d}v_b/\mathrm{d}t>0}} - i_c\Big|_{\substack{v_c=0 \\ \mathrm{d}v_c/\mathrm{d}t>0}} \right] \\ B_\gamma^{bc} &= -\frac{1}{3\sqrt{2}V}\left[i_b\Big|_{\substack{v_b=0 \\ \mathrm{d}v_b/\mathrm{d}t>0}} + i_c\Big|_{\substack{v_c=0 \\ \mathrm{d}v_c/\mathrm{d}t>0}} - i_a\Big|_{\substack{v_a=0 \\ \mathrm{d}v_a/\mathrm{d}t>0}} \right] \\ B_\gamma^{ca} &= -\frac{1}{3\sqrt{2}V}\left[i_c\Big|_{\substack{v_c=0 \\ \mathrm{d}v_c/\mathrm{d}t>0}} + i_a\Big|_{\substack{v_a=0 \\ \mathrm{d}v_a/\mathrm{d}t>0}} - i_b\Big|_{\substack{v_b=0 \\ \mathrm{d}v_b/\mathrm{d}t>0}} \right]. \end{aligned} \tag{64}$$

The desired compensating susceptances are thus expressed in terms of the three line currents sampled at instants defined by the positive-going zero-crossings of the line-neutral voltages v_a, v_b, and v_c. (Note that an artificial neutral may need to be provided in the measuring circuit). Equation 64 can be used directly as the basis of a compensator control system. Since the positive-going zero-crossings of v_a, v_b, and v_c follow one another at intervals of $2\pi/3$ electrical radians, the signal defining each desired compensating susceptance can in principle be updated three times per cycle.

Desired Compensating Susceptances Expressed in Terms of Average Real or Reactive Power Quantities. Since $\mathbf{V}_a = \mathbf{V}$, $\mathbf{V}_b = \mathbf{h}^2\mathbf{V}$, and $\mathbf{V}_c = \mathbf{hV}$, we can include $\mathbf{V}_a$, $\mathbf{V}_b$, and $\mathbf{V}_c$ in Equation 59 in the following way:

$$
\begin{aligned}
B_\gamma^{ab} &= \frac{1}{3V^2}\left[\operatorname{Im}(\mathbf{V}_a\mathbf{I}^*_{a(\ell)}) + \operatorname{Im}(\mathbf{V}_b\mathbf{I}^*_{b(\ell)}) - \operatorname{Im}(\mathbf{V}_c\mathbf{I}^*_{c(\ell)})\right] \\
B_\gamma^{bc} &= \frac{1}{3V^2}\left[\operatorname{Im}(\mathbf{V}_b\mathbf{I}^*_{b(\ell)}) + \operatorname{Im}(\mathbf{V}_c\mathbf{I}^*_{c(\ell)}) - \operatorname{Im}(\mathbf{V}_a\mathbf{I}^*_{a(\ell)})\right] \qquad (65)\\
B_\gamma^{ca} &= \frac{1}{3V^2}\left[\operatorname{Im}(\mathbf{V}_c\mathbf{I}^*_{c(\ell)}) + \operatorname{Im}(\mathbf{V}_a\mathbf{I}^*_{a(\ell)}) - \operatorname{Im}(\mathbf{V}_b\mathbf{I}^*_{b(\ell)})\right].
\end{aligned}
$$

We now make use of the relation

$$\operatorname{Im}[\mathbf{VI}^*] = \frac{1}{T}\int_T v\left(-\frac{\pi}{2}\right) i\, dt, \qquad (66)$$

where $v(-\pi/2)$ represents the signal v phase-shifted by $-\pi/2$ electrical radians at the fundamental frequency, and T is the period $2\pi/\omega$. This expression can be interpreted as a reactive power averaged over one complete cycle. Thus the first "reactive power" term in Equation 65 is

$$\operatorname{Im}[\mathbf{V}_a\mathbf{I}^*_{a(\ell)}] = \frac{1}{T}\int_T v_a\left(-\frac{\pi}{2}\right) i_{a(\ell)}\, dt. \qquad (67)$$

The reactive power represented by this equation cannot be meaningfully associated with one phase or one branch of the load circuit because the voltages $\mathbf{V}_a$, $\mathbf{V}_b$, and $\mathbf{V}_c$ are, by definition, balanced, whereas the currents $\mathbf{I}_{a(\ell)}$, $\mathbf{I}_{b(\ell)}$, and $\mathbf{I}_{c(\ell)}$ are not. Now in practice the signals $v_a(-\pi/2)$, and so on, can be immediately derived from the diagram relating line-neutral voltages to line-line voltages (Section E in Appendix). Since $\mathbf{V}_{ab} = -j\sqrt{3}\,\mathbf{V}_c$, $\mathbf{V}_{bc} = -j\sqrt{3}\,\mathbf{V}_a$, and $\mathbf{V}_{ca} = -j\sqrt{3}\,\mathbf{V}_b$ we have

$$v_a\left(-\frac{\pi}{2}\right) = \frac{v_{bc}}{\sqrt{3}}$$

$$v_b\left(-\frac{\pi}{2}\right) = \frac{v_{ca}}{\sqrt{3}} \qquad (68)$$

$$v_c\left(-\frac{\pi}{2}\right) = \frac{v_{ab}}{\sqrt{3}},$$

so that Equation 65 finally becomes

$$B_\gamma^{ab} = \frac{1}{3\sqrt{3}\,V^2}\,\frac{1}{T}\int_T (v_{bc}i_{a(\ell)} + v_{ca}i_{b(\ell)} - v_{ab}i_{c(\ell)})\,dt$$

$$B_\gamma^{bc} = \frac{1}{3\sqrt{3}\,V^2}\,\frac{1}{T}\int_T (v_{ca}i_{b(\ell)} + v_{ab}i_{c(\ell)} - v_{bc}i_{a(\ell)})\,dt \qquad (69)$$

$$B_\gamma^{ca} = \frac{1}{3\sqrt{3}\,V^2}\,\frac{1}{T}\int_T (v_{ab}i_{c(\ell)} + v_{bc}i_{a(\ell)} - v_{ca}i_{b(\ell)})\,dt\,.$$

Like Equation 64, this equation also can be directly employed as the basis of a compensator control system, since all the mathematical operations on the right-hand side can be performed straightforwardly by electronic circuits. Although the integration or averaging period is shown as one period of fundamental frequency, it is not strictly necessary to reset the electronic integrator every cycle, and a *continuous* signal representing desired susceptance can thus be generated for each branch of the compensator. It is also not strictly necessary that the integrator time-constant be equal to T; for some applications (e.g., arc furnaces) it may be appreciably shorter in order to make the compensator rapid in its response. Equation 69, of course, gives no information about the *response* of the compensator under such rapidly changing conditions.

Compensator Represented as Separate Positive- and Negative-Sequence Admittance Networks. It is possible to consider the compensator split into two networks, one of which supplies the positive-sequence components of the compensating currents and the other the negative-sequence components.

In Equation 58, we can substitute for the sequence components from Equation 54. The result is

$$\begin{aligned} B_{\gamma}^{ab} &= B_{\gamma 1}^{ab} + B_{\gamma 2}^{ab} \\ B_{\gamma}^{bc} &= B_{\gamma 1}^{bc} + B_{\gamma 2}^{bc} \\ B_{\gamma}^{ca} &= B_{\gamma 1}^{ca} + B_{\gamma 2}^{ca} \end{aligned} \tag{70}$$

where $B_{\gamma 1}^{ab}$, and so forth, form the positive-sequence compensating network:

$$B_{\gamma 1}^{ab} = B_{\gamma 1}^{bc} = B_{\gamma 1}^{ca} = -\frac{1}{3}\,[B_{\ell}^{ab} + B_{\ell}^{bc} + B_{\ell}^{ca}]\,, \tag{71}$$

and $B_{\gamma 2}^{ab}$, and so forth, form the negative-sequence compensating network:

$$\begin{aligned} B_{\gamma 2}^{ab} &= \frac{1}{\sqrt{3}}\,(G_{\ell}^{ca} - G_{\ell}^{bc}) + \frac{1}{3}\,(B_{\ell}^{bc} + B_{\ell}^{ca} - 2B_{\ell}^{ab}) \\ B_{\gamma 2}^{bc} &= \frac{1}{\sqrt{3}}\,(G_{\ell}^{ab} - G_{\ell}^{ca}) + \frac{1}{3}\,(B_{\ell}^{ca} + B_{\ell}^{ab} - 2B_{\ell}^{bc}) \\ B_{\gamma 2}^{ca} &= \frac{1}{\sqrt{3}}\,(G_{\ell}^{bc} - G_{\ell}^{ab}) + \frac{1}{3}\,(B_{\ell}^{ab} + B_{\ell}^{bc} - 2B_{\ell}^{ca})\,. \end{aligned} \tag{72}$$

These networks are shown in Figure 15. The negative-sequence compen-

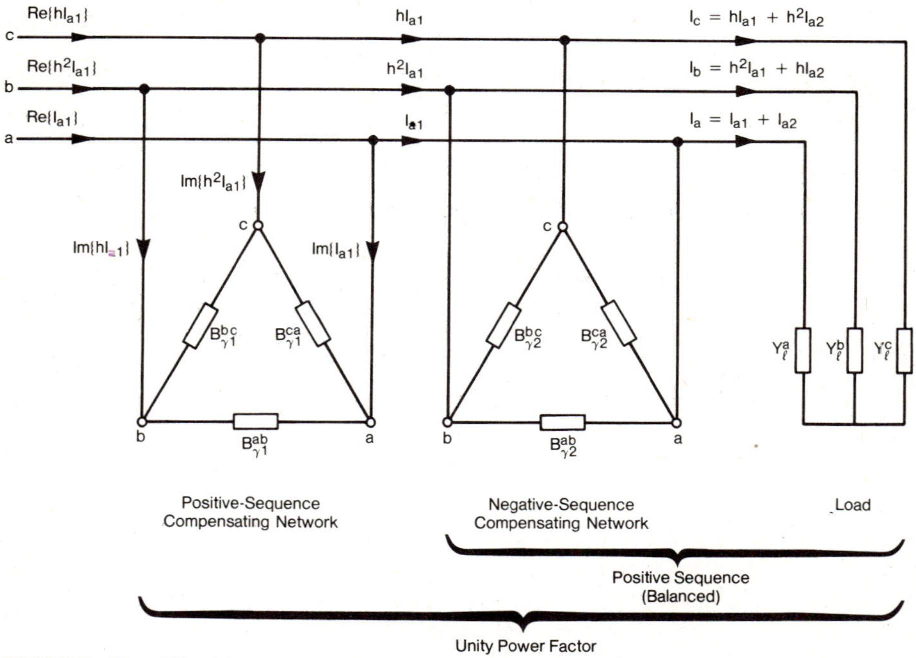

FIGURE 15. The ideal compensating network separated into positive- and negative-sequence components.

sating network carries the negative-sequence currents necessary to balance the load, so that no negative-sequence current flows to or from the supply. The positive-sequence compensating network provides the reactive power compensation necessary to correct the power-factor (on all three phases simultaneously) to unity. As a check on Equations 70, 71, and 72, the sum of the positive- and negative-sequence compensating susceptances agrees with Equation 47.

1.10. CONCLUSION

In order to deal in a practical way with the major functions of power-factor correction, voltage regulation, and load balancing, a theory of load compensation characterizes the compensator variously as a controlled source of reactive power, as a voltage-regulating feedback device, and as a network of desired susceptances.

It has been shown that the purely reactive three-phase compensator can fulfill all the three major functions of power-factor correction, voltage regulation, and phase balancing. The essential compensator characteristics necessary to achieve this have been derived. A reactive compensator with a voltage-regulating control system cannot maintain unity power-factor all the time, but can be biased so that unity *average* power-factor is obtained.

The symmetrical-component analysis of load compensation enables a more general approach to be formulated. It leads to the derivation of ways in which "susceptance-demand" control signals can be physically obtained in an actual compensator. It also provides a basis on which the performance of different types of compensator can be compared under unbalanced conditions.

APPENDIX

A. Wye-Delta Impedance Transformation

In terms of the impedances and admittances marked on Figure A1,

$$\mathbf{Z}_a = \frac{\mathbf{Z}_{ab}\mathbf{Z}_{ca}}{\mathbf{Z}_{ab} + \mathbf{Z}_{bc} + \mathbf{Z}_{ca}} \qquad \text{etc.}$$

$$\mathbf{Y}_{ab} = \frac{\mathbf{Y}_a\mathbf{Y}_b}{\mathbf{Y}_a + \mathbf{Y}_b + \mathbf{Y}_c} \qquad \text{etc.}$$

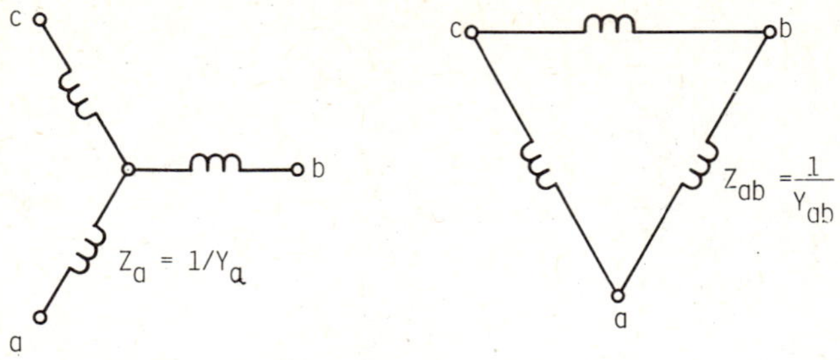

FIGURE A1. Wye-delta impedance transformation.

B. Solution for Compensating Reactive Power to Achieve a Given Terminal Voltage

The problem is to find Q_s from Equation 13 which will make $E = V$. Then $Q_\gamma = Q_s - Q_\ell$.

Rearranging Equation 13,

$$aQ_s^2 + bQ_s + c = 0$$

where

$$a = R_s^2 + X_s^2$$
$$b = 2V^2X_s$$
$$c = (V^2 + R_sP_\ell)^2 + X_s^2P_\ell^2 - E^2V^2\,;$$

therefore

$$Q_s = \frac{-b \pm \sqrt{b^2-4ac}}{2a}$$

C. Solution for Voltage When Real and Reactive Power are Known

From Equation 12,

$$\Delta \mathbf{V} = \frac{PR_s + QX_s}{V} + \mathrm{j}\,\frac{PX_s - QR_s}{V}$$

$$= \frac{A}{V} + \mathrm{j}\,\frac{B}{V}$$

From Equation 13,

$$E^2 = \left(V + \frac{A}{V}\right)^2 + \left(\frac{B}{V}\right)^2$$

$$= V^2 + 2A + \frac{A^2 + B^2}{V^2}$$

This can be rearranged as a simple quadratic equation in V^2.

D. The Symmetrical Components Transformation

The transformation is written with the multiplying factor $1/\sqrt{3}$ in order to ensure power invariance. The transformation is therefore unitary, that is, $C^{-1} = C'^{*}$. In matrix terms, the forward transformation is

$$\begin{bmatrix} \mathbf{I}_{a0} \\ \mathbf{I}_{a1} \\ \mathbf{I}_{a2} \end{bmatrix} = \frac{1}{\sqrt{3}} \begin{bmatrix} 1 & 1 & 1 \\ 1 & \mathbf{h} & \mathbf{h}^2 \\ 1 & \mathbf{h}^2 & \mathbf{h} \end{bmatrix} \begin{bmatrix} \mathbf{I}_a \\ \mathbf{I}_b \\ \mathbf{I}_c \end{bmatrix}$$

where $\mathbf{h} = e^{j2\pi/3}$. The inverse transformation is

$$\begin{bmatrix} \mathbf{I}_a \\ \mathbf{I}_b \\ \mathbf{I}_c \end{bmatrix} = \frac{1}{\sqrt{3}} \begin{bmatrix} 1 & 1 & 1 \\ 1 & \mathbf{h}^2 & \mathbf{h} \\ 1 & \mathbf{h} & \mathbf{h}^2 \end{bmatrix} \begin{bmatrix} \mathbf{I}_{a0} \\ \mathbf{I}_{a1} \\ \mathbf{I}_{a2} \end{bmatrix}$$

E. Expressions for Average Real and Reactive Power in Terms of Instantaneous Voltages and Currents

Taking $\mathbf{V}_a = \mathbf{V}$ as reference phasor, $\mathbf{I}_a = I_{aR} + jI_{aX}$ represents an inductive (lagging) load current in phase "a." The apparent power is

$$\mathbf{S}_a = \mathbf{V}_a\mathbf{I}_a^* = VI_{aR} - jV_aI_{aX} = P_a + jQ_a ,$$

that is,

$$P_a = \text{Re}\,[\mathbf{V}_a\mathbf{I}_a^*]\,;\; Q_a = -\,\text{Im}\,[\mathbf{V}_a\mathbf{I}_a^*]\,.$$

The instantaneous value of voltage in phase "a" is

$$v_a = \sqrt{2}\,V\,\text{Re}[e^{j\omega t}] = \sqrt{2}\,V\cos\omega t\,.$$

Likewise

$$\begin{aligned} i_a &= \sqrt{2}\,\text{Re}\,[(I_{aR} + jI_{aX})\,e^{j\omega t}] \\ &= \sqrt{2}[I_{aR}\cos\omega t - I_{aX}\sin\omega t]\,. \end{aligned}$$

Therefore the instantaneous power in phase "a" is

$$\begin{aligned} v_a i_a &= 2V[I_{aR}\cos^2\omega t - I_{aX}\sin\omega t\cos\omega t] \\ &= V[I_{aR}(\cos 2\omega t + 1) - I_{aX}\sin 2\omega t]\,. \end{aligned}$$

Integrating over one complete cycle of period T, and dividing by T, the average power is

$$P_a = \frac{1}{T}\int_T v_a i_a \, dt = VI_{aR} .$$

Let $v_a(-\pi/2)$ represent v_a phase-shifted through $-90°$ or $\pi/2$ radians. Then

$$v_a(-\pi/2) = \sqrt{2}\, V \cos(\omega t - \pi/2) = \sqrt{2}\, V \sin \omega t .$$

Note also that

$$v_a(-\pi/2) = \sqrt{2}\, V \,\text{Re}\, [e^{j(\omega t - \pi/2)}] = \sqrt{2}\, V \,\text{Re}[-je^{j\omega t}] .$$

Taking the instantaneous product with i_a,

$$\begin{aligned} v_a(-\pi/2)\, i_a &= 2V\, [I_{aR} \cos \omega t \sin \omega t - I_{aX} \sin^2 \omega t] \\ &= V\, [I_{aR} \sin 2\omega t - I_{aX}(1 - \cos 2\omega t)] \end{aligned}$$

Integrating over one complete cycle of period T and dividing by T, the average reactive power is

$$Q_a = \frac{1}{T}\int_T v_a(-\pi/2)\, i_a \, dt = VI_{aX} .$$

The following relationships have been used in Section E:

$$V_{ab} = -j\sqrt{3}\, V_c$$
$$V_{bc} = -j\sqrt{3}\, V_a$$
$$V_{ca} = -j\sqrt{3}\, V_b$$

REFERENCES

1. L. E. Bock and A. H. Moore, "Application of Capacitors in Arc Furnace Power Supply Systems," *Industrial Power Syst.*, 18–23. (September 1976).
2. D. E. Steeper and R. P. Stratford, "Reactive Compensation and Harmonic Suppression for Industrial Power Systems Using Thyristor Converters," *Trans. IEEE* **IA-12**, 232–257. (May/June 1976).
3. P. H. Ashmole, "Technical Summary of Reactive Compensation in Power Systems Industrial Applications," IEE Digest 1979/52, University of Birmingham, England. (11–12 September 1979).
4. L. Gyugyi, R. A. Otto, and T. H. Putman, "Principles and Application of Static Thyristor-Controlled Shunt Compensators," *Trans. IEEE, Power Appar. Syst.* **97**, 1935–1945. (September/October 1978).

Chapter 2

THE THEORY OF STEADY-STATE REACTIVE POWER CONTROL IN ELECTRIC TRANSMISSION SYSTEMS

T. J. E. MILLER

CONVENTIONS AND SYMBOLS

Reactive Power

In accordance with the widely used convention,

1. Reactive power at a generating station is
   positive if generated
   negative if absorbed.
2. Reactive power at a load is
   negative if generated
   positive if absorbed.
3. The receiving end of a transmission line is always treated as a load.

Per-Unit and Ordinary Units

Most equations are given in terms of positive-sequence phase-neutral quantities. The per-unit system is based on rated voltage V_0 and the surge impedance Z_0. Base power is therefore $3V_0^2/Z_0$ watts if V_0 is phase-neutral voltage; or V_0^2/Z_0 if V_0 is phase-phase voltage. In both cases base power is three-phase power.

Lower-case letters are sometimes used for per-unit voltages, currents, reactances, power, and so on. Inductances and capacitances in lower case are usually per-mile values. They may be in per-unit or in ordinary units.

Symmetrical and Radial Lines

The term *radial line* is used to describe a single transmission line with synchronous machines controlling the voltage at the sending end and no voltage control at the receiving end. The *symmetrical* line has identical synchronous machines at both ends, constraining the terminal voltages to be equal.

Emf and Voltage

The symbol E is used for an emf or *controlled* voltage.
The symbol V is used for voltage more generally.

Phasors are denoted by boldface type, for example, **E**

Primes indicate compensated or *virtual* quantities, for example, P_0'

Frequency is assumed to be 60 Hz.

Symbols

a	Line length, mi
B	Shunt susceptance, S
c	Capacitance per mile, F/mi (positive-sequence equivalent, phase-neutral)
E	(controlled) voltage
I	current, A
k_m	Mid-point compensation factor
k_{se}	Degree of series compensation
k_{sh}	Degree of shunt compensation
l	Inductance per mile, H/mi (positive-sequence equivalent, phase-neutral)
P	Power, W
P_0	Natural load or *Surge-Impedance load* (*SIL*), W
P_{max}	Maximum transmissible power, W
Q	Reactive power, VAr
s	Series compensation factor
V	Voltage

X	Reactance, Ω
Y	Admittance, S
Z	Impedance, Ω
Z_0	Characteristic or Surge Impedance, Ω
β	Wavenumber, radian/mi
δ	Transmission Angle, radian or degree
θ	Electrical length of line, radian
Γ	Propagation constant

Subscripts

r	Receiving end
s	Sending end
c	Capacitive
l	Inductive
γ	Compensator
0	Natural or characteristic

2.1. INTRODUCTION

2.1.1. Historical Background

The economics of ac power transmission have always forced the planning engineer to transmit as much power as possible through a given transmission line. Today, however, additional constraints loom much larger than they did in the past. First, the dependence of load centers on the continuity of electrical supplies has become more critical (as witnessed by events during the North-East power blackout of 1965 and the New York City blackout of 1977[(6)]). This means that the security, or reliability, of transmission circuits has needed to be continuously improved. Modern compensation methods have helped to make these improvements possible. Second, there has been extensive development of remote hydroelectric resources, such as the El Chocon-Cerros Colorados complex in Argentina, 1000 km from Buenos Aires, and the James Bay scheme in Québec, 1000 km from Montréal and Québec City.[(16)] Both these ac schemes are characterized not only by the long distances, but also by the large amounts of power to be transmitted (over 11,000 MW in the case of the James Bay scheme). The development of compensation schemes has helped to make ac transmission technically and economically competitive even in an age when the dc transmission alternative has made great strides also.

A third planning constraint has been the difficulty of acquiring right-of-way for new transmission circuits (the so-called corridor crisis). Increased pressure to maximize the utilization of both new and existing lines has helped to motivate the development and application of compensation systems.

This chapter begins with an account of the fundamental requirements in ac power transmission, *stability* and *voltage control.* The principles of operation of the main types of compensation are then studied in a generalized way, assuming that the compensation is uniformly *distributed* along a single transmission line. Later, in Sections 2.4, 2.5, and 2.6, detailed attention is given to *lumped* shunt, series, and dynamic shunt compensation respectively. The effects of each of these types of compensation on voltage control, reactive power requirements, and the steady-state stability limit are systematically explored.

2.1.2. Fundamental Requirements in ac Power Transmission

Bulk transmission of electrical power by ac is possible only if the following two fundamental reqirements are satisfied:

1. *Major synchronous machines must remain stably in synchronism.*

The *major* synchronous machines in a transmission system are the generators and synchronous condensers, all of which are incapable of operating usefully other than in synchronism with all the others.†

The central concept in the maintenance of synchronism is *stability.* Stability is the tendency of the power system (and of the synchronous machines in particular) to continue to operate steadily in the intended mode.‡ It is also a measure of the inherent ability of the system to recover from extraneous disturbances (such as faults, lightning, and changes of load), as well as from planned disturbances (such as switching operations).

One of the limits to the utilization of a transmission line is that for a given length of line the stability tends to become less as the transmitted power is increased. If the power could be gradually increased (with no extraneous disturbances), a level would be reached at which the system

† Synchronous motors are usually (but not always) smaller than even the smallest generators. They are not usually considered individually in the study and planning of the bulk transmission system, even though they can cause severe local disturbances as a result of loss of synchronism. Sometimes the major synchronous machines in one power station or in one region are grouped together and treated as one machine, in order to simplify analysis of the system as a whole. Each group is called a *dynamic equivalent* group, and must remain in synchronism with all the other connected groups.

‡ The intended or normal mode is that in which power and reactive power flows have their intended values, while voltages and currents and the mechanical phase angles between synchronous-machine rotors are all constant.

would suddenly become unstable. The synchronous machines at the two ends of the line would pull out of step, that is, lose synchronism. This level of power transmission is the *steady-state stability limit,* so called because it is the maximum steady power that can (in theory) be transmitted stably. The steady-state stability limit is not a hard and fast number fixed forever by the design of the synchronous machines and the transmission equipment. It can be considerably modified by several factors. Among the most important are the excitation of the synchronous machines (and therefore the line voltage); the number and connection of the transmission lines; the number and types of synchronous machines connected (which frequently change with the time of day); the pattern of real and reactive power flows in the system; and, of central interest here, the connection and characteristics of compensation equipment.

It is not practical to operate a transmission system too near to its steady-state stability limit; there must be a margin in the power transfer to allow for disturbances (such as load changes, faults, and switching operations). In determining an appropriate margin, the concepts of *transient* and *dynamic stability* are useful. A transmission system is said to be *dynamically* stable if it recovers normal operation following a specified *minor* disturbance. The degree of dynamic stability can be expressed in terms of the rate of damping of the transient components of voltages, currents, and the load angles of the synchronous machines. The rate of damping, or settling, is the principal interest in a dynamic stability study. Accordingly, modern calculations are usually based on small-perturbation theory and eigenanalysis.

A third major concern in the stability of power transmission systems is whether the system will recover normal operation following a major disturbance, such as a fault severe enough to trip a major circuit, or failure of a major item of plant, such as a generator, overhead line, or transformer. This is the so-called question of *transient stability.* A system has transient stability if it can recover normal operation following a specified *major* disturbance. Whether recovery is possible depends, among other factors, on the level of power transmission immediately before the disturbance occurred. The *transient stability limit* is the highest level of prior power transmission for which the system has transient stability following the specified disturbance.†

2. *Voltages must be kept near to their rated values.*

The second fundamental requirement in ac power transmission is the maintenance of correct voltage levels. Modern power systems are not very tolerant of abnormal voltages, even for short periods.

† For a fuller discussion of these aspects of power system stability, see References 7 through 9.

Undervoltage, which is generally associated with heavy loading and/or a shortage of generation, causes degradation in the performance of loads, particularly induction motors. In heavily loaded systems, undervoltage may be an indication that the load is approaching the steady-state stability limit. Sudden undervoltages can result from the connection of very large loads.

Overvoltage is a dangerous condition because of the risk of flashover or the breakdown of insulation. Saturation of transformers subjected to overvoltage can produce high currents rich in harmonics, and in the presence of sufficient capacitance there is a risk of ferroresonance as well as of harmonic resonances. Overvoltages arise from several causes. The reduction of load during certain parts of the daily load cycle causes a gradual voltage rise. Uncontrolled, this overvoltage would shorten the useful life of insulation even if the breakdown level were not reached. Sudden overvoltage can result from disconnection of loads or other equipment, while overvoltages of extreme rapidity and severity can be caused by line switching operations, faults, and lightning. In long-distance transmission systems, the Ferranti effect (overvoltage at light load) would limit the power transfer and the transmission distance if no compensating measures were taken.

2.1.3. Engineering Factors Affecting Stability and Voltage Control

The design of virtually every item of plant in a transmission system has a bearing on at least one of the fundamental requirements discussed above. The general study of power system voltage control and stability is too vast a subject for this chapter, which is concerned only with compensation techniques. In view of this, the broad picture of power system voltage control and stability is briefly summarized in Tables 1 and 2. (See also the References at the end of this chapter).

In Table 1 the main problems or applications of compensating equipment are grouped under the two fundamental transmission requirements discussed previously, and it can be seen that most of the special-purpose compensating equipments have a role to play under several headings. This makes the general subject of the deployment of compensating equipment a rather complicated one, and the literature is correspondingly extensive. In this chapter emphasis is laid on the theory of what can be achieved using (mainly) series capacitors, shunt capacitors and reactors, polyphase saturated reactors, and thyristor-controlled compensators. In the next section the basic compensation requirements are defined by looking at the *uncompensated* transmission line.

PROBLEM, APPLICATION, OR PURPOSE		EXISTING EQUIPMENT OR METHOD: Increase transmission voltage	EXISTING EQUIPMENT OR METHOD: Increase No. of lines in parallel	EXISTING EQUIPMENT OR METHOD: Transformer tapchanging	EXISTING EQUIPMENT OR METHOD: Slow AVR control	EXISTING EQUIPMENT OR METHOD: Fast AVR control	EXISTING EQUIPMENT OR METHOD: Fast turbine valving	EXISTING EQUIPMENT OR METHOD: Rapid line-switching operations, reclosing of circuit-breakers	EXISTING EQUIPMENT OR METHOD: Braking resistors	SPECIAL-PURPOSE COMPENSATING EQUIPMENT: Shunt reactor (switched/unswitched; linear/nonlinear)	SPECIAL-PURPOSE COMPENSATING EQUIPMENT: Shunt capacitor	SPECIAL-PURPOSE COMPENSATING EQUIPMENT: Series reactor	SPECIAL-PURPOSE COMPENSATING EQUIPMENT: Series capacitor	SPECIAL-PURPOSE COMPENSATING EQUIPMENT: Synchronous condenser	SPECIAL-PURPOSE COMPENSATING EQUIPMENT: Polyphase saturated reactor*	SPECIAL-PURPOSE COMPENSATING EQUIPMENT: Thyristor controlled reactor*	SPECIAL-PURPOSE COMPENSATING EQUIPMENT: Thyristor switched capacitor	SPECIAL-PURPOSE COMPENSATING EQUIPMENT: Short-circuit limiting coupling (or Fault current limiter)
FUNDAMENTAL REQUIREMENT #1	Improve steady-state stability	•	•		•	•				•	•		•	•	•	•	•	
	Improve dynamic stability					•								•		•	•	
	Improve transient stability	•				•	•	•	•		•		•	•	•	•	•	
FUNDAMENTAL REQUIREMENT #2	Limit rapid voltage decline					•		•			•			•	•	•	•	
	Limit slow voltage decline			•	•						•			•	•	•	•	
	Limit rapid voltage increase					•		•		•				•	•	•	•	
	Limit slow voltage increase			•	•					•				•	•	•	•	
	Limit fast-wavefront overvoltages due to lightning, switching, etc.									•				•	•	•		
OTHER REQUIREMENTS	Reactive power support at dc converter terminals										•			•	•	•	•	
	Increase short-circuit level												•	•				
	Decrease short-circuit level											•						•

*in conjunction with shunt capacitor where necessary

TABLE 2
Advantages and Disadvantages of Different Types of Compensating Equipment for Transmission Systems

Compensating Equipment	Advantages	Disadvantages
Switched shunt reactor	Simple in principle and construction	Fixed in value
Switched shunt capacitor	Simple in principle and construction	Fixed in value Switching transients
Series capacitor	Simple in principle Performance relatively insensitive to location	Requires overvoltage protection and subharmonic filters Limited overload capability
Synchronous condenser	Has useful overload capability Fully controllable Low harmonics	High maintenance requirement Slow control response Performance sensitive to location Requires strong foundations
Polyphase saturated reactor[a]	Very rugged construction Large overload capability No effect on fault level Low harmonics	Essentially fixed in value Performance sensitive to location Noisy
Thyristor-controlled reactor[a] (TCR)	Fast response Fully controllable No effect on fault level Can be rapidly repaired after failures	Generates harmonics Performance sensitive to location
Thyristor-switched capacitor (TSC)	Can be rapidly repaired after failures No harmonics	No inherent absorbing capability to limit overvoltages Complex buswork and controls Low frequency resonances with system Performance sensitive to location

[a] With shunt capacitors where necessary.

2.2. UNCOMPENSATED TRANSMISSION LINES

2.2.1. Electrical Parameters

A transmission line is characterized by four *distributed* circuit parameters: its series resistance r and inductance l, and its shunt conductance g and capacitance c, the lower-case symbols indicating per-mile values. All four parameters are functions of the line design, that is, of the conductor size, type, spacing, height above ground, frequency, and temperature. They also vary according to the number of nearby parallel lines, and different values are obtained for positive-sequence and zero-sequence currents. Some typical values are given in Table 3.

The characteristic behavior of the line is dominated by the series inductance and the shunt capacitance. Series resistance has a secondary but not insignificant influence, and has a separate importance in determining losses. In this chapter it is largely ignored. Positive-sequence nominal values are assumed, and shunt conductance is ignored. Balanced conditions are assumed except where stated, and one phase of the positive-sequence equivalent circuit is used.

Figure 1 shows a lumped-parameter equivalent circuit of one phase of a transmission line, having identical synchronous machines connected at both ends. Such a line is called *symmetrical.*

2.2.2. Fundamental Transmission Line Equation

The fundamental equation governing the propagation of energy along a transmission line is the wave equation

$$\frac{\mathrm{d}^2\mathbf{V}}{\mathrm{d}x^2} = \Gamma^2\mathbf{V} \text{ with } \Gamma^2 = (r + \mathrm{j}\omega l)(g + \mathrm{j}\omega c)\,. \tag{1}$$

Frequency is assumed fixed, and $\mathbf{V}$ is the phasor voltage $\hat{v}\mathrm{e}^{\mathrm{j}\omega t}/\sqrt{2}$ at any point on the line. The phasor current $\mathbf{I}$ satisfies the equation also. (For derivation see Reference 10). Since x is distance along the line measured from any convenient reference point, the equation describes the variation of the voltage $\mathbf{V}$ and the line current $\mathbf{I}$ along the line, and it implies that both will have a wavelike or sinusoidal variation.

Solution of the transmission line wave equation: Standing Waves. If the line is assumed lossless, the general solution of Equation 1 (for voltage and current) is

$$\mathbf{V}(x) = \mathbf{V}_{\mathrm{r}} \cos\beta(a - x) + \mathrm{j}Z_0\mathbf{I}_{\mathrm{r}} \sin\beta(a - x) \tag{2a}$$

TABLE 3
Typical EHV-UHV Transmission Line Parameters[a,b]

	Nominal Voltage (kV)									
	345		500			765		1100		1500
	Conductor Configuration									
Characteristics	Hor.	Delta	Hor.	Vert.	Delta	Hor.	Delta	Hor.	Delta	Hor.
ωl (Ω/mi)	0.59	0.59	0.60	0.49	0.51	0.53	0.53	0.48	0.47	0.47
r (Ω/mi)	0.060	0.060	0.025	0.026	0.019	0.020	0.020	0.0079	0.0076	0.0072
ωc (μS/mi)	7.27	7.29	7.23	8.80	8.35	8.02	8.01	8.94	8.92	9.21
β (rad/mi $\times 10^{-3}$)	2.07	2.07	2.08	2.08	2.06	2.06	2.06	2.07	2.05	2.08
Surge impedance Z_0 (Ω)	285	283	287	235	247	258	257	232	231	225
Natural load P_0 (MW)	417	420	870	1060	1010	2270	2280	5220	5250	10000
Charging MVA (MVA/mi)	0.866	0.868	1.81	2.20	2.09	4.70	4.70	10.8	10.8	20.7
Line current at SIL (A)	700	700	1000	1230	1170	1710	1720	2740	2750	3850

[a] Reproduced by kind permission of J.J. LaForest, Electric Utility Systems Engineering Department.

[b] This table contains typical positive-sequence characteristics of transmission lines from 345 to 1500 kV, calculated at nominal voltage, 60 Hz.

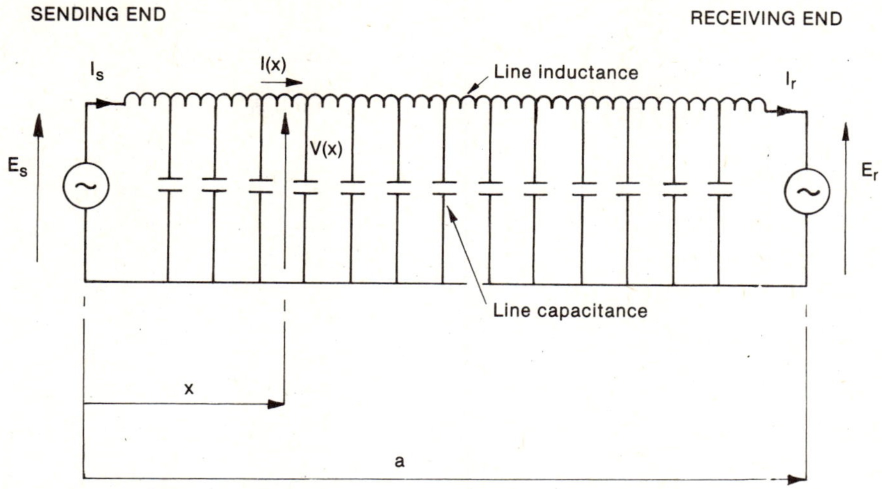

FIGURE 1. Lumped-element representation of a long transmission line.

$$\mathbf{I}(x) = \mathrm{j}\left(\frac{\mathbf{V}_r}{Z_0}\right) \sin\beta\,(a - x) + \mathbf{I}_r \cos\beta\,(a - x)\,, \tag{2b}$$

where β is derived from the propagation constant Γ by putting $r = g = 0$. This gives $\Gamma = \mathrm{j}\beta$ and

$$\beta = \omega\sqrt{lc}\ . \tag{3}$$

The form of Equation 2 shows the expected sine wave variation of **V** and **I** along the line, each quantity having two terms or components. **V** and **I** are said to form *standing waves* because of the sinusoidal variation of both their real and imaginary parts along the line.

The quantity 1 shilling fraction $\sqrt{lc}$ is the propagation velocity of electromagnetic effects along the line. For overhead high-voltage transmission lines it has a value somewhat less than the velocity of light, $u = 3 \times 10^8$ m/sec = 186,000 mi/sec. Since also $\omega = 2\pi f$, Equation 3 gives

$$\beta = \frac{2\pi f}{u} = \frac{2\pi}{\lambda}\,, \tag{4}$$

where λ is the wavelength. β is the wave number, that is, the number of complete waves per unit of line length.

At 60 Hz λ = 3100 mi and β can be expressed as one wavelength per 3100 mi, that is, 360° per 3100 mi, or 0.116° per mi, or 2.027×10^{-3} rad/mi. The quantity βa is the *electrical length* of the line expressed in radians or in wavelengths: symbol θ.

2.2.3. Surge Impedance and Natural Loading

The constant Z_0 in Equation 2 is the *surge impedance* (sometimes called the characteristic impedance):

$$Z_0 = \sqrt{\frac{l}{c}}. \tag{5}$$

Its value depends on the line design (see Section 2.2.1 and Table 3). For high-voltage overhead lines, the positive-sequence value typically lies in the range 200–400 Ω.

If losses are neglected the line is characterized entirely by its length and by the two parameters Z_0 and β. Since these values are roughly comparable for all lines, the behavior of all lines is fundamentally the same, and differences only arise according to the length, the voltage, and the level of power transmission.

The surge impedance is the apparent impedance of an infinitely long line, that is, the ratio of voltage to current at any point along it. A line of finite length terminated at one end by an impedance Z_0 is electrically indistinguishable from an infinite line, so that if $\mathbf{V}_r/\mathbf{I}_r = Z_0$ then from Equation 2 the apparent impedance at any point is

$$\mathbf{Z}(x) = \frac{\mathbf{V}(x)}{\mathbf{I}(x)} = \frac{Z_0\mathbf{I}_r[\cos\beta(a-x) + \mathrm{j}\sin\beta(a-x)]}{\mathbf{I}_r[\cos\beta(a-x) + \mathrm{j}\sin\beta(a-x)]} = Z_0 \tag{6}$$

which is independent of x. More importantly,

$$\mathbf{V}(x) = \mathbf{V}_r[\cos\beta(a-x) + \mathrm{j}\sin\beta(a-x)] = \mathbf{V}_r e^{\mathrm{j}\beta(a-x)} \tag{7a}$$

$$\mathbf{I}(x) = \mathbf{I}_r[\cos\beta(a-x) + \mathrm{j}\sin\beta(a-x)] = \mathbf{I}_r e^{\mathrm{j}\beta(a-x)}, \tag{7b}$$

that is, both **V** and **I** have constant amplitude along the line. The line is said to have a *flat voltage profile.* While **V** and **I** are in phase with each other all along the line, both are rotated in phase. The phase angle between sending-end and receiving-end quantities is implicit in Equation 7; it is $\theta = \beta a$ rad. For a 200-mi line at 60 Hz the angle is 0.405 rad or 23.2°. The phasor relationships are shown in Figure 2.

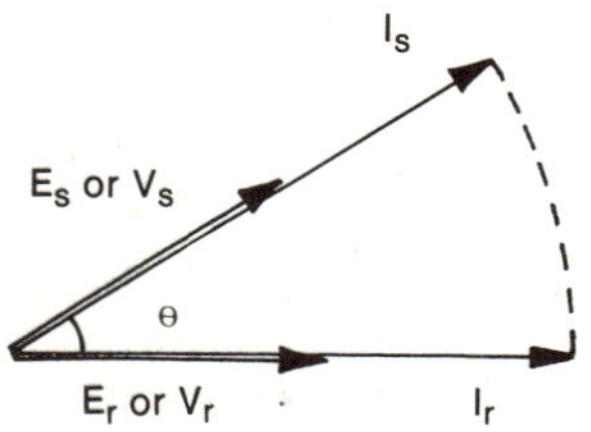

FIGURE 2. Phasor diagram of naturally loaded line.

A line in this condition is said to be *naturally loaded.* The natural load (or *surge-impedance load,* SIL) is

$$P_0 = \frac{V_0^2}{Z_0}, \tag{8}$$

where V_0 is the nominal or rated voltage of the line. If V_0 is the line-to-neutral voltage, Equation 8 gives the per-phase value of surge-impedance power; if V_0 is the line-to-line voltage, P_0 is the three-phase value. The natural load is an important reference quantity which will be used extensively below.

An advantage of operating the line at the natural load is that because of the flat voltage profile, the insulation is uniformly stressed at all points.

The natural load of the uncompensated line increases with the *square* of the voltage (Equation 8). This helps to explain why transmission voltages have increased as the level of transmitted power has grown. Table 3 shows the natural load for some common line voltages.

The surge impedance Z_0 is a real number. Therefore, at the natural load the power factor — that is, the cosine of the angle between $\mathbf{V}$ and $\mathbf{I}$ — is unity at all points along the line, including the ends. This is apparent from Equation 6. It means that at the natural load *no reactive power has to be absorbed or generated at either end.* The reactive power generated in the shunt capacitance of the line is exactly absorbed by the series inductance. This important condition can be further explained as follows. In any short element of the line the reactive power per unit length generated by the shunt capacitance is $V^2 b = V^2 \omega c$, while the reactive power per unit length absorbed by the series inductance is $I^2 \omega l$. For reactive-power balance in this element of line,

$$V^2 \omega c = I^2 \omega l ,$$

i.e.,

$$\frac{V}{I} = \sqrt{\frac{l}{c}} = Z_0 . \tag{9}$$

This must be true at all points along the line, including the sending and the receiving end. Therefore, reactive power balance is achieved at the natural loading, with $P_0 = V^2/Z_0$. This is the only value of transmitted power that gives a flat voltage profile and unity power factor at both ends of the line.

In the sense that P_0 is the natural *power* of the line, the "natural" *reactive* power is zero.

2.2.4. The Uncompensated Line on Open-Circuit

Voltage and Current Profiles. A lossless line that is energized by generators at the sending end and is open-circuited at the receiving end is described by Equations 2a and b with $\mathbf{I}_r = 0$, so that

$$\mathbf{V}(x) = \mathbf{V}_r \cos\beta(a - x) \tag{10a}$$

and

$$\mathbf{I}(x) = j\left(\frac{\mathbf{V}_r}{Z_0}\right) \sin\beta(a - x)\ . \tag{10b}$$

The voltage and current at the sending end are given by these equations with $x = 0$:

$$\mathbf{E}_s = \mathbf{V}_r \cos\theta\ ; \tag{11a}$$

$$\mathbf{I}_s = j\left(\frac{\mathbf{V}_r}{Z_0}\right) \sin\theta = j\left(\frac{\mathbf{E}_s}{Z_0}\right) \tan\theta\ . \tag{11b}$$

$\mathbf{E}_s$ and $\mathbf{V}_r$ are in phase, which is consistent with the fact that there is no power transfer. (See Section 2.2.6). The phasor diagram is shown in Figure 3.

The line *voltage profile* expressed by Equation 10a can be written more conveniently in terms of $\mathbf{E}_s$:

$$\mathbf{V}(x) = \mathbf{E}_s \frac{\cos\beta(a - x)}{\cos\theta}\ . \tag{12a}$$

Similarly the *current profile* is given by

$$\mathbf{I}(x) = j\,\frac{\mathbf{E}_s}{Z_0}\,\frac{\sin\beta(a - x)}{\cos\theta}\ . \tag{12b}$$

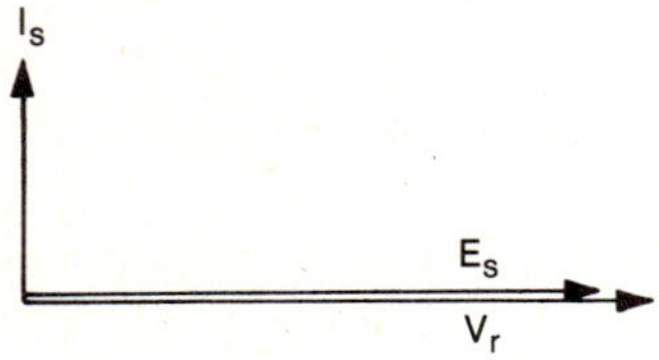

FIGURE 3. Phasor diagram of 200-mi line open-circuit at the receiving end.

These profiles are shown in Figure 4 for a line 200 miles in length, for which at 60 Hz $\theta = 0.405$ radian $= 23.2°$. With $E_s = 1.0$ pu the receiving-end voltage is $V_r = 1.088$ pu, that is a rise of 8.8%. This rise is called the *Ferranti effect.*

A rise of 8.8% is not enough to cause severe problems for insulation or for voltage regulating equipment. But at 400 mi the open-circuit voltage would be 1.579 pu which is unacceptable, if not dangerous. At 775 mi (one quarter-wavelength) the voltage rise would be infinite; operation of such a line is completely impractical without some means of compensation.

In practice the open-circuit voltage rise will be greater than is indicated by Equation 11a, which assumes that the sending-end voltage is fixed. Following a sudden open-circuiting of the line at the receiving end, the sending-end voltage tends to rise immediately to the *open-circuit voltage* of the sending-end generators, which exceeds the terminal voltage by approximately the voltage drop due to the prior current flowing in their short-circuit reactances. In spite of its practical importance, this complication will not be considered further, except to note that typically it is desirable to limit the open-circuit voltage rise to about 25% at the sending end and about 40% at the receiving end under worst conditions—that is,

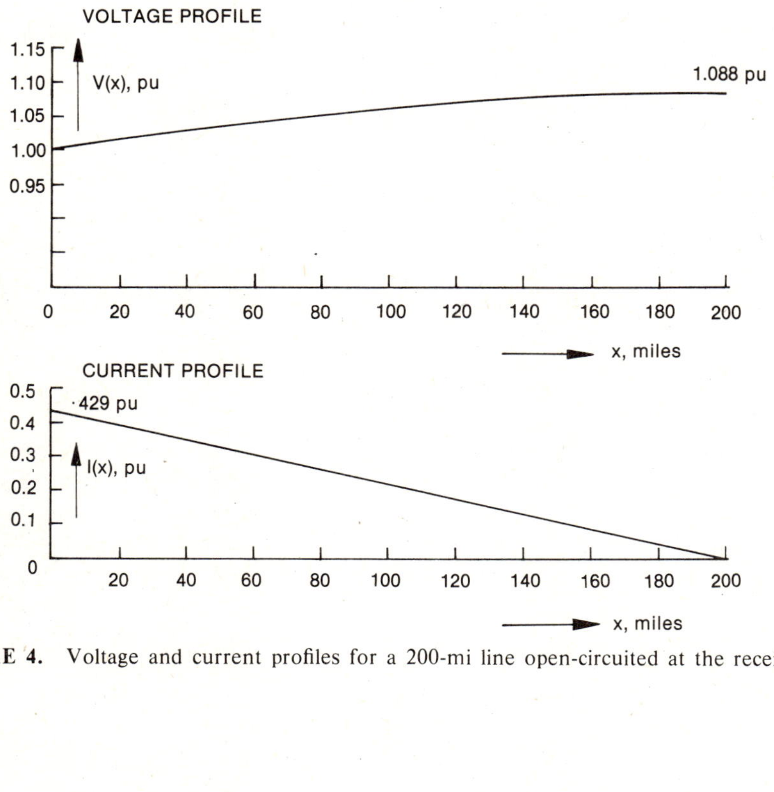

FIGURE 4. Voltage and current profiles for a 200-mi line open-circuited at the receiving end.

with all parallel lines connected, minimum generators connected, and before the excitation on the generators has been reduced to bring the voltage down to a safe level.

The magnitude of $\mathbf{I}_s$ in Figure 3 is 0.429 pu. This means that the *line-charging current* flowing in the sending-end generators is 42.9% of the current corresponding to the natural load.

The Symmetrical Line at No-Load. Akin to the open-circuited line energized from one end is the symmetrical line at no load. This is a line with identical synchronous machines at both ends, but no power transfer. Suppose that the terminal voltages are controlled to have the same *magnitude,* that is, $E_s = E_r$. From Equations 2a and b, with $x = 0$,

$$\mathbf{E}_s = \mathbf{E}_r \cos\theta + jZ_0\mathbf{I}_r \sin\theta \; ; \tag{13a}$$

$$\mathbf{I}_s = j\left(\frac{\mathbf{E}_r}{Z_0}\right) \sin\theta + \mathbf{I}_r \cos\theta \, . \tag{13b}$$

With no power transfer the electrical conditions are the same at both ends. Therefore by *symmetry*†

$$\mathbf{I}_s = -\mathbf{I}_r \, . \tag{14}$$

From Equation 13b,

$$-\mathbf{I}_r = j\,\frac{\mathbf{E}_r}{Z_0}\,\frac{\sin\theta}{1+\cos\theta} = j\frac{\mathbf{E}_r}{Z_0}\tan\frac{\theta}{2} \, . \tag{15}$$

Substituting this value for $\mathbf{I}_r$ in Equation 13a gives

$$\mathbf{E}_s = \mathbf{E}_r \tag{16}$$

and therefore

$$\mathbf{I}_s = j\,\frac{\mathbf{E}_s}{Z_0}\tan\frac{\theta}{2} \, . \tag{17}$$

Equation 16 shows that $\mathbf{E}_r$ and $\mathbf{E}_s$ are in phase, which again is consistent with the fact that there is no power transfer. The current at each end is line-charging current. Comparison of Equations 15 and 17 with Equation 11b shows that the line is equivalent to two equal halves connected back-to-back. Half the line-charging current is supplied from each end.[1] The phasor diagram is shown in Figure 5 for $a = 200$ mi, with $E_s = E_r = V_0 = 1.0$ pu.

† The negative sign in Equation 14 arises because of the convention in which positive current flows away from the sending end and towards the receiving end.

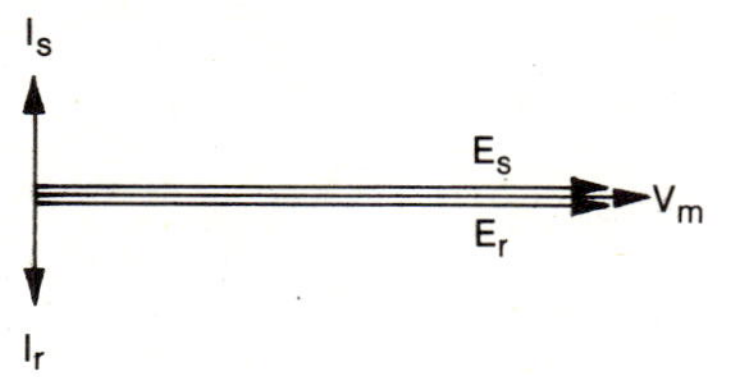

FIGURE 5. Phasor diagram of 200-mi symmetrical line.

By symmetry the midpoint current is zero. The midpoint voltage is, therefore, equal to the open-circuit voltage of a line having half the total length:

$$V_m = \frac{E_s}{\cos(\theta/2)} \ . \tag{18}$$

The voltage and current profiles for the symmetrical line at no load can be derived from Equations 12a and b with a replaced by $a/2$:

$$\mathbf{V}(x) = \mathbf{E}_s \frac{\cos \beta(a/2 - x)}{\cos(\theta/2)} \tag{19a}$$

and

$$\mathbf{I}(x) = \mathrm{j}\, \frac{\mathbf{E}_s}{Z_0} \frac{\sin \beta(a/2 - x)}{\cos(\theta/2)} \tag{19b}$$

for $x \leqslant a/2$. For the other half of the line, that is, $a/2 \leqslant x \leqslant a$,

$$\mathbf{V}(x) = \mathbf{V}(a - x) \tag{19c}$$

and

$$\mathbf{I}(x) = -\,\mathbf{I}(a - x) \ . \tag{19d}$$

The profiles are shown in Figure 6. It is interesting to compare these with Figure 4.

If $E_s \neq E_r$ the current and voltage profiles are no longer symmetrical and the highest voltage is no longer at the midpoint, but is nearer to the end of the line which has the higher terminal voltage. The currents in the synchronous machines are also unequal.

Underexcited Operation of Generators Due to Line-Charging. With $\mathbf{I}_r = 0$ the charging reactive power at the sending end is given by

$$Q_s = \mathrm{Im}\,(\mathbf{E}_s \mathbf{I}_s^*) \ . \tag{20}$$

The line-charging current is given by Equation 11b, so that if E_s is equal to the rated voltage of the line,

$$Q_s = -\,P_0 \tan\theta \ . \tag{21}$$

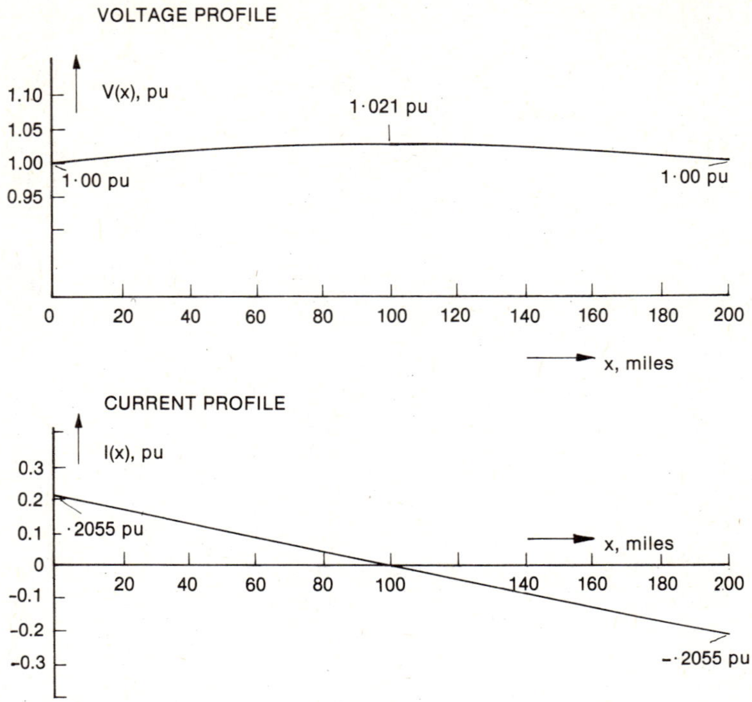

FIGURE 6. Voltage and current profiles for a 200-mi symmetrical line.

The charging current leads the line terminal voltage by 90° and flows in the generators. For the 200-mi line $I_s = 0.429$ pu and so Q_s is nearly 43% of the natural load expressed in MVA. At 400 kV the generators would have to absorb 172 MVAr.

The reactive power absorption capability of synchronous generators is limited for two reasons. First, underexcited operation increases the heating of the ends of the stator core. Second, the reduced field current reduces the internal emf of the generators, and this impairs stability (see later). The absorption limit is typically not more than 0.45 pu of the MVA rating. In the 200-mi example line, if the total MVA rating of the synchronous generators is equal to the natural load, the charging reactive power at 0.43 pu would be just within the limit. But if (for economy) half the generators were disconnected, the load being small or zero, the remainder would have to absorb 0.86 pu of their MVA capacity, which is definitely above the limit.

Aside from using compensation, there are two main ways in which this problem can sometimes be alleviated. First, if the line is made up of two or more parallel circuits, one or more of the circuits can be switched off under light-load or open-circuit conditions. This is permissible only if the consequent reduction in security of supply at the receiving end is acceptable. Second, if the generator absorption is limited by stability and not by core-end heating, the absorption limit can be increased by using a rapid-response excitation system which restores the stability margins when the steady-state field current is low.

The underexcited operation of generators can set a more stringent limit to the maximum length of an uncompensated line than the open-circuit voltage rise. Let the total generator rating be P_g and let their maximum reactive power absorption be $q_u P_g$. This must not be less than the line-charging reactive power given by Equation 21. It follows that the generating capacity must satisfy the relation

$$P_g \geqslant \frac{P_0 \tan\theta}{q_u} . \tag{22}$$

If, for instance, $q_u = 0.3$ and the sending-end generating capacity is $P_g = P_0$, the maximum length of uncompensated line is only 144 mi. Alternatively a line 200 mi long would require $P_g \geqslant 1.43\ P_0$ if q_u were limited to 0.3. It would generally be wasteful to have so much excess generating capacity connected (or even installed) merely in order to satisfy the line-charging requirement. It is better to satisfy this requirement by means of compensation. Shunt reactors, synchronous condensers, or static compensators can be connected at the receiving end or at points along the line. Their ratings and points of connection should ideally be optimized and coordinated with other equipment to achieve satisfactory control of line voltage under all conditions, as well as to relieve the generators of excessive reactive power absorption.

2.2.5. The Uncompensated Line Under Load: Effect of Line Length, Load Power, and Power Factor on Voltage and Reactive Power

Radial Line with Fixed Sending-end Voltage. A load $P + jQ$ at the receiving end of a transmission line draws the current

$$\mathbf{I}_r = \frac{P - jQ}{\mathbf{V}_r} . \tag{23}$$

From Equation 2a with $x = 0$, if the line is assumed lossless the sending- and receiving-end voltages are related by

$$\mathbf{E}_s = \mathbf{V}_r \cos\theta + j\mathbf{Z}_0 \sin\theta \, \frac{P - jQ}{\mathbf{V}_r} . \tag{24}$$

If $\mathbf{E}_s$ is fixed, this quadratic equation can be solved for $\mathbf{V}_r$. The solution shows how $\mathbf{V}_r$ varies with the load and its power factor and with the line length. A typical result is shown in Figure 7, for which a = 200 mi. The magnitude V_r is plotted against the normalized load power P/P_0 for five different power factors, with $E_s = V_0 = 1.0$ pu.

Several fundamentally important properties of ac transmission are evident from Figure 7. For each load power factor there is a *maximum transmissible power.* (See Section 2.5). For any value of P below the maximum there are two possible solutions for V_r (i.e., two roots of Equation 24). Normal operation of the power system is always at the upper value, within narrow limits around 1.0 pu. When $P = Q = 0$, Equation 24 reduces to Equation 11a for the open-circuit condition. Also apparent in Figure 7 is the flat voltage profile achieved at unity power factor when $P = P_0$, that is, $V_r = E_s$.

The load power factor has a strong influence on the receiving-end voltage. Loads with lagging power factor, with unity, or with very high leading power factor, tend to reduce V_r as the load P increases. With leading power factors (except those very near unity), the tendency is to increase V_r until P reaches a much higher value. Leading power factor loads generate reactive power which supplements the line-charging reactive power and tends to support the line voltage.

The effect of the line length can be determined by redrawing Figure 7 for different values of a. Figures 8a through c show the results for three different power factors with a = 100, 200, 300, 400, and 500 mi. It appears from Figure 8 that uncompensated lines between about 100 and 200 mi long can be operated at normal voltage provided that the load power factor is high. Because of their large voltage variations, longer lines are impractical at all power factors unless some means of voltage

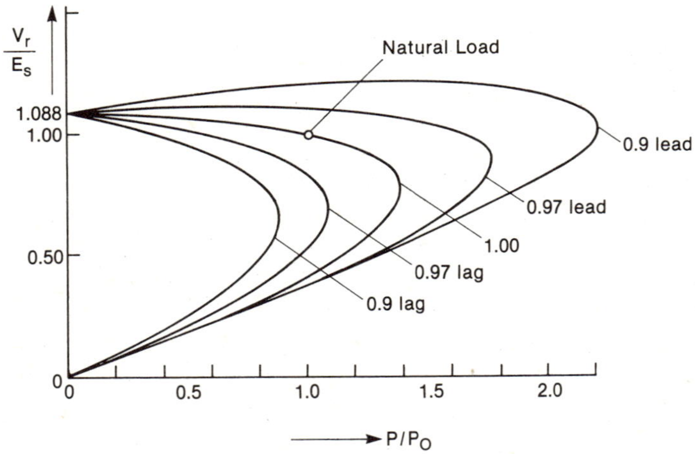

FIGURE 7. Receiving-end voltage magnitude as a function of load (P) and load power factor for a 200-mi lossless radial line.

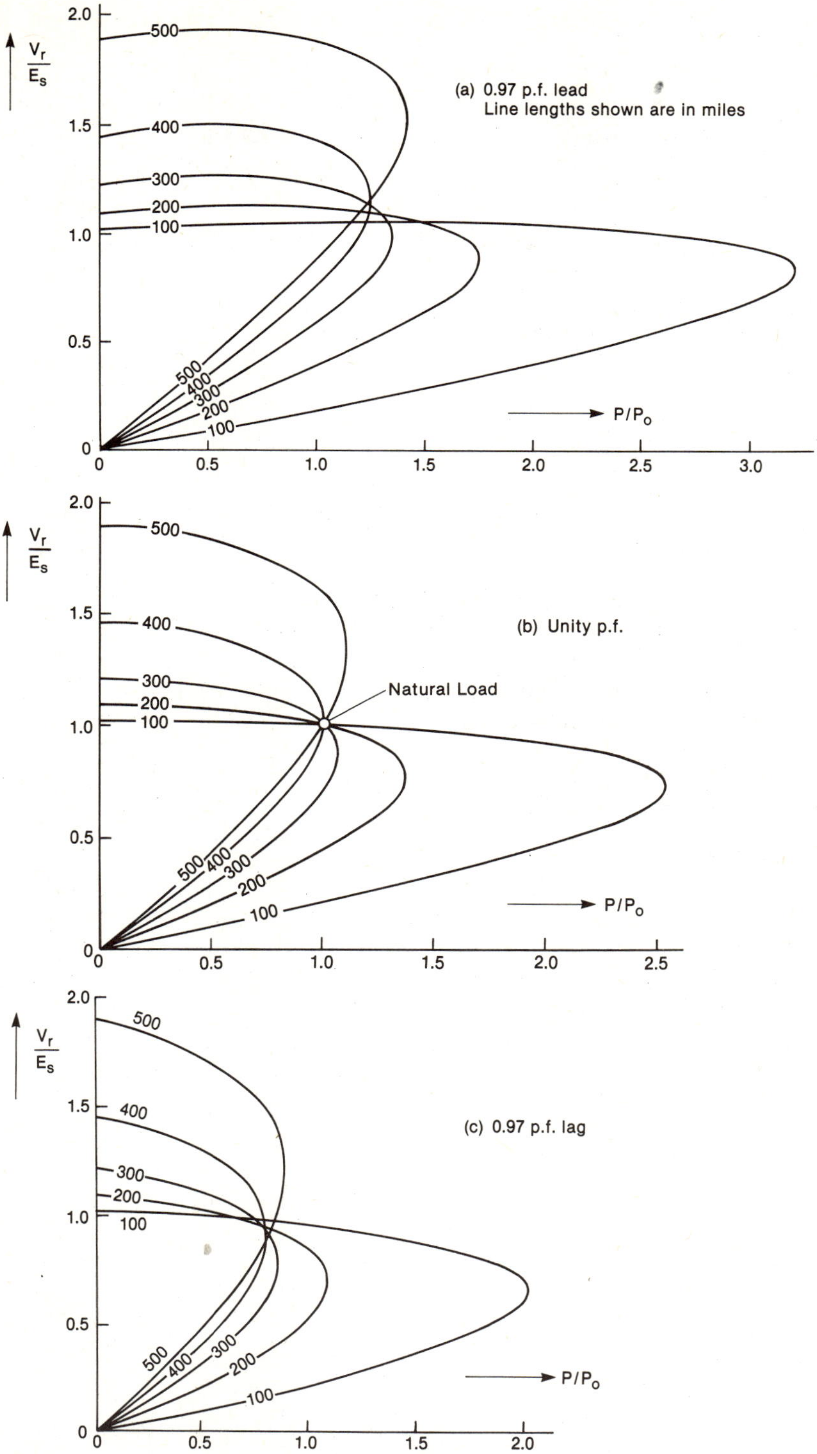

FIGURE 8. Receiving-end voltage as a function of line length, load, and load power factor (radial line).

control or compensation is provided. Even though $V_r = E_s =$ 1.0 pu at the natural load $(P + jQ = P_0)$, if the line length is longer than about 200 mi V_r is extremely sensitive to any *variation* in P. If a is greater than 390 mi or $\lambda/8$ (i.e., $\theta > 45°$) then at the natural load the receiving-end voltage is the *lower* of the two roots of Equation 24: See Figure 8*b* for $a = 400$ and 500 mi, with $P/P_0 = 1$. In virtually all cases such operation would be unstable.

Symmetrical Line. In Section 2.2.4 the no-load behavior of the symmetrical line was deduced in terms of two open-circuited half-length lines connected back-to-back. The symmetrical line under load can be treated in the same way. Although the symmetrical line is a special case, the treatment provides a physical understanding which is helpful in dealing with more complex cases.[1]

By definition the symmetrical line has $E_s = E_r$. Under load $\mathbf{E}_s$ leads $\mathbf{E}_r$ in phase, and by symmetry the midpoint voltage is midway in phase between them: See Figure 9. By symmetry again, the power factor angle at one end must be the negative of that at the other end, while the power factor at the midpoint is unity. This being so, it is possible and convenient to use Figure 8*b* to describe how V_m varies with the transmitted power. Provided that $E_s = E_r = 1.0$ pu, the line length is replaced by $a/2$ and V_m can be read off Figure 8*b*. For example, the midpoint voltage variations on a symmetrical 200-mi line are the same as the receiving-end voltage variations on a 100-mi line with a unity power-factor load. At 200 mi a marked improvement results from having synchronous machines at both ends. A symmetrical 500-mi line, however, would still have unacceptably large voltage variations at the midpoint (equal to the receiving-end variations on a 250-mi line).

Reactive Power Requirements. The reactive power requirements of the line are determined by the voltage and the level of power transmission. It is important to know what these requirements are, because they deter-

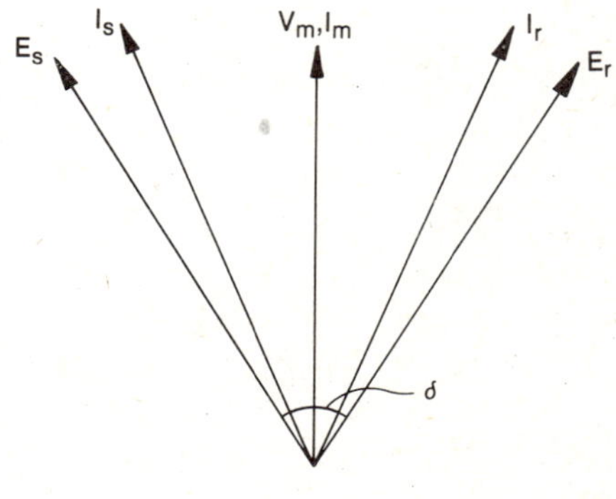

FIGURE 9. Phasor diagram of symmetrical line with $P > P_0$. Note that the receiving end has a leading power factor and that both ends are supplying reactive power to the line.

mine the reactive power ratings of the terminal synchronous machines as well as of any compensating equipment. Note that the terminal power factor is the resultant of all circuits connected at that end of the line. Any inductive load connected, for example, at the sending end will assist the synchronous generators to absorb the line-charging reactive power. In general, in the absence of compensating equipment, the synchronous machines must absorb or generate the difference between the reactive power of the line and that of the local load.

The equations for the sending-end half of the symmetrical line are

$$\mathbf{E}_s = \mathbf{V}_m \cos\frac{\theta}{2} + jZ_0\mathbf{I}_m \sin\frac{\theta}{2} \tag{25a}$$

$$\mathbf{I}_s = j\frac{\mathbf{V}_m}{Z_0}\sin\frac{\theta}{2} + \mathbf{I}_m \cos\frac{\theta}{2}. \tag{25b}$$

At the midpoint,

$$P_m + jQ_m = \mathbf{V}_m\mathbf{I}_m^* = P, \tag{26}$$

where P is the transmitted power. Note that $Q_m = 0$; that is, no reactive power flows past the midpoint. The real and reactive power which must be supplied at the sending end are given by

$$P_s + jQ_s = \mathbf{E}_s\mathbf{I}_s^*. \tag{27}$$

Substituting for $\mathbf{E}_s$ and $\mathbf{I}_s$ from Equation 25 and treating $\mathbf{V}_m$ as reference phasor, $P_m = V_m I_m$ and

$$P_s + jQ_s = P + j\frac{\sin\theta}{2}\left[Z_0 I_m^2 - \frac{V_m^2}{Z_0}\right]. \tag{28}$$

Since the line is assumed lossless, the result $P_s = P$ is expected; likewise $P_r = P$ at the receiving end. The expression for Q_s can be rearranged as follows. Making use of the relations $P_0 = V_0^2/Z_0$ and $P_m = V_m I_m$,

$$Q_s = P_0\frac{\sin\theta}{2}\left[\left(\frac{P}{P_0}\right)^2\left(\frac{V_0}{V_m}\right)^2 - \left(\frac{V_m}{V_0}\right)^2\right]. \tag{29}$$

This equation shows how the midpoint voltage is related to the reactive power requirement of the symmetrical line. By symmetry, Equation 29 applies to both ends of the line, and each end supplies half the total. Because of the reactive power sign convention, this is written $Q_s = -Q_r$.

Where $P = P_0$, that is, at the natural load, if $V_m = 1.0$ pu Equation 29 gives the familiar result: $Q_s = 0$. In this condition $Q_r = 0$ also, and

$E_s = E_r = V_m = V_0 = 1.0$ pu. At no load, that is, $P = 0$, if the terminal voltages are both adjusted so that $E_s = E_r = V_0 = 1.0$ pu, then $I_m = 0$ and from Equation 29.

$$Q_s = - P_0 \tan \frac{\theta}{2} . \tag{30}$$

This is identical in form to Equation 21. It shows that when $E_s = E_r$ and $P = 0$, the sending-end reactive power is the line-charging reactive power for half the line. The receiving-end generators absorb an equal amount from the other half.

If the terminal voltages are continuously adjusted so that the midpoint voltage $V_m = V_0 = 1.0$ pu at all levels of power transmission, then from Equation 29

$$Q_s = P_0 \frac{\sin \theta}{2} \left[\left(\frac{P}{P_0} \right)^2 - 1 \right] = - Q_r . \tag{31}$$

In addition, from Equations 25 and 26 it can be shown that for $V_m = V_0$,

$$E_s = V_m \sqrt{1 - \sin^2 \frac{\theta}{2} \left[1 - \left(\frac{P}{P_0} \right)^2 \right]} = E_r . \tag{32}$$

These two equations illustrate the general behavior of the symmetrical line. If $P < P_0$, the midpoint voltage is higher than the terminal voltages. If $P > P_0$, the reverse is true, and if $P = P_0$ the voltage profile is flat. When $P < P_0$ there is an excess of line-charging reactive power; that is, Q_s is negative and Q_r is positive, indicating absorption at both terminals. When $P > P_0$ there is an overall deficit of reactive power in the line. The excess or deficit can be corrected by means of compensation, as is seen in Section 2.3.

It should be noted that the reactive power requirement is determined by the *square* of the power transmitted. As an example, consider an uncompensated symmetrical line 200 mi long with $P = 1.5\ P_0$. Then $\sin \theta = 0.394$ and $Q_s = -Q_r = 0.246\ P_0$. For every megawatt of transmitted power, a total reactive power of $2 \times 0.246/1.5 = 0.329$ MVAr has to be supplied from the ends.

Alternative useful equations for the reactive power requirements are given later in Section 2.2.6.

2.2.6. The Uncompensated Line Under Load: Maximum Power and Stability Considerations

Symmetrical Line. If the load at the receiving end of a lossless transmission line is $P + jQ$, then the terminal voltages are related by Equation 24:

$$\mathbf{E}_s = \mathbf{E}_r \cos\theta + jZ_0 \frac{P - jQ}{\mathbf{E}_r} \sin\theta . \tag{33}$$

This equation is valid for synchronous and nonsynchronous loads alike. Here the load is assumed to be synchronous and $\mathbf{E}_r$ is written instead of $\mathbf{V}_r$. If $\mathbf{E}_r$ is taken as reference phasor, $\mathbf{E}_s$ can be written as

$$\mathbf{E}_s = E_s e^{j\delta} = E_s (\cos\delta + j\sin\delta) , \tag{34}$$

where δ is the phase angle between $\mathbf{E}_s$ and $\mathbf{E}_r$ (see Figure 9). δ is called the *load angle* or the *transmission angle.* Equating the real and imaginary parts of Equations 33 and 34,

$$E_s \cos\delta = E_r \cos\theta + Z_0 \frac{Q}{E_r} \sin\theta , \tag{35}$$

$$E_s \sin\delta = Z_0 \frac{P}{E_r} \sin\theta . \tag{36}$$

Equation 36 can be rearranged in the form

$$P = \frac{E_s E_r}{Z_0 \sin\theta} \sin\delta . \tag{37}$$

This equation is important because of its simplicity and its wide-ranging validity. The equation is true when $E_s \neq E_r$ and is valid for synchronous and nonsynchronous loads alike. Its only major shortcoming is that it neglects losses. A more familiar form is obtained when, for an electrically short line, $\sin\theta$ is replaced by $\theta = \beta a = \omega a\sqrt{lc}$. Then $Z_0\theta = \omega a\sqrt{lc} \,.\, \sqrt{l/c} = \omega a l = X_l$, the series reactance of the line, and

$$P \simeq \frac{E_s E_r}{X_l} \sin\delta . \tag{38}$$

Equation 37 shows that if E_s and E_r are fixed, the power transmitted can be expressed in terms of only one variable: the transmission angle δ. If $E_s = E_r = V_0$, then

$$P = \frac{P_0}{\sin\theta} \sin\delta \; . \tag{39}$$

Figure 10 shows this relationship for a 200-mi line at 60 Hz, for which $\theta = 0.405$ rad and $\sin\theta = 0.394$. The graph is ordinarily plotted with P as ordinate; but here it is rotated clockwise by 90° to reflect the fact that P is the independent and δ the dependent variable. It also reflects the similarity with the voltage-versus-power characteristic of Figure 8*b*, which for convenience is reproduced as Figure 11 with the correct line-lengths for the symmetrical line. It can be seen that as the load is increased from zero the load angle increases. If E_s and E_r are held constant the voltage profile "sags," the midpoint voltage experiencing the greatest decrease along the line.

As indicated in Section 2.2.5 there is a maximum power that can be transmitted. It is useful to have a physical understanding of this phenomenon. Let the sending-end synchronous machines be thought of as an equivalent synchronous generator, and the receiving-end machines as an equivalent synchronous motor. The load angle δ is then a measure of the relative mechanical position of the rotors of these two machines as

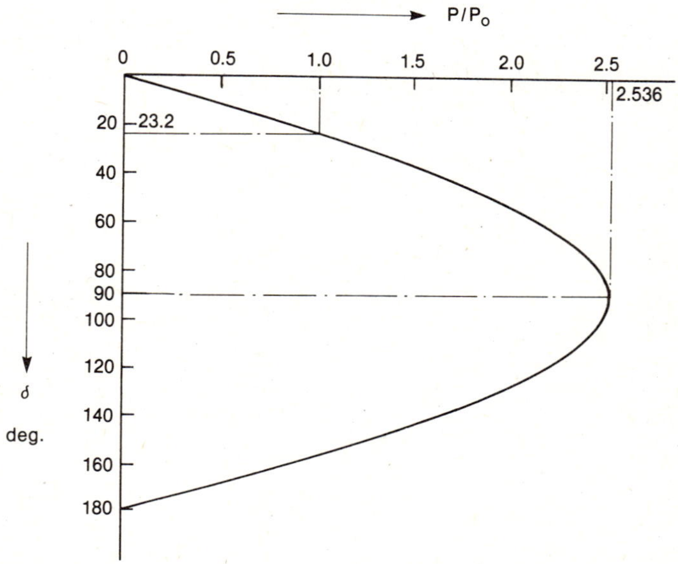

FIGURE 10. Power transmission angle characteristic for 200-mi line.

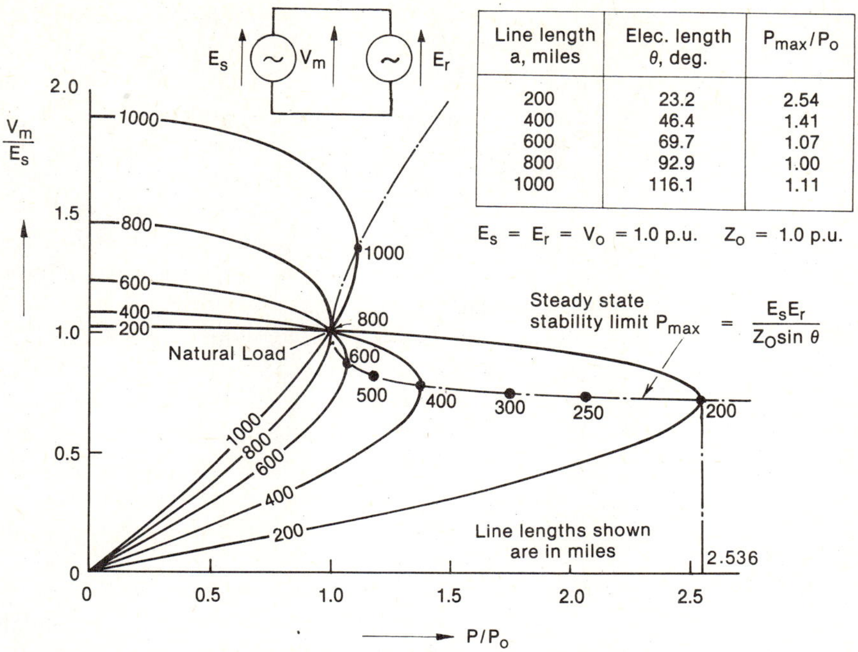

Line length a, miles	Elec. length θ, deg.	P_{max}/P_o
200	23.2	2.54
400	46.4	1.41
600	69.7	1.07
800	92.9	1.00
1000	116.1	1.11

FIGURE 11. Variation of midpoint voltage with power transmitted along a symmetrical line.

they rotate in synchronism. The terminal voltages are assumed to be held constant by excitation control. The load power can be increased by increasing the torque on the shaft of the motor. This causes the motor to slow down, so that if the generator is assumed to continue at constant speed, the load angle increases. According to Figure 10 and Equation 37, the increased load angle is accompanied by an increase in the transmitted power, which causes the motor to speed up again until a new steady state is attained at the new power level. V_m will now be smaller than it was before, and I_m larger. For successive power increments this process cannot continue indefinitely because there comes a point at which the fractional reduction in V_m exceeds the fractional increase in I_m, and their product $P = V_m I_m$ decreases with any further increase in the transmission angle, however small. This point occurs when $\delta = 90°$, and if $E_s = E_r = V_0$ then $V_m = V_0/\sqrt{2}$. If the load torque is increased slightly, δ increases but the transmitted power now decreases, so that if the generator maintains constant speed the motor will slow down still more and will lose synchronism. The system is unstable, and if this condition is arrived at by the gradual process just described, the power is said to have exceeded the steady-state stability limit $P_0/\sin\theta$. At 200 mi this is $P_0/0.394 = 2.54\ P_0$. The theoretical steady-state stability limit for other line lengths is given in the table in Figure 11 and its locus is plotted also. For lines less than $\lambda/4$ in length (775 mi at 60 Hz), the steady-state stability limit decreases rapidly with increasing line length.

Because of frequent minor disturbances in the power transmitted in any real system, as well as occasional major disturbances caused by faults and switching operations, it is not practical to operate an uncompensated line too near to its steady-state stability limit. A margin is necessary, and, based on experience, a general rule is that the load angle on any uncompensated line should not exceed about 30°, corresponding to a power transmission of half the steady-state limit. If this empirical rule is followed, then the maximum electrical length over which the natural load can be transmitted without compensation is given by Equation 39 with $P = P_0$ and $\delta = 30°$, that is, $\theta = 30°$ or $a = 260$ mi at 60 Hz. A smaller power can be stably transmitted over a longer distance, but in the absence of compensation the maximum permissible line length is still limited by the no-load value of V_m or the reactive power ratings of the synchronous machines (whether absorbing at no load or generating at full load).

It appears from Figure 11 that if the uncompensated line length exceeds one quarter-wavelength, then a flat voltage profile is unattainable at *any* stable level of power transmission. The fact that the steady-state stability limit increases with line length for $a > \lambda/4$ is not of practical interest because of the high voltages, the impractical reactive power requirements, and the high voltage sensitivity associated with the upper parts of the curves for $a > \lambda/4$. Even without these difficulties it would be practically impossible to "maneuver" a line into such an operating condition without passing through an unstable range.

Radial Line with Nonsynchronous Load. Figures 7 and 8 show that there is a maximum power that can be transmitted over a line, even when the load is nonsynchronous. The value of the maximum power can be calculated simply, as follows, for a unity power factor load.

From elementary circuit theory, the maximum power that can be drawn by a unity power factor load from a supply represented as an open-circuit voltage in series with a short-circuit impedance, is given by

$$P_{max} = \frac{E_0 I_s}{2\,(1 + \cos\phi_s)}\,. \tag{40}$$

E_0 is the open-circuit voltage, I_s is the short-circuit current, and ϕ_s is the phase angle between them when the supply is short-circuited. If the supply is regarded as the sending-end emf together with the transmission line up to the receiving-end terminals, then from Equations 2a and b

$$\mathbf{E}_0 = \frac{\mathbf{E}_s}{\cos \theta} \tag{41}$$

and

$$\mathbf{I}_s = \frac{\mathbf{E}_s}{(jZ_0 \sin \theta)} . \tag{42}$$

Since $\mathbf{E}_0$ leads $\mathbf{I}_s$ by 90°, $\phi_s = 90°$, $\cos \phi_s = 0$, and from Equation 40,

$$P_{max} = \frac{E_s^2}{Z_0 \sin 2\theta} . \tag{43}$$

This equation describes a locus on which lie all the maximum-power points in Figure 8*b*.

Since the load is nonsynchronous, the angle δ cannot be interpreted in terms of the relative angular positions of the rotors of an equivalent machine at the receiving-end and the sending-end generator. The question of the maintenance of synchronism is therefore not an issue.

If the system is operating on the upper part of one of the curves in Figure 8*b*, an increase in power transmission can be caused by a reduction in the effective resistance of the load. In practice this might be done by switching on more lighting load. Alternatively, the load torque on induction motors might be increased, causing them to slow down: the increased slip then reduces the effective resistance of the motors. Reduced load resistance draws more current from the supply, and at unity power factor the voltage decreases (Figure 8*b*). Up to the point of maximum power the product of voltage and current increases, and the system is stable. At the point of maximum power, any further reduction of the effective load resistance produces a reduction in transmitted power.

Effect of Generator Reactances. The internal reactances of the synchronous machines at the end(s) of the line add to the series impedance and alter the phase angle between the internal emf's. This is best illustrated by an example.

Figure 12 shows a symmetrical 200-mi line carrying the natural load P_0. The power system at each end of the line is represented by an equivalent synchronous machine with its step-up transformer. The ratings of these machines are each assumed equal to P_0 and their transient reactances are $x'_d = 0.25$ pu.† The transformer reactances on the same

† The transient reactance is used on the assumption that the generators are fitted with fast-acting voltage regulators. See Reference 9.

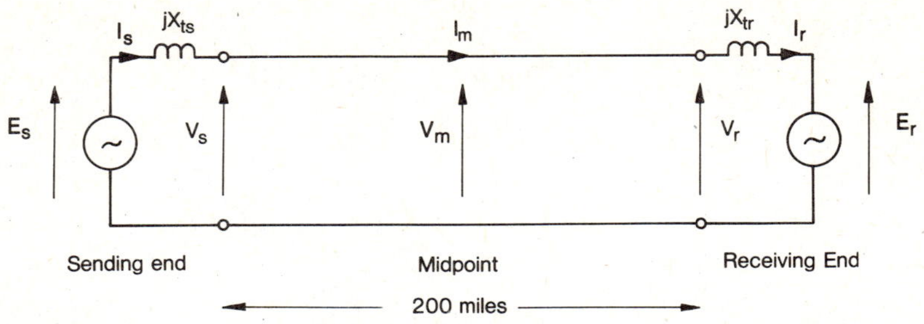

FIGURE 12. Symmetrical line transmitting the natural load P_0: effect of generator reactances at the ends.

per-unit base are assumed to be $x_T = 0.1$ pu so that $x_{ts} = x_{tr} = x_d' + x_T = 0.25 + 0.1 = 0.35$ pu.

The phasor diagram is shown in Figure 13. The excitation of the synchronous machines is adjusted so that $V_s = V_r = V_0 = 1.0$ pu, so that the voltage profile is flat along the line and $V_m = 1.0$ pu. The line angle is 23.2°, as calculated in Section 2.2.3, or from Equation 37. The power factor is unity at both ends, so that if the synchronous machine resistances are neglected the voltage drops across x_{ts} and x_{tr} are in phase quadrature with $\mathbf{I}_s$ and $\mathbf{I}_r$, and both are equal to 0.35 pu, since $I_s = I_r = 1.0$ pu. The total angle δ is 64.2°—nearly 2.8 times that of the line alone.

In the general case of a symmetrical line (with $x_{ts} = x_{tr} = x_t$), the relationship between P and δ (i.e., the phase angle between $\mathbf{E}_s$ and $\mathbf{E}_r$) can be calculated from Equations 25a and b with $P = V_m I_m$ and

$$\mathbf{E}_s = \mathbf{V}_s + jx_{ts}\mathbf{I}_s \,. \tag{44}$$

The result is:

$$P = \frac{E^2}{\left(Z_0 - \dfrac{x_t^2}{Z_0}\right)\sin\theta + 2x_t\cos\theta}\sin\delta\,, \tag{45}$$

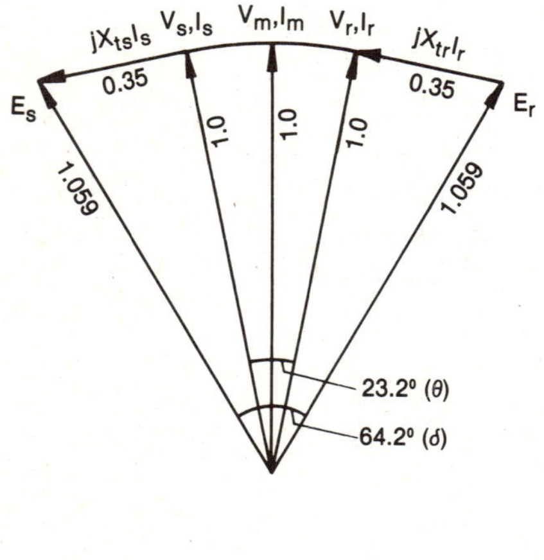

FIGURE 13. Symmetrical line of Figure 12, phasor diagram. (Indicated values are in pu)

where $E_s = E_r = E$. The form of this relationship is exactly as shown in Figure 10, and it reduces to Equation 37 when $x_t = 0$. The maximum power is modified considerably by x_t. The effect depends on the line length and is illustrated in Figure 14, which shows the maximum power as a function of x_t for various line lengths. In practice, uncompensated line lengths do not exceed 100–200 mi, and x_t will generally be less than Z_0. The effect of the generator reactance is therefore to widen the phase angle δ for a given level of power transmission, or to lower the power transmission corresponding to a given angle. (For very long lines there is an upturn in the maximum power when x_t is high. This phenomenon is not of practical interest because operation under such conditions involves very high voltages and reactive-power levels.)

The power transmission relationship can be expressed with E replaced by V_m in Equation 45. The result is

$$P = V_m^2 \frac{\cos\dfrac{\theta}{2} - \dfrac{x_t}{Z_0}\sin\dfrac{\theta}{2}}{Z_0\sin\dfrac{\theta}{2} + x_t\cos\dfrac{\theta}{2}}\tan\frac{\delta}{2}. \tag{46}$$

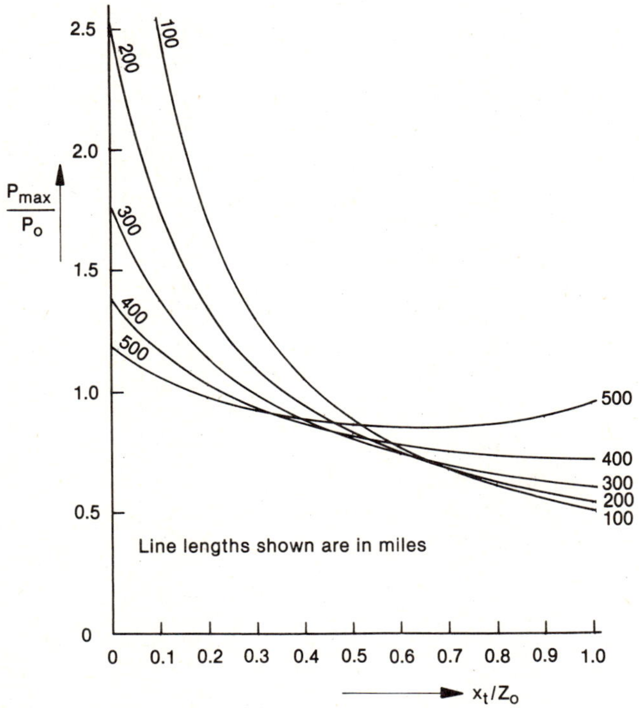

FIGURE 14. Effect of generator reactances on maximum transmissible power for different line lengths.

If $x_t = 0$, this reduces to

$$P = \frac{V_m^2}{Z_0 \tan \frac{\theta}{2}} \tan \frac{\delta}{2} . \tag{47}$$

Alternative Expression for Reactive Power Requirements. In Section 2.2.5, Equation 29 for the reactive power required at the ends of the line assumes that the line is symmetrical and that V_m is known. A useful alternative formula is obtained from Equation 35 for the receiving-end reactive power:

$$Q_r = \frac{V_r(V_s \cos \delta - V_r \cos \theta)}{Z_0 \sin \theta} . \tag{48}$$

A similar procedure can be used to derive the following formula for the sending-end reactive power:

$$Q_s = - \frac{V_s(V_r \cos \delta - V_s \cos \theta)}{Z_0 \sin \theta} . \tag{49}$$

These expressions are valid when the line is not symmetrical, i.e., $V_s \neq V_r$. If $V_s = V_r$ the line is symmetrical and

$$Q_s = - \frac{V_s^2 (\cos \delta - \cos \theta)}{Z_0 \sin \theta} = -Q_r . \tag{50}$$

If $P < P_0$ and $V_s = 1.0$ pu, δ is less than θ, $\cos \delta > \cos \theta$, and Q_s is negative while Q_r is positive. This implies that reactive power is being absorbed at both ends of the line. If $P > P_0$, reactive power is generated at both ends; whereas if $P = P_0$, $Q_s = Q_r = 0$. If $P = 0$, $\cos \delta = 1$ and Equation 50 reduces to Equation 30. The terminal reactive power requirements represented by Equation 50 are illustrated in Figure 15.

For an electrically short line, $\cos \theta \rightarrow 1$ and $Z_0 \sin \theta \rightarrow X_l$, the series reactance of the line, so that Equation 50 reduces to

$$Q_s = \frac{V_s^2 (1 - \cos \delta)}{X_l} = -Q_r . \tag{51}$$

with $V_s = V_r$.

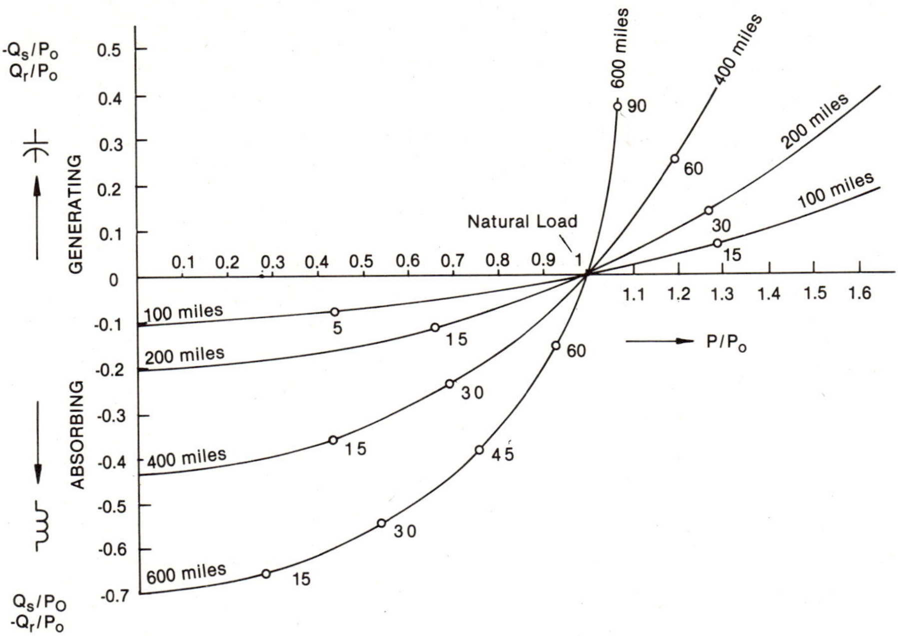

FIGURE 15. Terminal reactive power requirements of symmetrical line, as a function of power transmitted and line length. (Figures show transmission angle δ°)

2.3. COMPENSATED TRANSMISSION LINES

2.3.1. Types of Compensation: Virtual-Z_0, Virtual-θ, and "Compensation by Sectioning"

In this chapter, *compensation* means the modification of the electrical characteristics of a transmission line in order to increase its power transmission capacity while satisfying the fundamental requirements for transmission stated in Section 2.1. With this general objective, a *compensation system* ideally performs the following functions:

1. It helps produce a substantially flat voltage profile at all levels of power transmission;
2. It improves stability by increasing the maximum transmissible power;
3. It provides an economical means for meeting the reactive power requirements of the transmission system.

A figure of merit used to gauge the effectiveness of a compensated system is the product of line length and maximum transmissible power. In

Section 2.2 it was shown, for instance, that at 60 Hz without compensation it is not practical to transmit even the natural load over a distance greater than about 200 mi. Compensated lines enable the transmission of the natural load over greater distances, and shorter compensated lines can carry loads greater than the natural load.

A *flat voltage profile* can be achieved if the effective surge impedance of the line is modified so as to have a *virtual* value, Z_0', for which the corresponding *virtual natural load* V_0^2/Z_0' is equal to the *actual* load. The uncompensated surge impedance is $Z_0 = \sqrt{l/c}$. At fundamental frequency this can be written $\sqrt{x_l x_c}$, implying that if the series and/or the shunt reactances x_l and/or x_c are modified (for example by appropriate connection of capacitors or reactors), then the line can be made to have a virtual surge impedance Z_0' and a virtual natural load P_0' for which

$$P_0' = \frac{V_0^2}{Z_0'} = P, \tag{52}$$

where P is the actual power to be transmitted and V_0 is the rated voltage of the line. Since the load P varies, sometimes suddenly, the ideal compensation would be capable of variation also—and without delay. Compensation which can be said to have the objective of modifying Z_0 (or P_0) will be termed *surge-impedance compensation* or Z_0*-compensation.*

Controlling the virtual surge impedance to match a given load (Equation 52) is not sufficient by itself to ensure the *stability* of transmission over longer distances. This is made clear by Figure 11, which shows that without compensation, even under *ideal* conditions (i.e., with no disturbances) the natural load cannot be transmitted stably over distances greater than $\lambda/4$(= 775 mi at 60 Hz). In practice stability is a limiting factor at distances much shorter than this, in the absence of compensation.

Both of the fundamental line parameters Z_0 and θ influence stability through their influence on the transmission angle δ. (See Equation 37). Once a line is compensated in such a way as to satisfy Equation 52—to achieve a flat voltage profile—Z_0' is determined, and the only way to improve stability is to reduce the effective value of θ. Two alternative compensation strategies have been developed to achieve this. One is to apply *series capacitors* to reduce X_l and thereby reduce θ, since $\theta = \beta a = \sqrt{X_l/X_c}$ at fundamental frequency. This strategy might be called *line-length compensation* or θ*-compensation.* The other approach is to divide the line into shorter sections which are more or less independent of one another (except that they all transmit the same power). This might be called *compensation by sectioning.* It is achieved by connecting constant-voltage compensators at intervals along the line. The maximum transmissible power is that of the weakest section, but since this is necessarily shorter than the whole line, an increase in maximum power and, therefore, in stability can be expected.

All three types of compensation may be used together in a single transmission line.

2.3.2. Passive and Active Compensators

It is helpful to distinguish between *passive* and *active* compensators. *Passive* compensators include shunt reactors and capacitors and series capacitors. These devices may be either permanently connected, or switched; but in their usual forms they are incapable of continuous (i.e., stepless) variation. They operate by modifying the natural inductance and capacitance and their operation is essentially static. Apart from switching, they are uncontrolled.

Passive compensators are used only for surge-impedance compensation and line-length compensation. For example, shunt reactors are used to compensate for the effects of distributed line capacitance, particularly in order to limit voltage rise on open circuit or at light load. They tend to increase the virtual surge impedance and reduce the virtual natural load P_0'. Shunt capacitors may be used to augment the natural capacitance of the line under heavy loading. They generate reactive power which tends to boost the voltage. They tend to reduce the virtual surge impedance and to increase P_0'. Series capacitors are used for line-length compensation. Usually a measure of surge-impedance compensation is necessary in conjunction with series capacitors, and this may be provided by an active compensator.

Active compensators are usually shunt-connected devices which have the property of tending to maintain a substantially constant voltage at their terminals. They do this by generating or absorbing precisely the required amount of corrective reactive power in response to any small variation of voltage at their point of connection. They are usually capable of continuous (i.e., stepless) variation and rapid response. Control may be inherent, as in the saturated-reactor compensator; or by means of a control system, as in the synchronous condenser and thyristor-controlled compensators.

Active compensators may be applied either for surge-impedance compensation or for compensation by sectioning. In Z_0-compensation they are capable of all the functions performed by fixed shunt reactors and capacitors and have the additional advantages of continuous variability with rapid response. Compensation by sectioning is fundamentally different in that it is possible *only* with active compensators, which must be capable of virtually immediate response to the smallest variation in power transmission or voltage, that is, their operation is essentially dynamic (in the control engineer's sense). All active compensators except the saturated-reactor type are also capable of acting as passive compensators. Table 4

TABLE 4
Classification of Compensators by Function and Type

Function	Passive	Active
Surge-impedance compensation (Virtual-Z_0 compensation) voltage control, reactive power management	Shunt reactors (linear or nonlinear) Shunt capacitors	Synchronous machines Synchronous condensers Saturated-reactor compensators Thyristor-switched capacitors Thyristor-controlled reactors
Line-length compensation (Virtual-θ compensation) voltage control, reactive power management, stability improvement	Series capacitors	—
Compensation by sectioning Dynamic shunt compensation, stability improvement on longer lines	—	Synchronous condensers Saturated-reactor compensators Thyristor-switched capacitors Thyristor-controlled reactors

summarizes the classification of the main types of compensator according to their usual functions. (Most of these types of compensator had been proposed by the mid-1920's, and several of them can be found in the papers of E.F.W. Alexanderson, who in 1925 discussed the use of shunt reactors controlled by thyratrons, foreshadowing the modern thyristor-controlled reactor compensator.†)

Rapid-response excitation systems used on synchronous machines also have a strong and important compensating effect in a power system. Fitted to the generators at either end of a line, they modify the effective series reactance of the transmission line as a whole, and contribute improvements in both voltage control and stability. They have the effect of reducing the synchronous machine effective reactance to the *transient* reactance x_d'.[9]

The application of reactive power compensation must be done as economically as possible. In some cases the management of reactive

† E.L. Owen, private communication.

power in a power system can be enhanced by modifications to the design of existing (or planned) plant; sometimes this is a cheaper way of improving performance than installing compensation equipment. For example, feedback signals can be used in the automatic voltage regulators of synchronous machines to enhance the stability and enable an increase in power transmission. As another example, shunt reactors and capacitors can often be advantageously relocated after a period of evolutionary change in the loading pattern of the system. More usually, however, compensating equipment is introduced precisely because it is the least expensive way of satisfying the reactive power requirements. This is typically the case when the alternatives are an increase in the number of transmission lines, in the ratings of planned synchronous generators, or in the system voltage.

Other applications and functions of compensators on transmission systems include the management of reactive power flows in order to minimize losses; the damping of power oscillations; the provision of reactive power at dc converter terminals. These are not considered in this chapter.

Both passive and active compensators are in use today and so are all the compensation strategies; virtual-Z_0, virtual-θ, and compensation by sectioning. Although most of the fundamental concepts have a long pedigree, modern activity is considerable. In equipment development, activity is concentrated on the static reactive power controller† or static compensator, to improve its efficiency, reliability, and response characteristics. In the analytical field, attention is focussed on the optimal deployment of compensators, the relative merits of series and shunt compensation schemes for long lines, and the modeling of compensators in power systems on the digital computer.

The remaining sections of this chapter attempt to develop the theory of compensation to the point where the state of the art can be understood in all the main compensation strategies and applications. For further reading the references listed at the end of the chapter should prove helpful.

2.3.3. Uniformly Distributed Fixed Compensation

Modified Line Parameters: Virtual Z_0, θ, and P_0. Compensators are normally connected at the ends of a line or at discrete points along it. In spite of their lumped or concentrated nature, it is useful to derive certain basic relationships for the ideal case of *uniformly distributed* compensation because these relationships are simple and independent of the characteristics of any particular type of compensator. They also give considerable

† This includes the thyristor-controlled reactor (TCR), the thyristor-controlled leakage transformer (TCT), the thyristor-switched capacitor (TSC), and hybrid forms.

physical insight, and help to determine the fundamental nature of the type of compensation required, without reference to extensive computer studies. The formulas derived are in most cases approximately true for practical systems with concentrated compensation because the spacing between compensators is limited by the same factors that limit the maximum length of uncompensated line.

The surge impedance Z_0 of an uncompensated line can be written

$$Z_0 = \sqrt{\frac{l}{c}} = \sqrt{\frac{\mathrm{j}\omega l}{\mathrm{j}\omega c}} = \sqrt{x_l x_c}\,. \tag{53}$$

If a uniformly distributed shunt compensating inductance $l_{\gamma\mathrm{sh}}$ (H/mile) is introduced, the effective value of the shunt capacitive admittance per mile becomes

$$(\mathrm{j}\omega c)' = \mathrm{j}\omega c + \frac{1}{\mathrm{j}\omega l_{\gamma\mathrm{sh}}}$$

$$= \mathrm{j}\omega c(1 - k_{\mathrm{sh}}), \tag{54}$$

where k_{sh} is the *degree of shunt compensation:*

$$k_{\mathrm{sh}} = \frac{1}{\omega^2 l_{\gamma\mathrm{sh}} c} = \frac{x_c}{x_{\gamma\mathrm{sh}}} = \frac{b_{\gamma\mathrm{sh}}}{b_c}\,. \tag{55}$$

Here $x_{\gamma\mathrm{sh}}$ and $b_{\gamma\mathrm{sh}}$ are the reactance and susceptance per mile of the shunt compensating inductance. Substituting for $(\mathrm{j}\omega c)'$ in Equation 53, the surge impedance has the effective or virtual value:

$$Z_0' = \frac{Z_0}{\sqrt{1 - k_{\mathrm{sh}}}}\,. \tag{56}$$

If shunt capacitance $c_{\gamma\mathrm{sh}}$ is added instead of shunt inductance, then k_{sh} is negative and has the value

$$k_{\mathrm{sh}} = \frac{c_{\gamma\mathrm{sh}}}{c} = \frac{x_c}{x_{\gamma\mathrm{sh}}} = \frac{b_{\gamma\mathrm{sh}}}{b_c}\,, \tag{57}$$

where $x_{\gamma\mathrm{sh}}$ and $b_{\gamma\mathrm{sh}}$ are the reactance and susceptance per mile of the shunt compensating capacitance.

Shunt inductive compensation, therefore, increases the virtual surge impedance, whereas shunt capacitive compensation reduces it.

In a similar way the effect of uniformly distributed series capacitance $c_{\gamma\mathrm{se}}$ on l can be shown to give:

$$Z_0' = Z_0\sqrt{1 - k_{\mathrm{se}}}\,, \tag{58}$$

where k_{se} is the *degree of series compensation, given by*

$$k_{se} = \frac{1}{\omega^2 l c_{\gamma se}} = \frac{x_{\gamma se}}{x_l} = \frac{b_l}{b_{\gamma se}}. \tag{59}$$

$x_{\gamma se}$ and $b_{\gamma se}$ are the reactance and susceptance per mile of the series compensating capacitance. The parameters k_{sh} and k_{se} are a useful measure of the reactive power ratings required of the compensating equipment.

Combining the effects of shunt and series compensation,

$$Z_0' = Z_0\sqrt{\frac{1-k_{se}}{1-k_{sh}}}. \tag{60}$$

Corresponding to the virtual surge impedance Z_0' is a virtual natural load P_0' given by V_0^2/Z_0', so that

$$P_0' = P_0\sqrt{\frac{1-k_{sh}}{1-k_{se}}}. \tag{61}$$

The wavenumber β is also modified and has the virtual value

$$\beta' = \beta\sqrt{(1-k_{sh})(1-k_{se})}\ . \tag{62}$$

The electrical length θ is modified according to this equation also:

$$\theta' = \theta\sqrt{(1-k_{sh})(1-k_{se})} \tag{63}$$

where $\theta = a\beta$ and $\theta' = a\beta'$. These relationships are shown graphically in Figures 16, 17, and 18.

All the equations in Section 2.2 are valid for the line with uniformly distributed compensation, if the virtual surge impedance Z_0' and the virtual wavenumber β' (or $\theta' = a\beta'$) are substituted for the uncompensated values. This means, for example, that Figure 11 can be used to determine the midpoint voltage of a compensated line under load; Equation 37 can be used to determine the maximum transmissible power and the load angle; and Equation 29 can be used to determine the reactive power requirements at the ends of the line.

Effect of Distributed Compensation on Voltage Control. For any fixed degree of series compensation, additional capacitive shunt compensation increases θ' and P_0' and decreases Z_0', while inductive shunt compensation has the reverse effect. 100% inductive shunt compensation (i.e., $k_{sh} = 1$) reduces θ' and P_0' to zero and increases Z_0' to infinity: this implies a flat voltage profile at zero load, suggesting the use of shunt reactors to cancel the Ferranti effect. Under heavy loading, a flat voltage

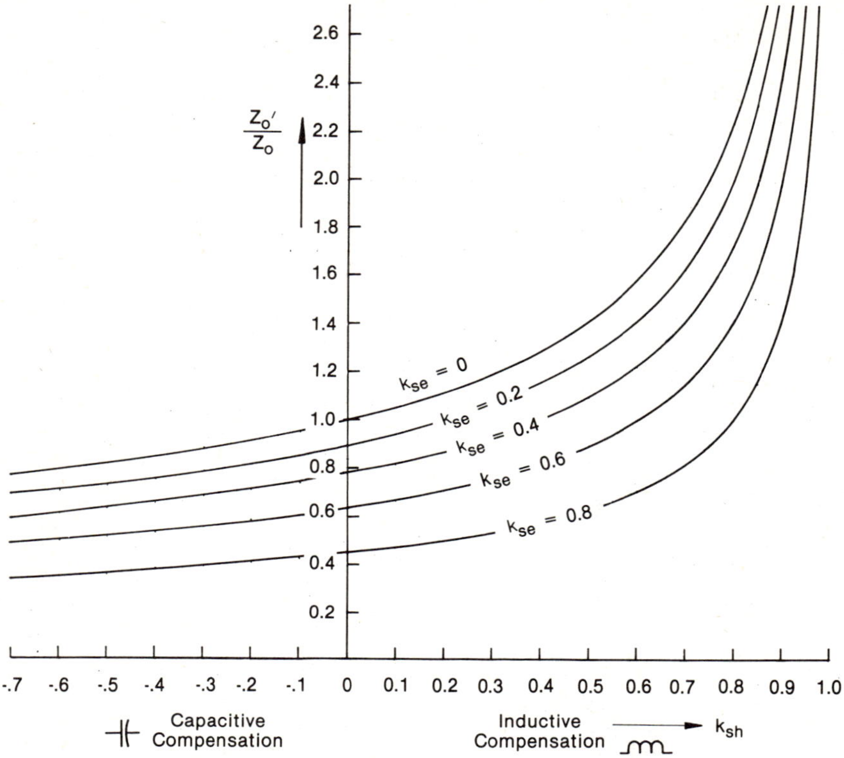

FIGURE 16. Virtual surge impedance Z_0' as a function of k_{sh} and k_{se}. (Uniformly distributed compensation)

profile can be restored by replacing the reactors with shunt capacitors. For example, in order to transmit 1.2 P_0 with a flat voltage profile without series capacitors ($k_{se} = 0$) would require 0.45 pu of distributed shunt capacitive compensation, (from Figure 17); that is, $k_{sh} = -0.45$.

The effect of series capacitive compensation ($k_{se} > 0$) is to decrease Z_0' and θ' and to increase P_0'. Series capacitive compensation can, in principle, be used instead of shunt capacitors to give a flat voltage profile under heavy loading. For example, to transmit 1.2 P_0 with a flat voltage profile without shunt compensation ($k_{sh} = 0$) would require about 0.30 pu of *distributed* series compensation, according to Figure 17; that is, $k_{se} = 0.30$. In reality the lumped nature of series capacitors makes them unsuitable for line voltage control. Their natural application is rather in stabilization, by reducing the virtual line length θ'.

At no-load the midpoint voltage of a compensated symmetrical line is given by Equation 18:

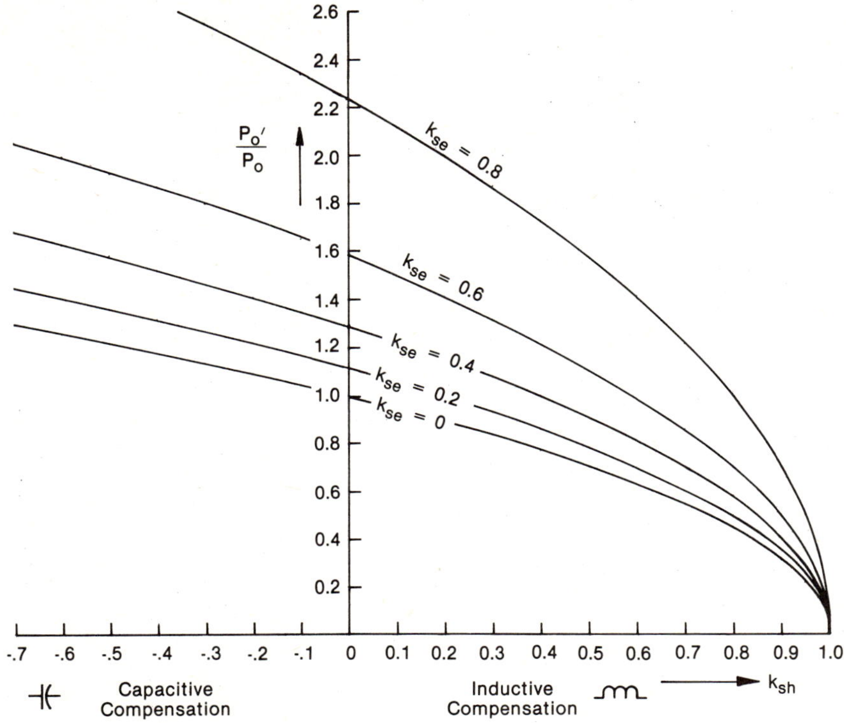

FIGURE 17. Virtual natural load P_0' as a function of k_{se}, k_{sh}.

$$V_m = \frac{E_s}{\cos \dfrac{\theta'}{2}} \tag{64}$$

It can be seen through their effect on θ' that both series capacitive and shunt inductive distributed compensation tend to reduce the Ferranti voltage rise, whereas shunt capacitive compensation tends to aggravate it.

Effect of Distributed Compensation on Line-Charging Reactive Power. At no-load the line-charging reactive power which has to be absorbed by the terminal synchronous machines is given by

$$Q_s = -P_0' \tan \theta' \tag{65}$$

for a radial line, and

$$Q_s = -P_0' \tan \frac{\theta'}{2} = -Q_r \tag{66}$$

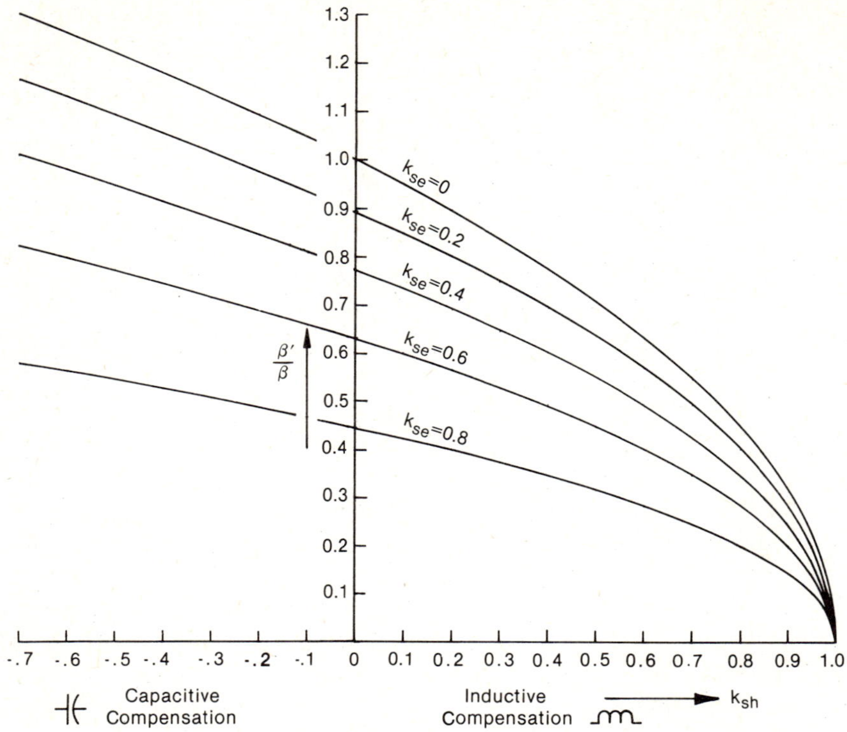

FIGURE 18. Virtual wavenumber β'/β as a function of k_{sh} and k_{se}. (Uniformly distributed compensation)

for a symmetrical line. With a significant degree of either shunt inductive or series capacitive compensation, θ' will tend to be small enough so that $\tan\theta' \simeq \theta'$ and $\tan(\theta'/2) \simeq \theta'/2$. If P_0' and θ are now substituted from Equations 61 and 63 into Equation 66, the factor $\sqrt{1-k_{se}}$ cancels, and for the symmetrical line

$$Q_s = -\frac{P_0\theta}{2}(1-k_{sh}) = -Q_r\,. \tag{67}$$

In the absence of shunt inductive compensation ($k_{sh}=0$) the series-compensated line generates roughly as much line-charging reactive power at no-load as a completely uncompensated line of the same length. If the line is long enough to justify series compensation in the first place, the reactive power absorption required in the terminal synchronous machines at no-load will be excessive. Moreover, the tendency of the synchronous machines to run underexcited at all but the heaviest loads degrades the stability which the capacitors are intended to enhance. This problem can be relieved by means of additional shunt inductive compensation (see Equation 67). Series compensation schemes have virtually always included synchronous condensers and/or shunt reactors for this purpose. Static reactive power controllers can also be used with advantage instead of synchronous condensers.

Effect of Distributed Compensation on Maximum Power. The power transmission Equation 37 can be very approximately written:

$$P \simeq P_0' \frac{\delta}{\theta'} . \tag{68}$$

When $P = P_0'$, $\delta = \theta'$ and the equation is exactly true. The first general objective of a compensation scheme is to produce a high value of P_0', the power level at which the voltage profile is flat. If the system is operated with P near P_0', then δ will necessarily be near θ'. The compensation scheme must now acquire a second objective, which is to ensure that θ' is small enough so that the transmission is stable; that is, that P is not too close to the steady-state stability limit. These objectives will be recognized as the fundamental requirements of transmission (Section 2.1.2).

From Figure 17, a high P_0' can be obtained with series capacitors and/or *capacitive* shunt compensation. On the other hand a low β' (and θ') can be obtained with series capacitors and/or *inductive* shunt compensation, (Figure 18). Only series capacitive compensation contributes to both objectives.

Of course, not all transmission systems requiring compensation require it for *both* objectives. Short lines may require voltage support, equivalent to an increase in P_0', even though their electrical length is much less than 90°. This may be provided by shunt capacitors, provided that θ' does not become excessive as a result. It is common for shunt-compensated lines not exceeding about 200 mi in length to be loaded above the uncompensated natural load. On the other hand, lines longer than about 300–500 mi cannot be loaded even up to the natural load because of the excessive uncompensated electrical length. In these cases the reduction of θ' is the first priority.

The effect of uniformly distributed compensation on the maximum transmissible power (i.e., the steady-state stability limit) is determined from Equation 37 as follows. If the terminal voltages are held constant at V_0, the maximum power is given by

$$P_{max}' = \frac{V_0^2}{Z_0' \sin \theta'} . \tag{69}$$

Combining this with Equations 8, 61, and 63,

$$\frac{P'_{max}}{P_0} = \frac{1}{\sqrt{\dfrac{1-k_{se}}{1-k_{sh}}}\sin\left[\theta\sqrt{(1-k_{se})(1-k_{sh})}\right]} \quad . \tag{70}$$

The form of this equation suggests that a given degree of series compensation has a more pronounced effect on P'_{max} than does the same degree of shunt compensation, because the factors $(1 - k_{sh})$ in the denominator produce opposing influences. This is borne out by numerical examples. The effect of series compensation by itself ($k_{sh} = 0$) is shown in Figure 19 where P'_{max}/P_0 is plotted as a function of k_{se} for various line lengths. The improvement in P'_{max}/P_0 is marked for higher values of $k_{se}(>0.5)$. Very high values of k_{se} can lead to resonance problems, and in practice it is rare to find $k_{se}>0.8$.

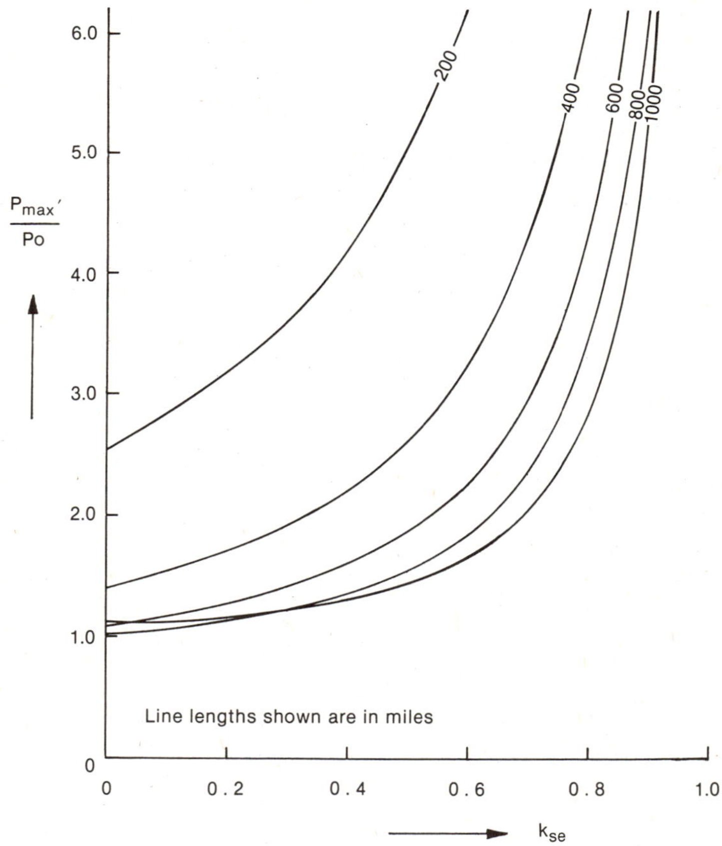

FIGURE 19. Effect of series compensation on maximum transmissible power. (No shunt compensation)

An example of the use of Figures 16 through 19 and their associated equations is as follows. A line 600 miles long has uncompensated values of $\theta = 69.7°$ and $P_{max} = 1.066\ P_0$. The natural load would be unacceptably close to the stability limit. To operate the line with a transmission angle of $\delta = 30°$, the power transfer would have to be no more than about one-half the natural load. With $P = 0.5\ P_0$, Equation 32 shows that in order to maintain a midpoint voltage of 1.0 pu, the terminal voltages have to be $E_s = E_r = 0.869$ pu. These voltages are unacceptably low. Furthermore, from Equation 31 the terminal synchronous machines have to absorb 0.704 kVAr of line-charging reactive power for every kW of power transmitted, operating at a leading (underexcited) power factor of 0.818. With 80% series compensation, Figure 19 gives $P'_{max} = 4.321\ P_0$, Figure 18 gives $\theta' = 0.447\ \theta = 31.2°$, and Figure 17 gives $P'_0 = 2.236\ P_0$. If the power transmitted is now equal to the virtual natural load P'_0, the transmission angle is $\sin^{-1}(P'_0/P'_{max}) = \sin^{-1}(2.236/4.321) = 31.2° = \theta'$ and the voltage profile is flat. The terminal power factors are both unity and no reactive power has to be supplied or absorbed at the ends of the line. This is clearly a marked improvement in the power transmission characteristics of the line.

At no-load the uncompensated midpoint voltage of this line would rise to 1.218 pu (Equation 18). With series compensation the midpoint voltage would rise to only 1.038 pu. The internal reactance of the terminal synchronous machines would tend to cause this voltage to be higher, and it may be desirable to employ shunt reactors to limit it, as well as to relieve the generators of some of the reactive power absorption. Adding uniformly distributed shunt reactance, with $k_{sh} = 0.5$, $\theta' = 22°$, $P'_0 = 1.581\ P_0$, and P'_{max} becomes $4.214\ P_0$. The maximum power, or steady-state stability limit, is not much affected by the additional shunt reactive compensation, but the virtual natural load is appreciably reduced and so is the electrical length. At the virtual natural load the transmission angle is 22.0°, while the midpoint voltage at no-load is 1.019 pu—comparable to that of an uncompensated symmetrical line of length 190 mi.

Ideally it is desirable to compensate a line in such a way that the normal actual load is equal to the virtual natural load P'_0. As already indicated, for stability assessment it is useful to be able to calculate the ratio P'_{max}/P'_0 for the compensated line; from Equations 70 and 61 this is given by

$$\frac{P'_{max}}{P'_0} = \frac{1}{\sin\left[\theta\sqrt{(1-k_{se})(1-k_{sh})}\right]}. \tag{71}$$

The effect of compensation is to improve the ratio P'_{max}/P'_0, that is, to decrease the transmission angle at the virtual natural load. The same result can of course be deduced from Figure 18, since at the virtual natural load $\delta = \theta' = \beta' a$.

A special case arises when $k_{sh} = 1$, that is, with 100% shunt reactive compensation. Equation 70 reduces to

$$\frac{P'_{max}}{P_0} = \frac{1}{\theta(1 - k_{se})}. \tag{72}$$

This relationship is not markedly different from Figure 19. Again, the maximum power is not appreciably affected by shunt compensation. However, with $k_{sh} = 1$, $P'_0 = 0$, that is, a flat voltage profile is obtained only at zero load. The line with 100% shunt compensation behaves exactly as a series inductance or reactance $X_l = \omega a l$. With series (capacitive) compensation the reactance is modified to $X'_l = X_l(1 - k_{se})$.

2.3.4. Uniformly Distributed Regulated Shunt Compensation

This section develops the theory of uniformly distributed *regulated* shunt compensation, which is an ideal system of compensation to which the closest approach in practice is compensation by sectioning, described in detail in Section 2.6.

A line operating at the natural load has a flat voltage profile and, from Equation 39, the transmission angle is then equal to the electrical length of the line, that is, $\delta = \theta$. Considering shunt compensation only ($k_{se} = 0$), suppose that k_{sh} could be continuously *regulated* in such a way that $P'_0 = P$ at all times. (See Equation 61.) Then $\theta' = \delta$ at all times. Therefore,

$$\frac{P}{\delta} = \frac{P'_0}{\theta'}. \tag{73}$$

Substituting for P'_0 and θ' from Equations 61 and 63, respectively,

$$\frac{P}{\delta} = \frac{P_0\sqrt{1 - k_{sh}}}{\theta\sqrt{1 - k_{sh}}} = \frac{P_0}{\theta} = \text{constant}. \tag{74}$$

This implies a *linear* relationship between P and δ, as shown in Figure 20a, suggesting that the maximum transmissible power is *infinite*.

The constant in Equation 74 is the slope of the $P - \delta$ characteristic and is given by

$$\frac{P_0}{\theta} = \frac{V_0^2}{Z_0 \beta a} = \frac{V_0^2}{X_l}. \tag{75}$$

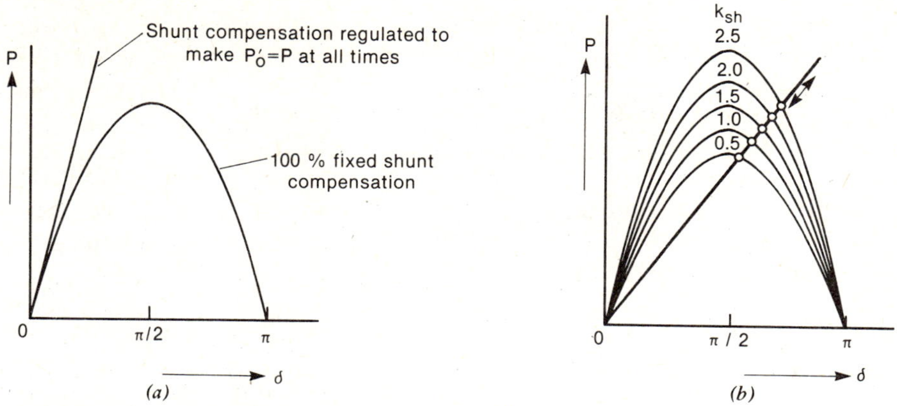

FIGURE 20. (*a*) Effect of regulated shunt compensation on power transmission characteristic. (*b*) Composition of Figure 20*a* from constant-k_{sh} curves.

At zero power the $P-\delta$ line is, therefore, tangent to the sinusoidal $P-\delta$ characteristic of the line with 100% *fixed* shunt compensation ($k_{sh}=1$).

The performance of the *regulated* shunt compensation can be further understood in terms of Figure 20*b*. For every fixed value of k_{sh} there is a sinusoidal $P-\delta$ curve, with P'_{max} given by Equation 69 with $k_{se}=0$. Some of these curves are drawn in Figure 20*b*. As the power varies, the regulating mechanism (whatever it is) adjusts k_{sh} so that the operating point continually shifts from one curve to another in such a way that it always lies on the straight line with positive slope.

It is essential to realize that at large transmission angles the stability is maintained only if k_{sh} can be varied rapidly enough to keep up with any change of P that might occur. If k_{sh} were varied only slowly, a rapid change of P would cause the operating point to move along the sinusoidal $P-\delta$ curve corresponding to the current value of k_{sh} and if $\delta > \pi/2$ rad the system would not be stable. A line with ideal regulated shunt compensation is said to be *dynamically stabilized.*

The regulation of k_{sh} must not only be sufficiently rapid, it must also be *continuous.* If there were any "slack" or dead band in the response of k_{sh} to a change of P, the operating point would move along the "current" sinusoidal $P-\delta$ curve, and when the compensation reacted it would have to over-correct. There would be a tendency towards sustained limit-cycling or hunting of the operating point.

The equations used so far do not suggest how such a compensation scheme could be realized in practice. In practice any compensating devices would not be uniformly distributed, but would be lumped at discrete intervals along the line. The simplest and probably the only practical way to control the compensation is not to try to control k_{sh} directly, but to design the compensation as *constant voltage regulators,* since by continuously forcing the voltage to be constant and equal to V_0 at several points along the line the condition $P_0' = P$ is automatically satisfied whatever the value of P. Such constant-voltage compensators are *active* compensators and can be synchronous condensers, saturated-reactor compensators, or thyristor-controlled compensators.

Reactive Power Required for Compensation. The total reactive power which must be absorbed or supplied by the compensating equipment is easily calculated because the line voltage is constant. It is given by

$$Q_\gamma = -(I^2\omega l - V^2\omega c)a = P_0\theta\left[1 - \left(\frac{P}{P_0}\right)^2\right]. \tag{76}$$

This can also be expressed in terms of the current value of k_{sh}. From Equations 76 and 61 with $k_{se} = 0$ and $P = P_0'$,

$$Q_\gamma = P_0\theta k_{sh}\,. \tag{77}$$

This equation is valid also for fixed compensation if $P = P_0'$.

The reactive power required is capacitive if $P > P_0'$ and it increases with the square of the transmitted power. There will be a transmitted power above which it is uneconomic to provide the necessary reactive power, and at this level alternatives may need to be considered [e.g., a higher transmission voltage, the use of series as well as shunt compensation, or the use of high voltage direct current (HVDC)]. There are other practical limits to the performance of the regulated shunt compensated scheme. One is the speed of response of the compensators. A second is the problem that if the regulating function fails in one compensator (or if it reaches a reactive-power limit and continues as a fixed shunt susceptance), the stability of the entire system may be seriously impaired. Perhaps the most serious limit is that the dynamic characteristic is straight only while the compensation equipment is within its capacitive current rating. In order to significantly improve the transmission characteristic while maintaining adequate stability margins, a very large amount of compensating capacitance may be required (see Section 2.6).

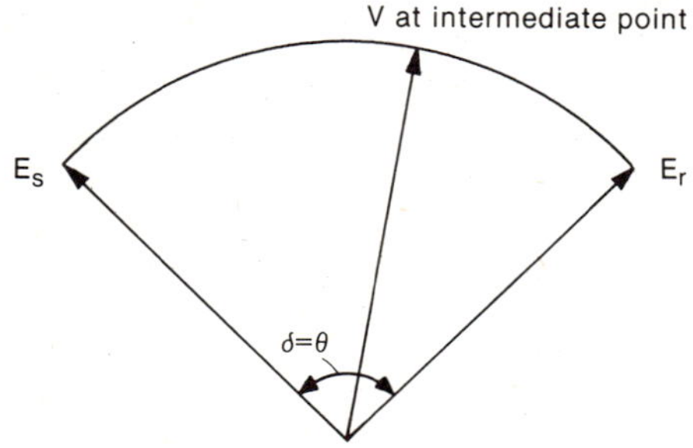

FIGURE 21. Phasor diagram of line with regulated uniformly distributed shunt compensation.

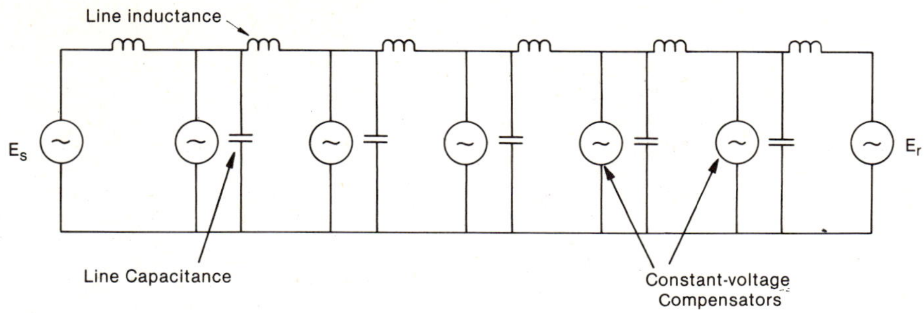

FIGURE 22. Approximate equivalent circuit of line with regulated uniformly distributed shunt compensation.

The phasor diagram of a system with ideal regulated shunt compensation is shown in Figure 21 and Figure 22 is an approximate lumped equivalent circuit. It is of interest to note that the concept of distributed shunt compensation was proposed in 1921 by F. G. Baum(3) as a means of securing constant *voltage* along a long line, but the stability issue was not fully understood at that time and no mention was made of the dynamic nature of this compensation scheme. Baum's compensators would have been synchronous condensers. A truly dynamically compensated line was not built until more than half a century later. One of the largest of these is the 735 kV James Bay Transmission scheme between James Bay and Montréal, Québec, Canada.(16)

2.4. PASSIVE SHUNT COMPENSATION

2.4.1. Control of Open-Circuit Voltage with Shunt Reactors

Shunt compensation with reactors increases the virtual surge impedance (Figure 16) and reduces the virtual natural load, that is, the load at which a flat voltage profile is achieved. With $k_{sh} = 1.0$ the voltage profile is flat at *no-load* (or on open-circuit).

In practice, shunt compensating reactors cannot be uniformly distributed. Instead, they are connected at the ends of the line and at intermediate points—usually at intermediate switching substations. A typical arrangement for a double-circuit line is shown in Figure 23. On a long radial line, the switching stations may occur typically at intervals of between 50 and 250 mi.

In the case of very long lines, at least some of the shunt reactors are permanently connected to the line (as shown in Figure 23) in order to give maximum security against overvoltages in the event of a sudden rejection of load or open-circuiting of the line. On shorter lines, or on sections of line between unswitched reactors, the overvoltage problem is less severe and the reactors may be switched frequently to assist in the hour-by-hour management of reactive power as the load varies. Shunt

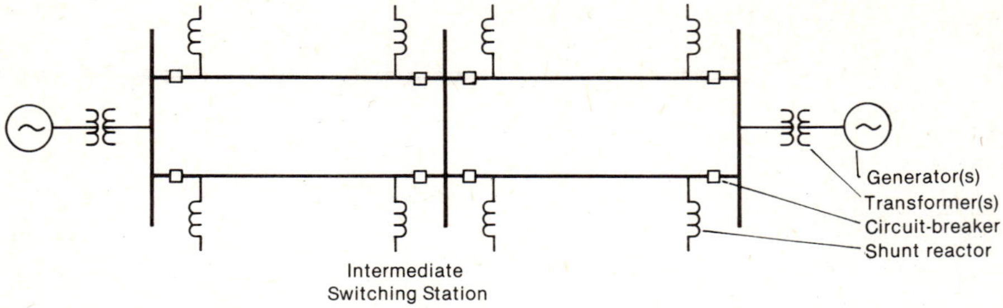

FIGURE 23. Arrangement of shunt reactors on a long-distance high-voltage ac transmission line.

capacitors are usually switched. If there is a sudden load-rejection or open-circuiting of the line, it may be necessary to disconnect them very quickly, to prevent them from increasing the voltage still futher, and also to reduce the likelihood of ferroresonance where transformers remain connected.

Required Reactance Values of Shunt Reactors. The calculation of the optimum ratings and points of connection of shunt reactors and capacitors is generally done by means of extensive load-flow studies, taking into account all possible system configurations.† A simple approach to the problem for a single line is, nevertheless, instructive.

Consider the simple circuit in Figure 24, which has a single shunt reactor of reactance X at the receiving end and a pure voltage source $\mathbf{E}_s$ at the sending end. The receiving-end voltage is given by

$$\mathbf{V}_r = jX\mathbf{I}_r \,. \tag{78}$$

From Equation 2a,

$$\mathbf{E}_S = \mathbf{V}_r \cos \beta a + jZ_0\mathbf{I}_r \sin \beta a$$

$$= \mathbf{V}_r \left[\cos \theta + \frac{Z_0}{X} \sin \theta\right]. \tag{79}$$

$\mathbf{E}_S$ and $\mathbf{V}_r$ are, therefore, in phase, which is consistent with the fact that no real power is being transmitted. For the receiving-end voltage to be equal to the sending-end voltage, X must be given by

$$X = Z_0 \frac{\sin \theta}{1 - \cos \theta} \,. \tag{80}$$

† A comparative study of several shunt and series compensation schemes was published by Iliceto and Cinieri.[(2)]

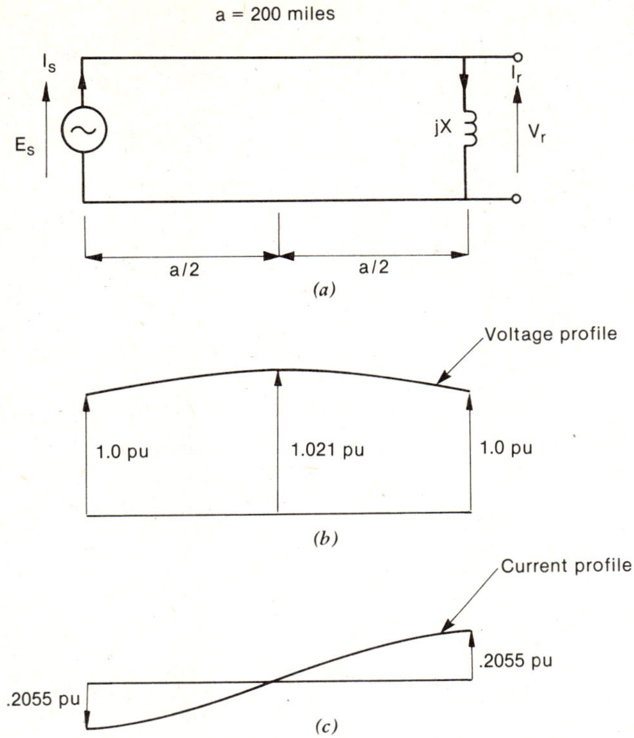

FIGURE 24. Voltage and current profiles of a shunt-compensated line at no-load. (a = 200 mi)

The sending-end current is given by Equation 2*b* as

$$\mathbf{I}_S = j\,\frac{\mathbf{E}_S}{Z_0}\sin\theta + \mathbf{I}_r\cos\theta\,. \tag{81}$$

Making use of Equations 78, 79, and 80, this can be rearranged to give

$$\mathbf{I}_S = j\,\frac{\mathbf{E}_S}{Z_0}\,\frac{1-\cos\theta}{\sin\theta} = j\,\frac{\mathbf{E}_S}{X} = -\mathbf{I}_r \tag{82}$$

since $\mathbf{E}_S = \mathbf{V}_r$. This means that the generator at the sending end behaves exactly like the shunt reactor at the receiving end in that both absorb the same amount of reactive power:

$$Q_S = -Q_r = \frac{E_S^2}{X} = \frac{E_S^2}{Z_0}\left[\frac{1-\cos\theta}{\sin\theta}\right]. \tag{83}$$

The charging current divides equally between the two halves of the line. The voltage profile is symmetrical about the midpoint, and is shown in Figure 24 together with the line-current profile. In the left half of the line the charging current is negative; at the midpoint it is zero; and in the right half it is positive. The maximum voltage occurs at the midpoint and is given by Equation 2a with $x = a/2$:

$$\mathbf{V}_{\mathrm{m}} = \mathbf{V}_{\mathrm{r}} \left[\cos \frac{\theta}{2} + \frac{Z_0}{X} \sin \frac{\theta}{2} \right] = \frac{\mathbf{E}_{\mathrm{s}}}{\cos (\theta/2)} . \tag{84}$$

Note that $\mathbf{V}_{\mathrm{m}}$ is in phase with $\mathbf{E}_{\mathrm{s}}$ and $\mathbf{V}_{\mathrm{r}}$, as is the voltage at all points along the line. For a 200-mi line with $E_{\mathrm{s}} = V_0 = 1.0$ pu, the midpoint voltage is 1.021 p.u. and the reactive power absorbed at each end is 0.2055 P_0. These values should be compared with the receiving-end voltage of 1.088 pu and the sending-end reactive-power absorption of $Q_{\mathrm{s}} = 0.429\ P_0$ in the absence of the compensating reactor. For continuous duty at no-load with a line voltage of 500 kV (phase-to-phase) the rating of the shunt reactor shown in Figure 24 would be 68.5 MVAr per phase, Z_0 being 250 Ω.

Equation 84 shows that with the shunt reactor the line behaves at no-load as though it were two separate open-circuited lines placed back-to-back and joined at the midpoint. The open-circuit voltage rise on each half is given by Equation 84.

Multiple Shunt Reactors along a Long Line. The analysis of the simple case shown in Figure 24 can be generalized to describe the behavior of a line divided into n sections by $n - 1$ shunt reactors spaced at equal intervals along the line, together with a shunt reactor at each end. Because of the standing-wave nature of the voltage and current variation along the line, the voltage and current profiles in Figure 24 could be reproduced in every section if every section were of length a and if the *terminal conditions were the same.* The terminal voltages at the ends of the section shown in Figure 24 are equal in magnitude and phase. The currents are equal but opposite in phase. The correct conditions could, therefore, be achieved by connecting shunt reactors of *twice* the susceptance given by Equation 80 at every junction between two sections. The concept of building up a long shunt-compensated line in this way is shown in Figure 25. The shunt reactors at the ends of the line are each of half the susceptance of the intermediate ones. If a is the total length of the composite line, replacing a by a/n in Equation 80 gives the required reactance of each of the intermediate reactors:

$$X = \frac{Z_0}{2} \left[\frac{\sin (\theta/n)}{1 - \cos (\theta/n)} \right] . \tag{85}$$

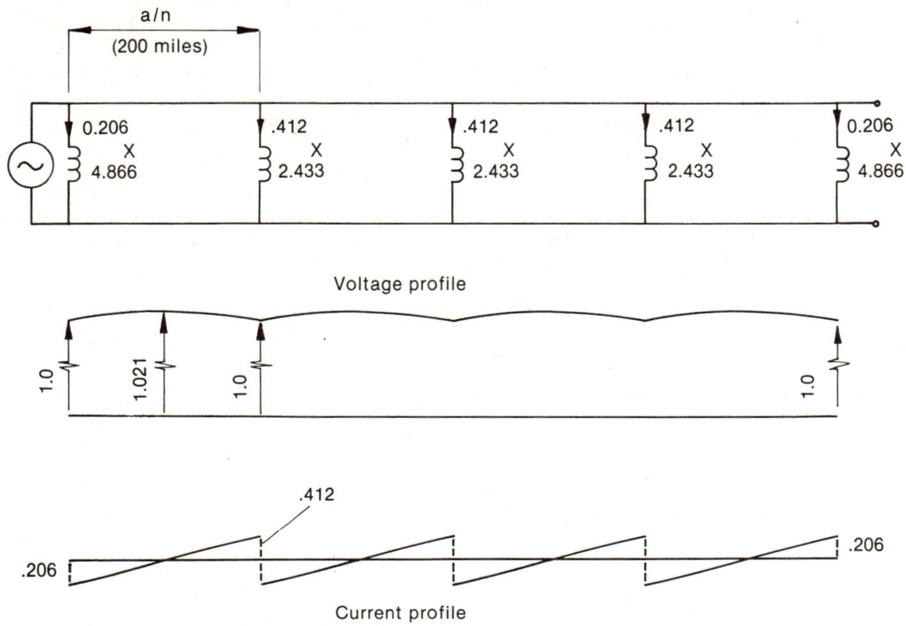

FIGURE 25. Line compensated with multiple shunt reactors at no-load.

In Figure 25, the sending-end generator supplies no reactive current. In practice it would supply, to a first approximation, only the losses.

Each intermediate reactor can be thought of as absorbing the line-charging current or reactive power from two half-sections of line, each of length $a/2n$ on either side of it, whereas the reactors at each end absorb the reactive power from the half-section on one side only. This again explains why the intermediate reactors have twice the susceptance.

In the shunt-compensated systems in Figures 24 and 25, the equivalent degree of compensation k_{sh} is approximately unity. The reactive power absorbed from each half of each 200-mi section is given exactly by Equation 83 as 0.2055 P_0. With *uniformly distributed* compensation and $k_{sh} = 1$, the reactive power to be absorbed from each 100-mi half-section is given by Equation 77 as 0.2027 P_0. The two figures differ by about 1.4% only.

The definition of k_{sh} given in Section 2.3.2 (Equation 55) can be used to define k_{sh} for *lumped* compensation if $b_{\gamma sh}$ is replaced by the total shunt compensating susceptance and b_c by the total shunt capacitance of the line. For example, a typical value of b_c at 500 kV is 8 μS/mi ($c =$ 0.0212 μF/mi) giving a line-charging reactive power of 2 MVAr/mi (three-phase). If $Z_0 = 250\ \Omega$, then from Equation 80 the receiving-end shunt reactor has $X = 1216\ \Omega = 4.866$ pu, $B = 8.22 \times 10^{-4}$ S, and a reactive power of 205.5 MVAr, corresponding to the charging reactive power of half the line, which is 100 × 2 = 200 MVAr. The lumped value of k_{sh} in this example is $8.22 \times 10^{-4}/(100 \times 8 \times 10^{-6}) = 1.028$. The use and interpretation of k_{sh}, therefore, has to be treated with care over longer line sections.

2.4.2. Voltage Control by Means of Switched Shunt Compensation

The voltage regulation diagram for the line of Figure 24 is shown in Figure 26 for three different power factors. The curves labeled "U" are for the uncompensated line; those labeled "L" apply when the shunt reactor discussed in Section 2.4.1 is connected; and those labeled "C" apply when a capacitor of equal reactance is connected instead of the reactor.

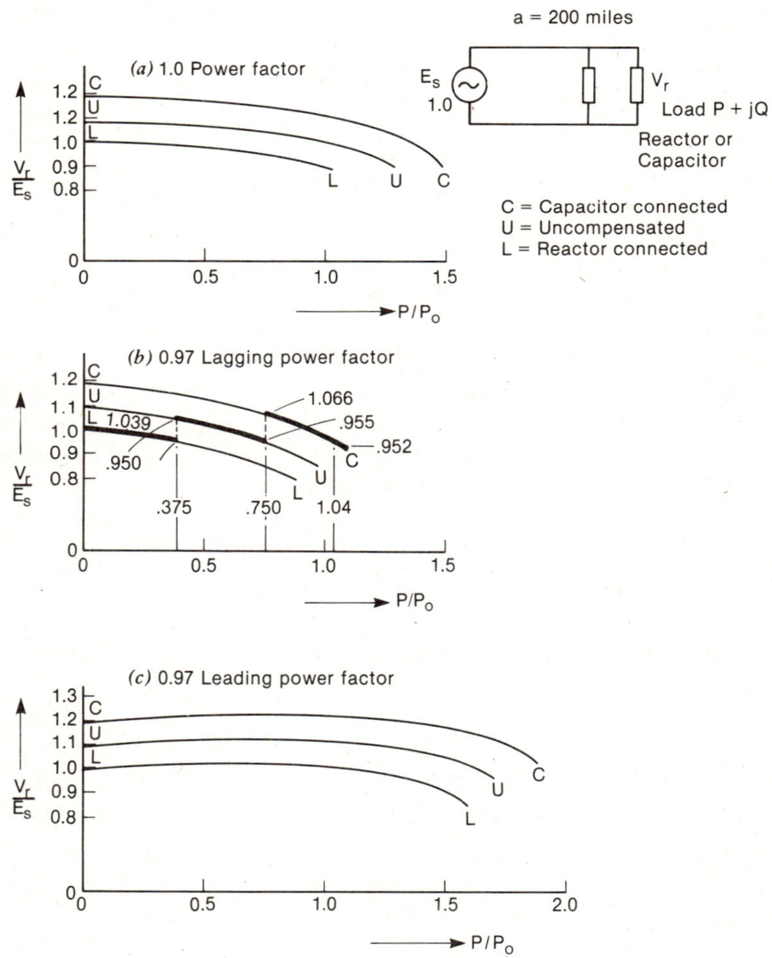

FIGURE 26. Voltage control by means of switched shunt compensation.

The fact that the "L" curve lies below the "U" curve in every case reflects the fact that on the uncompensated line the line-charging reactive power supports the voltage to a significant degree under load. This advantage is lost if the shunt reactor is permanently connected.

Figure 26*b* illustrates the principle by which the receiving-end voltage can be kept more or less constant as the load varies, by switching the reactor and the capacitor. In this example the load power factor is assumed to be 0.97 lagging. At zero load, the shunt reactor is connected and reduces the uncompensated open-circuit voltage from 1.088 pu to 1.0 pu. It remains connected until the power transfer reaches 0.375 P_0, at which level the receiving-end voltage has decreased to 0.95 pu. The reactor is then disconnected and the line remains uncompensated between $P = 0.375\ P_0$ and $P = 0.75\ P_0$, at which level the voltage has decreased to 0.955 pu. When $P = 0.75\ P_0$ the capacitor is connected, and it sustains the voltage above 0.95 pu until P reaches 1.04 P_0. The voltage control in this illustrative example is very coarse. In practice, the switching of shunt reactors and/or capacitors is coordinated with the control of tap-changing transformers and other voltage-regulating equipment to maintain the voltage within narrower limits than are indicated in Figure 26*b*.

2.4.3. The Midpoint Shunt Reactor or Capacitor

The shunt reactor or capacitor located at the midpoint of a symmetrical line is a special case of the line with $n - 1$ reactors considered in Section 2.4.1. It has $n = 2$. It is useful to study this case in detail because it can be meaningfully compared with the series-compensated line treated in Section 2.5 and the line compensated by sectioning in Section 2.6. Its main application is in the control of line voltage and power factor, as will be shown.

Each half of the line is represented by its π-equivalent circuit as in Figure 27*a*. The terminal synchronous machines are assumed to have constant voltage and zero internal reactance. The two capacitive shunt susceptances $B_c/4$ connected at the ends can be omitted from the analysis if it is remembered that the terminal synchronous machines absorb their reactive power at all times. To account for this, $\mathbf{I}_s$ is written $\mathbf{I}_s = \mathbf{I}_s' + j(B_c/4)\mathbf{E}_s$, and similarly for $\mathbf{I}_r$. In effect, the two extreme quarters of the line length have 100% compensation of their shunt capacitance. The central half of the line is compensated by a single shunt reactor or capacitor, which may be switched on or off, but whose susceptance cannot be varied continuously. The degree of compensation for the central half is given by

$$k_m = \frac{B_\gamma}{\frac{1}{2} B_c}. \tag{86}$$

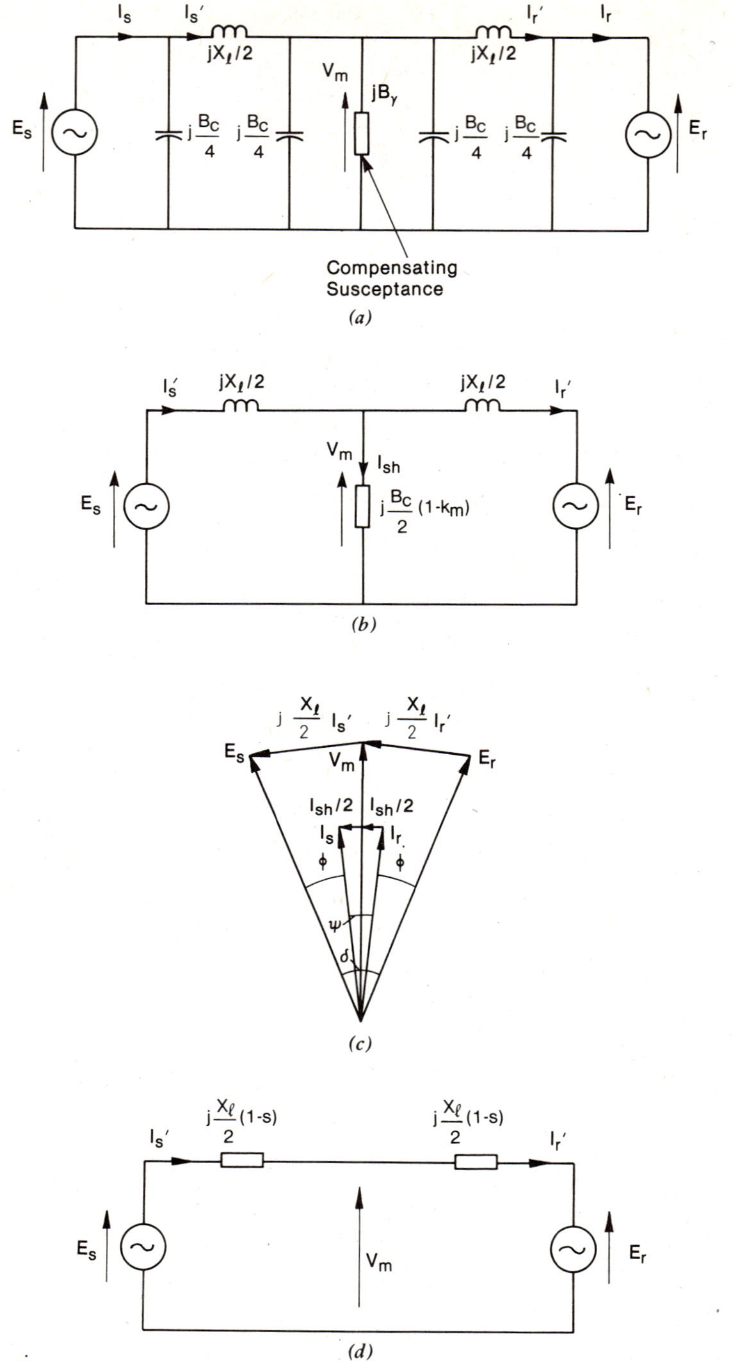

FIGURE 27. Analysis of line compensated with midpoint reactor or capacitor.

k_m is arbitrarily taken as positive if B_γ is inductive, negative if B_γ is capacitive, and is unity when $B_\gamma = B_c/2$, B_c being the shunt susceptance of the whole line.

At fundamental frequency, the equivalent circuit reduces to Figure 27*b*. The phasor diagram is given in Figure 27*c*. The basic relationships are

$$P = V_m I_m \,, \tag{87}$$

$$V_m = \mathbf{E}_r + j\,\frac{X_l}{2}\,\mathbf{I}_r' = \mathbf{E}_s - j\,\frac{X_l}{2}\,\mathbf{I}_s' \,, \tag{88}$$

and

$$\mathbf{I}_{sh} = \mathbf{I}_s' - \mathbf{I}_r' = 2(\mathbf{I}_s' - \mathbf{I}_m) = 2(\mathbf{I}_m - \mathbf{I}_r') = j\,\frac{B_c}{2}\,(1 - k_m)\mathbf{V}_m \,. \tag{89}$$

From these it can be shown that with $E_s = E_r = E$,

$$P = \frac{E^2}{X_l(1 - s)} \sin\delta \,, \tag{90}$$

where

$$s = \frac{X_l}{2}\,\frac{B_c}{4}\,(1 - k_m) \,. \tag{91}$$

The midpoint voltage can be expressed as

$$V_m = \frac{E\cos(\delta/2)}{1 - s} \,, \tag{92}$$

which shows that with $s < 1$ the midpoint compensation increases the midpoint voltage by the factor $1/(1 - s)$. This factor of increase tends to offset the voltage drop in the series reactance of the line and, therefore, mitigate the midpoint voltage sag under load. If $\cos(\delta/2)/(1 - s) > 1$ the midpoint voltage exceeds the terminal voltages even under heavy load.

If $k_m = 1$ the shunt capacitance is canceled along the entire line length, making the virtual natural load zero. The compensating susceptance is then a reactor of admittance $B_c/2$ (Equation 86). Only at no-load is the voltage profile flat. The line is practically reduced to its series impedance. If the terminal voltages are adjusted to keep $V_m = V_0 = 1$ pu while the power transfer is increased, then their values are given by the equation

$$E_s = E_r = V_0 \sqrt{1 + \left(\frac{x_l}{2}\right)^2 \left(\frac{P}{P_0}\right)^2}, \tag{93}$$

which can be deduced from the phasor diagram. Note that here we have written $x_l = X_l/Z_0 = \theta = \beta a$.

Among other equations of interest for the midpoint-compensated symmetrical line are the following. The terminal currents are given by

$$\mathbf{I}_s = \mathbf{I}_r^* = I_m \left(1 - \frac{B_c X_l}{8}\right) + jV_m \frac{B_c}{4} (2 - k_m - s). \tag{94}$$

The terminal reactive powers are given by

$$Q_s = -Q_r = \frac{P^2}{V_m^2} \frac{X_l}{2} \left\{1 - \frac{B_c X_l}{8}\right\}$$

$$- V_m^2 \frac{B_c}{4} (1 - s)(2 - k_m - s). \tag{95}$$

This expression includes the reactive power of the extreme capacitive branches of the π-equivalent circuits, which the terminal synchronous machines must absorb at all times.

The reactive power of the midpoint compensating susceptance is given by

$$Q_\gamma = E^2 \frac{B_c k_m}{4(1 - s)^2} (1 + \cos \delta). \tag{96}$$

Note that Q_γ is positive for a reactor ($k_m > 0$) and negative for a capacitor ($k_m < 0$).

An example illustrates the features of the midpoint-compensated line. For a 200-mi line $B_c/Y_0 = X_l/Z_0 = \theta = 0.4054$ pu.† For 100% compensation of the line capacitance $B_\gamma = B_c/2 = 0.2027$ per-unit of Y_0. At 500 kV with $Z_0 = 250\Omega$ the required shunt reactor would have a reactance of

$$X_\gamma = \frac{2}{B_c} Z_0 = \frac{2}{0.4054} \times 250 = 1233\Omega \tag{97}$$

† The short-section π-equivalent circuit is used in this example.

(phase-neutral), and its reactive-power rating would be $[500/\sqrt{3}]^2/1233 = 67.6$ MVA per phase. This agrees with the approximate figure obtained by reckoning 2 MVA/mi for the three phases over the central 100-mi section of the line (see Table 3). At no-load the terminal synchronous machines must each absorb $\frac{1}{2} \times 3 \times 67.6 = 101.4$ MVA, which is about one-tenth of the natural load. With the reactors connected, $k_m = 1$ and $s = 0$. The voltage profile is approximately flat, with $E_s = E_r = V_m = 500$ kV (phase-to-phase). Without the reactor, the midpoint voltage would be $V_m = 1.021$ pu $= 510.5$ kV (see Figure 5*b*).

With the midpoint reactors connected, at the natural load $P = P_0 = 1000$ MW (three-phase), and from Equation 90, $\delta = 23.92°$. The midpoint voltage is given by Equation 92: with $E = 1.0$ pu $= 500$ kV, $V_m = \cos(\delta/2) = 0.978$ pu $= 489$ kV (phase-to-phase). The voltage profile is as shown in Figure 28*b*. The terminal reactive powers are given by Equation 95, that is, 212 MVAr generation at each end, giving a terminal power factor of 0.978 lagging at the sending end and 0.978 leading at the receiving end.

If the reactor is replaced by a capacitor of equal susceptance, $k_m = -1$ and from Equation 91

$$s = \frac{0.4054\ Z_0}{2} \times \frac{0.4054\ Y_0}{4}\ [1 - (-1)] = 0.0411\,. \qquad (98)$$

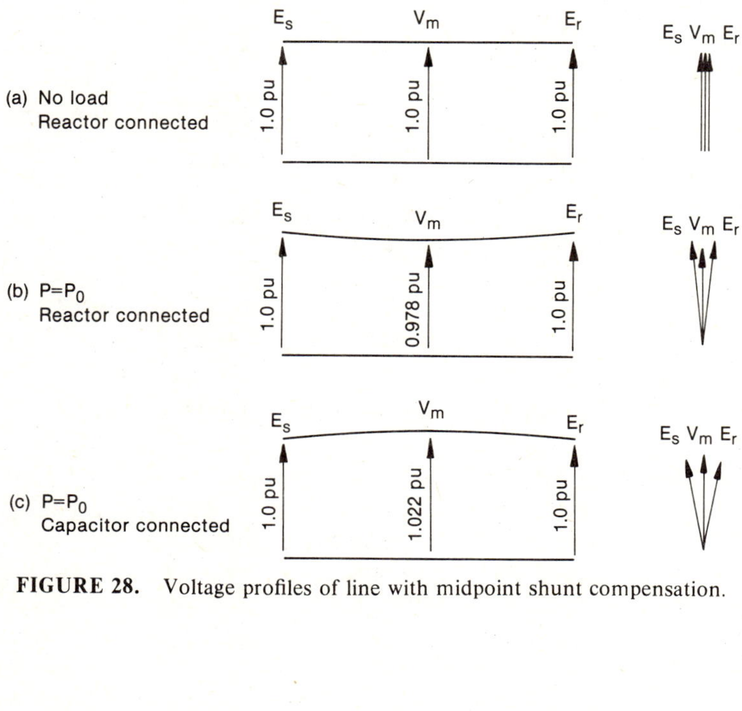

FIGURE 28. Voltage profiles of line with midpoint shunt compensation.

At the natural load, from Equation 90, $\delta = 22.88°$ and from Equation 92, $V_m = 1.022$ pu. The terminal reactive powers are now each 9 MVAr absorption at each end. As would be expected with such a low value of *s*, the stabilizing influence of the midpoint capacitor is insignificant, although the midpoint *voltage support* is worthwhile.

In the preceding example the midpoint susceptance is much more effective in controlling the line voltage and the terminal power factors than in reducing the transmission angle. It is for voltage and power factor control that this type of compensation is normally used. It appears from Equation 90 that a significant effect on the transmission angle would be achieved only if *s* were an order of magnitude larger than in the example: this would be the case if the line length were very much longer and/or if the capacitive compensating susceptance were extremely large. However, since the midpoint *voltage* is also subject to the factor $1/(1 - s)$, with so much compensation it would either become unacceptably high, or become excessively sensitive to the switching of the capacitor, or both. For these reasons the single passive shunt susceptance is not a practical way of increasing the maximum transmissible power of a long line. For shunt stabilization of a long line, the susceptance has to be large and *dynamically controlled,* as for example in the thyristor-controlled or saturated-reactor compensators (see Section 2.6).

The form of Equation 90 suggests a parallel between the midpoint-compensated line and the series-compensated line because the coefficient of $\sin\delta$ is of the same form as Equation 109 with k_{se} replaced by *s*. In this sense *s* can be interpreted as an equivalent degree of series compensation. The series-capacitor effect in the transmission-angle characteristic is weak and is obtained only with $s < 1$. This requires $k_m < 1$, which says that if B_γ is inductive it must not be so large as to compensate all the shunt capacitance of the central half of the line. In other words, the *shunt capacitance of the line has a stabilizing influence on the power transmission*—which can be enhanced if B_γ is made capacitive ($k_m < 0$), or weakened if B_γ is made inductive ($k_m > 0$).

2.5. SERIES COMPENSATION[11-13]

2.5.1. Objectives and Practical Limitations

As discussed in Section 2.3, the central concept in series compensation is to cancel part of the reactance of the line by means of series capacitors. This increases the maximum power, reduces the transmission angle at a given level of power transfer, and increases the virtual natural load. The line reactance, however, being effectively reduced, now absorbs less of the line-charging reactive power, often necessitating some form of shunt-inductive compensation.

As a means of reducing the "transfer reactance" between the ends of a line, the series capacitor finds two main classes of application:

1. It can be used to increase the power transfer on a line of any length. Sometimes a series capacitor is used to increase the load share on one of two or more parallel lines—especially where a higher-voltage line overlays a lower-voltage line in the same corridor.
2. It can be used to enable power to be transmitted stably over a greater *distance* than is possible without compensation. On long lines shunt inductive compensation is usually also necessary in order to limit the line voltage.

A practical upper limit to the degree of series compensation is of the order of 0.8. With $k_{se} = 1$ the effective line reactance would be zero, so that the smallest disturbance in the relative rotor angles of the terminal synchronous machines would result in the flow of large currents. The circuit would also be series resonant at the fundamental frequency, and it would be difficult to control transient voltages and currents during disturbances.

The capacitor *reactance* is determined by the desired steady-state and transient power transfer characteristics, as well as by the location of the capacitor on the line. Its location is decided partly by economic factors and partly by the severity of fault currents (which are a function of the capacitor location). The voltage rating is determined by the severity of the worst anticipated fault currents through the capacitor and any bypass equipment—"severity" being a function of duration as well as magnitude.

Clearly it is not practicable to distribute the capacitance in small units along the line. In practice lumped capacitors are installed at a small number of locations (typically one or two) along the line. This makes for an uneven voltage profile,[(2)] as will be seen.

2.5.2. Symmetrical Line with Midpoint Series Capacitor and Shunt Reactors

The case studied in this section is a lossless, symmetrical line with a midpoint series capacitor on either side of which are connected two equal shunt reactors (see Figure 29*a*).† The capacitor is shown split into two equal series parts for convenience in analysis. The purpose of the shunt reactors is to control the line voltage; this is illustrated by an example in Section 2.5.3.

† Other configurations are studied in Reference 2.

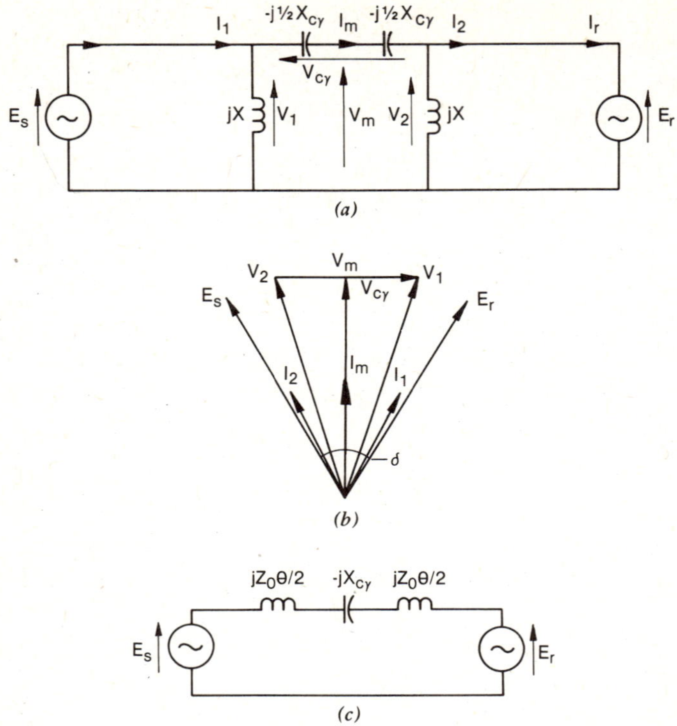

FIGURE 29. (*a*) Symmetrical line with midpoint series capacitor and shunt reactors. (*b*) General phasor diagram for Figure 29*a*. (*c*) Equivalent circuit with perfect shunt compensation.

Power-Transfer Characteristics and Maximum Transmissible Power. The general phasor diagram is shown in Figure 29*b*. Considering the left-hand (sending-end) line section, the conditions at its two ends are related by the fundamental Equation 2:

$$\mathbf{E}_s = \mathbf{V}_1 \cos \frac{\theta}{2} + jZ_0\mathbf{I}_1 \sin \frac{\theta}{2} \tag{99a}$$

$$\mathbf{I}_s = j\frac{\mathbf{V}_1}{Z_0} \sin \frac{\theta}{2} + \mathbf{I}_1 \cos \frac{\theta}{2}. \tag{99b}$$

The receiving-end half of the line behaves similarly. The capacitor reactance is $X_{c\gamma} = 1/\omega C_\gamma$ and the voltage across the capacitor is given by

$$\mathbf{V}_{c\gamma} = \mathbf{V}_1 - \mathbf{V}_2 = -j\mathbf{I}_m X_{c\gamma}. \tag{100}$$

By symmetry, $P = V_m I_m$, $E_s = E_r$, and

$$\mathbf{V}_m = \mathbf{V}_1 - \tfrac{1}{2}\mathbf{V}_{c\gamma} = \mathbf{V}_2 + \tfrac{1}{2}\mathbf{V}_{c\gamma}. \tag{101}$$

ents $\mathbf{I}_1$ and $\mathbf{I}_2$ are given by

$$\mathbf{I}_m = \mathbf{I}_1 + \frac{j\mathbf{V}_1}{X} = \mathbf{I}_2 - j\frac{\mathbf{V}_2}{X}. \tag{102}$$

Using these relationships, and taking $\mathbf{V}_m$ as reference phasor, it is possible to follow the same procedure as was used in Section 2.2.6 to derive the basic power-transfer characteristic (Equation 37). This yields the following results:

$$P = \frac{E_s V_m}{Z_0 \sin\frac{\theta}{2} - \frac{X_{c\gamma}}{2}\left[\cos\frac{\theta}{2} + \frac{Z_0}{X}\sin\frac{\theta}{2}\right]} \sin\frac{\delta}{2} \tag{103}$$

and

$$E_s \cos\frac{\delta}{2} = V_m\left[\cos\frac{\theta}{2} + \frac{Z_0}{X}\sin\frac{\theta}{2}\right] = E_r \cos\frac{\delta}{2}. \tag{104}$$

If V_m is substituted from Equation 104 into Equation 103, the following general result is obtained for the symmetrical line, if $E_s = E_r$:

$$P = \frac{E_s E_r}{\left[Z_0 \sin\theta - \frac{X_{c\gamma}}{2}(1 + \cos\theta)\,\mu_x\right]\mu_x} \sin\delta, \tag{105}$$

where

$$\mu_x = 1 + \frac{Z_0}{X}\frac{\sin\theta}{1 + \cos\theta} = 1 + \frac{Z_0}{X}\tan\frac{\theta}{2}. \tag{106}$$

In the absence of the shunt reactors, $\mu_x = 1$. With fixed terminal voltages, $E_s = E_r = E$, the transmission angle δ can be determined from Equation 105 for any level of power transmission below the maximum. Once δ is known, V_m can be determined from Equation 104. Then $\mathbf{V}_1$, $\mathbf{V}_2$, $\mathbf{V}_{c\gamma}$, and other quantities can be found from Equations 99 through 101.

Special Cases:

1. The most familiar form of the series-compensated power transfer characteristic is obtained from Equation 105 by ignoring the shunt capaci-

tance of the line and removing the shunt reactors. Then $Z_0 \sin\theta$ is replaced by X_l and $\mu_x = 1$, so that with $E_s = E_r = E$,

$$P = \frac{E^2}{X_l - X_{c\gamma}} \sin\delta . \tag{107}$$

If the degree of series compensation k_{se} is defined by

$$k_{se} = \frac{X_{c\gamma}}{X_l} = \frac{X_{c\gamma}}{\omega a l} , \tag{108}$$

then

$$P = \frac{E^2}{X_l(1 - k_{se})} \sin\delta , \tag{109}$$

which corresponds to Equation 72. This form of the power-transfer characteristic is useful because of its simplicity. The error resulting from the neglect of shunt capacitance is illustrated in the example of Section 2.5.3.

2. Another important special case arises when the shunt reactors are chosen to compensate the line capacitance perfectly as discussed in Section 2.4.1. Their reactances are then each

$$X = Z_0 \frac{\sin(\theta/2)}{1 - \cos(\theta/2)} . \tag{110}$$

From Equation 106 the shunt reactance factor μ_x reduces to

$$\mu_x = \sec\frac{\theta}{2} \tag{111}$$

and the power-transfer characteristic becomes (with $E_s = E_r = E$)

$$P = \frac{E^2}{2Z_0 \sin\frac{\theta}{2} - X_{c\gamma}} \sin\delta . \tag{112}$$

The perfect shunt compensation of the capacitance of each half of the line leaves only the series reactance in the equivalent circuit, which is shown in Figure 29*c*. This equivalent circuit does not take into account the fact that because the shunt-inductive compensation is concentrated, the line voltage profile is not perfectly flat, even at no-load.

The maximum transmissible power computed from Equation 105 can be compared with the value obtained with uniformly distributed shunt and series compensation (Equation 70). Reference 2 describes this comparison for the scheme of Figure 29*a*, as well as for several other capacitor locations. (See also Section 2.5.3).

Reactive Power Requirements at the Terminals. The reactive power at the sending-end terminal is given by

$$Q_s = \mathrm{Im}(\mathbf{E}_s \mathbf{I}_s^*)\,. \tag{113}$$

From Equations 99 through 102, it can be shown that

$$\mathbf{E}_s = \mu_x V_m \cos\frac{\theta}{2} + \mathrm{j}\frac{P}{V_m}\left[Z_0 \sin\frac{\theta}{2} - \frac{\mu_x X_{c\gamma}}{2}\cos\frac{\theta}{2}\right] \tag{114}$$

and

$$\mathbf{I}_s = I_m\left[\frac{\mu'_x X_{c\gamma}}{2Z_0}\sin\frac{\theta}{2} + \cos\frac{\theta}{2}\right] + \mathrm{j}\frac{\mu_x V_m}{Z_0}\sin\frac{\theta}{2}\,, \tag{115}$$

where

$$\mu'_x = 1 - \frac{Z_0}{X}\,\frac{1+\cos\theta}{\sin\theta} = 1 - \frac{Z_0}{X}\,\mathrm{ctg}\,\frac{\theta}{2}\,. \tag{116}$$

In the absence of the shunt reactors, $\mu'_x = 1$ $(= \mu_x)$. Substituting for $\mathbf{E}_s$ and $\mathbf{I}_s^*$ in Equation 113, the result is

$$Q_s = \frac{P_0}{2}\left(\frac{P}{P_0}\right)^2\left(\frac{V_0}{V_m}\right)^2\left\{\frac{X_{c\gamma}}{2Z_0}\left[(\mu'_x - \mu_x) - (\mu'_x + \mu_x)\cos\theta\right]\right.$$
$$\left. + \left[1 - \left(\frac{X_{c\gamma}}{2Z_0}\right)^2 \mu'_x\mu_x\right]\sin\theta\right\} - \frac{P_0}{2}\left(\frac{V_m}{V_0}\right)^2 \mu'_x\mu_x \sin\theta\,. \tag{117}$$

By symmetry the reactive power at the receiving end is $Q_r = -\,Q_s$. In all cases V_m is given by Equation 104, with δ derived from Equation 105.

Special Cases:

1. In the absence of the shunt reactors $\mu'_x = \mu_x = 1$ and the Equation 117 for Q_s and $-\,Q_r$ reduces to

$$Q_s = \frac{P_0}{2}\left\{ \sin\theta \left\{ \left(\frac{P}{P_0}\right)^2 \left(\frac{V_0}{V_m}\right)^2 \left[1 - \left(\frac{X_{c\gamma}}{2Z_0}\right)^2\right] - \left(\frac{V_m}{V_0}\right)^2 \right\} \right.$$

$$\left. - \cos\theta \left(\frac{P}{P_0}\right)^2 \left(\frac{V_0}{V_m}\right)^2 \frac{X_{c\gamma}}{Z_0} \right\}. \tag{118}$$

In the absence of the series capacitor and the shunt reactors, this equation reduces to Equation 29 for the uncompensated line.

2. If the shunt reactors are designed for perfect no-load compensation (Equation 110), then μ_x is given by Equation 111 and μ'_x by

$$\mu'_x = \frac{Z_0}{X} \csc \frac{\theta}{2}. \tag{119}$$

Then the following simplifications arise:

$$(\mu'_x - \mu_x) - (\mu'_x + \mu_x)\cos\theta = 2\left[1 - 2\cos\frac{\theta}{2}\right] \tag{120}$$

and

$$\mu'_x \mu_x = \frac{2Z_0}{X \sin\theta}. \tag{121}$$

The terminal reactive powers become

$$Q_s = \frac{P_0}{2}\left(\frac{P}{P_0}\right)^2 \left(\frac{V_0}{V_m}\right)^2 \left\{ \frac{X_{c\gamma}}{Z_0}\left(1 - 2\cos\frac{\theta}{2}\right) \right.$$

$$\left. + \left[1 - \left(\frac{X_{c\gamma}}{2Z_0}\right)^2 \frac{2Z_0}{X\sin\theta}\right] \sin\theta \right\} - \frac{P_0}{2}\left(\frac{V_m}{V_0}\right)^2 \frac{2Z_0}{X} \tag{122}$$

with $Q_r = -Q_s$. Note that at no-load ($P = 0$), Q_s reduces to $-V_m^2/X$; that is, the reactive power absorption at the sending end is identical to that in the left-hand shunt reactor. (See Section 2.4.1). Similarly, at no-load the receiving-end absorption equals the reactive power in the right-hand shunt reactor, both being equal to V_m^2/X. At no-load there is no current in the series capacitor and the two halves of the line behave as two back-to-back open-circuited lines each with a shunt reactor at its receiving end.

2.5.3. Example of a Series-Compensated Line

The following example shows the mechanism of series compensation in the steady state. In order to demonstrate the importance of the accompanying shunt reactors, the example is worked with and without them.

It is again emphasized that the midpoint is only one of many possible locations for the series capacitor, and is not necessarily always the most favorable, either technically or economically. For a comparative steady-state evaluation of different capacitor locations, the reader is referred to Reference 2.

Midpoint Series Capacitor Without Shunt Reactors. The example is for a 400-mi line, in order to facilitate comparisons with examples elsewhere in the chapter.† With $a =$ 400 mi, the electrical length is $\theta = \beta a = 0.8108$ rad or 46.5°. In the per-unit system based on V_0 and Z_0, the per-unit reactance is $X_l = 0.8108$ and the total shunt capacitive susceptance is $B_c = 0.8108$ pu. The series compensating capacitor is chosen so as to compensate 50% of the line reactance, so that $X_{c\gamma} = 0.5 \times 0.8108 = 0.4054$ pu. (At 500 kV with $Z_0 = 250\ \Omega$, $X_{c\gamma} = 101.4\ \Omega$.) Without midpoint shunt reactors, $X = \infty$ and $\mu_x = \mu'_x = 1$. Assuming that the terminal voltages are constant, $E_s = E_r = V_0 = 1.0$ pu. From Equation 104 the no-load midpoint voltage (with $\delta = 0$) is $1/\cos(\theta/2) = 1.0882$ pu. Since there is no current through the series capacitor, this voltage appears also on both sides of it. From Equation 105 the power transfer characteristic is given by

$$P = \frac{\sin \delta}{\sin 0.8108 - \tfrac{1}{2} \times 0.4054(1 + \cos 0.8108)}$$

$$= 2.6144 \sin \delta\,. \qquad (123)$$

If there were no compensation P_{max} would be 1.3796 pu (from Equation 37), so the series capacitor increases P_{max} by a factor of $2.6144/1.3796 = 1.8950$. If the compensation were uniformly distributed with $k_{se} = 0.5$, P_{max} would be 2.6072 P_0 (from Equation 70), which is only about 0.25% different from the value obtained here with lumped capacitors.

† Note that the full distributed-parameter representation has been used for both halves of the line, and not the equivalent-π circuits as in Sections 2.4 and 2.6.

The reactive power requirements at the terminals can be calculated from Equation 118, which gives

$$Q_s = -Q_r = \left[0.2079\frac{p^2}{v_m^2} - 0.3624\, v_m^2\right] P_0\,, \tag{124}$$

where $p = P/P_0$ and $v_m = V_m/V_0$. From Equations 100 and 101 the voltage on either side of the capacitor ($V_1 = V_2$) is given by

$$V_1 = \left| V_m - \mathrm{j}\,0.2027\,\frac{P}{V_m} \right| . \tag{125}$$

The voltage across the series capacitor is given by

$$\mathbf{V}_{c\gamma} = -\mathrm{j}I_m X_{c\gamma} = -\mathrm{j}\,0.4054\,\frac{P}{V_m}\,. \tag{126}$$

Table 5 shows the variation of these parameters as the power transfer increases from zero through 2.0 pu. Also shown is the variation of the transmission angle that would be obtained without compensation. For a given power transfer, this is about twice the value obtained with the series capacitor installed.

Although there is a marked reduction in transmission angle, the voltage V_1 ($= V_2$) on either side of the series capacitor is rather high, and there is very little reduction as the power increases. In addition there is appreciable absorption of reactive power at the line ends, even at the natural load P_0. This can be associated with the generally high voltage conditions along the line, together with the fact that the series capacitor itself generates reactive power.

Midpoint Series Capacitor with Shunt Reactors. The high voltage and reactive power absorption at the ends of the line can be corrected by means of shunt reactors. The value of X required on either side of the series capacitor is given by Equation 110 as $X = 4.8656\ Z_0$. From Equations 119 and 111, $\mu_x' = 0.5211$ and $\mu_x = 1.0882$. Substituting in Equation 105, or directly from Equation 112,

$$P = 2.6084 \sin\delta\,. \tag{127}$$

The maximum transmissible power is hardly affected by the addition of the shunt reactors. With uniformly distributed shunt compensation $k_{sh} = 1$ and from Equation 72, $P = P_0/0.8108(1-0.5) = 2.4667\ P_0$. The midpoint voltage V_m is given by Equation 104 as

$$V_m = 1.0\cos\frac{\delta}{2}\,, \tag{128}$$

TABLE 5
Performance of Midpoint Series Compensated Line without Shunt Reactors
(a = 400 mi, k_{se} = 0.5)

$p = \frac{P}{P_0}$	$v_m = \frac{V_m}{V_0}$	Terminal Reactive Power $\frac{Q_s}{P_0} = -\frac{Q_r}{P_0}$	Trans-Mission Angle δ (°)	$\frac{V_1}{V_0}$	$\frac{V_{c\gamma}}{V_0}$	δ Without Compen-sation (°)
0	1.0882	−0.4292	0	1.0882	0	0
0.25	1.0870	−0.4172	5.487	1.0880	0.0932	10.440
0.50	1.0832	−0.3809	11.026	1.0872	0.1871	21.249
0.75	1.0767	−0.3193	16.671	1.0859	0.2824	32.932
1.00	1.0673	−0.2303	22.488	1.0841	0.3798	46.456
1.25	1.0546	−0.1109	28.563	1.0816	0.4805	64.966
1.50	1.0378	0.0440	35.011	1.0784	0.5860	unstable
2.00	0.9866	0.5015	49.906	1.0688	0.8218	unstable

which indicates a flat voltage profile at no-load (when $\delta = 0$). The reactive power requirements are given by Equation 122 or 117 as

$$Q_s = -Q_r = P_0\left[0.1841\frac{p^2}{v_m^2} - 0.2055\ v_m^2\right]. \tag{129}$$

The voltage on either side of the series capacitor ($V_1 = V_2$) is the same as the voltage across the shunt reactors and is given again by

$$V_1 = \left| V_m - j\ 0.2027\ \frac{P}{V_m} \right|, \tag{130}$$

and the capacitor voltage by

$$\mathbf{V}_{c\gamma} = -j\ 0.4054\ \frac{P}{V_m}. \tag{131}$$

Table 6 shows the variation of these parameters as the power transfer increases from zero through 2 P_0.

The voltage on either side of the series capacitor is nearly 1.0 pu over practically the whole range of power transfer, implying that the shunt reactors could be permanently connected without disadvantage. The reactive absorption at the terminals is considerably reduced, and at high values of P the terminal power factors become lagging, which may be beneficial to transient stability because the internal generator voltages are increased.

Note that shunt reactors have not been connected at the line ends, although it is quite practical to do this. In the calculations the generators at the ends fulfill the shunt compensating function, and at no-load each absorbs exactly the same amount of reactive power as one of the central shunt reactors,—that is, the line-charging reactive power of one 100-mi section ($=0.2055\ P_0$).

TABLE 6
Performance of Midpoint Series Compensated Line with Shunt Reactors
(a = 400 mi, k_{se} = 0.5, Shunt Reactor X = 4.8656 Z_0)

$p = \dfrac{P}{P_0}$	$v_m = \dfrac{V_m}{V_0}$	Terminal Reactive Power $\dfrac{Q_s}{P_0} = -\dfrac{Q_r}{P_0}$	Transmission Angle δ (°)	$\dfrac{V_1}{V_0}$	$\dfrac{V_{c\gamma}}{V_0}$	δ Without Compensation (°)
0	1	−0.2055	0	1	0	0
0.25	0.9988	−0.1935	5.500	1.000	0.1015	10.440
0.50	0.9954	−0.1572	11.051	1.001	0.2036	21.249
0.75	0.9894	−0.0954	16.710	1.001	0.3073	32.932
1.00	0.9807	−0.0062	22.543	1.002	0.4134	46.456
1.25	0.9689	0.1135	28.635	1.004	0.5230	64.966
1.50	0.9534	0.2689	35.104	1.005	0.6378	unstable
2.00	0.9061	0.7282	50.063	1.011	0.8948	unstable

2.6. COMPENSATION BY SECTIONING[15–20] (DYNAMIC SHUNT COMPENSATION)

2.6.1. Fundamental Concepts

If a synchronous machine is connected at an intermediate point along a transmission line, it can maintain constant the voltage at that point, just as the terminal synchronous machines do at the ends. In so doing, it divides the line into two sections which are apparently quite independent. The voltage profile, the maximum transmissible power, and the reactive power requirements of each section can then be determined separately—the problems in each independent section being less severe than in the line as a whole. The maximum transmissible power in particular is now determined by the "weakest link in the chain," which is generally the longest section. For example, if a line is sectioned into two equal halves, then if shunt capacitance is neglected (or completely compensated by shunt reactors), the power transmission characteristic is given by Equation 38 for each half of the line separately. Thus, replacing δ by $\delta/2$ and X_l by $X_l/2$, with $E_s = E_r = E$,

$$P = 2\frac{E_m E}{X_l} \sin \frac{\delta}{2}, \tag{132}$$

where E_m is the midpoint voltage, held constant by the synchronous machine or by some other constant-voltage compensating device. The maximum transmissible power is doubled.

This scheme, called "compensation by sectioning," was proposed by F. G. Baum in 1921 (See Reference 3). Baum suggested that by connecting synchronous condensers at intervals of about 100 mi, a substantially flat voltage profile could be achieved at all levels of power transmission. The condensers would adjust the virtual natural load, P_0', to be equal to the actual load at all times. Baum calculated an example of an 800-mi line at 220 kV transmitting about 100 MW (approximately 0.8 P_0) *with a total transmission angle of about* 146°. Although Baum considered voltage regulation and reactive power requirements in detail, he did not consider stability. It was not until much later that the stability of long lines compensated in this way was adequately understood for practical use (see Section 2.3.4).

If losses are neglected the "compensating" current taken by the intermediate synchronous machine is purely reactive (i.e., in phase quadrature with the voltage at the point of connection), and the machine supplies or absorbs only reactive power to or from the line. In the steady state, therefore, the machine can maintain constant voltage at its point of connection *without requiring a mechanical prime mover.* In a given steady state there is a certain ratio between the compensating current $\mathbf{I}_\gamma$ and the voltage $\mathbf{V}$ at the point of connection. The ratio has the dimensions of a susceptance, which is capacitive if $\mathbf{I}_\gamma$ leads $\mathbf{V}$ and inductive if $\mathbf{V}$ leads $\mathbf{I}_\gamma$. This implies that the synchronous machine could, in the steady state, be replaced by a capacitor or a reactor.

However, if the power being transmitted along the line changed value, the voltage $\mathbf{V}$ would tend to change. In order to restore $\mathbf{V}$ to the constant value required to maintain the independence of the line sections, the capacitive or inductive susceptance would have to change also. This suggests that if the susceptance of a real capacitor or reactor could be modulated or controlled in such a way as to maintain constant voltage at its point of connection, the device would be functionally equivalent to the synchronous machine. Figure 30 illustrates the principle of modulating the susceptance in such a way as to maintain constant terminal voltage.

So far it has been implied that the shunt compensating device must maintain constant voltage *magnitude* at its point of connection. Under steady-state or very slowly varying conditions, the *static* compensator (i.e., a compensator having no moving parts) can be made to be functionally equivalent to an intermediate synchronous machine. Under more rapidly varying conditions, the inertia of the synchronous machine rotor influences the *phase* of the voltage at the point of connection, because of the exchange of kinetic energy between it and the system as the rotor accelerates or decelerates. The purely static compensator cannot exchange this energy with the system, and this leads to a different influence on the system under rapidly varying conditions.†

This section develops the theory of compensation by sectioning in the steady state and for very slowly varying conditions, that is, slowly enough for the rates of change of kinetic energy of rotating machines to be negligible. In spite of this, it is seen that in certain regimes compensation by sectioning is an essentially *dynamic* process (in the control engineer's sense).

2.6.2. Dynamic Working of the Midpoint Compensator

The *fixed* midpoint shunt susceptance was considered in Section 2.4.3. In this section the same representation of the transmission line is used, a π-

† It is true that the static compensator contains capacitors and inductors which can store energy, but because they carry alternating current there is no significant net storage of energy averaged over a period long enough for a significant change to take place in the kinetic energy of the synchronous machines in the system (typically a few cycles).

An exception to this is the static compensator constructed of a very large dc energy-storage coil connected to the line via a three-phase bridge rectifier.[21] To obtain a useful amount of stored energy, with acceptable losses, the storage coil would probably need to be superconducting.

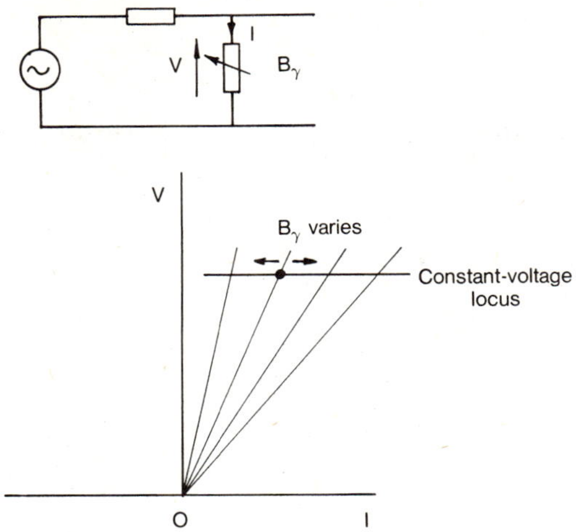

FIGURE 30. Principle of maintaining constant ac voltage at the terminals of a controlled susceptance.

equivalent circuit for each half, as in Figure 27*a*. It is assumed here, however, that the compensating susceptance is continuously controlled in such a way as to keep the midpoint voltage constant with a value E_m. The analysis applies equally well with a synchronous condenser instead of a controlled susceptance.

Now the midpoint voltage is related to the terminal voltages in Equation 92, so it follows that with $E_s = E_r = E$,

$$1 - s = \frac{E}{E_m} \cos \frac{\delta}{2}. \tag{133}$$

At this stage it is not of interest to interpret s as an equivalent degree of series compensation. Instead it is used to obtain the compensating susceptance B_γ necessary to satisfy Equation 133. Thus, (substituting from Equations 86 and 91).

$$B_\gamma = -\frac{4}{X_l}\left[1 - \frac{E}{E_m} \cos \frac{\delta}{2}\right] + \frac{B_c}{2}, \tag{134}$$

where B_c is the total shunt capacitive susceptance of the whole line and X_l is its total reactance, as in Figure 27*a*. This equation tells how B_γ must vary with the transmission angle δ in order to maintain the midpoint voltage equal to E_m. Naturally, through δ, B_γ varies with the power being transmitted.

Since s now *varies* with δ (Equation 133), the power transmission characteristic is modified. From Equation 90,

$$P = \frac{E^2}{X_l(1-s)} \sin \delta = \frac{E_m E}{X_l \cos(\delta/2)} \sin \delta , \tag{135a}$$

that is,

$$P = 2\frac{E_m E}{X_l} \sin \frac{\delta}{2} . \tag{135b}$$

This is identical to Equation 132. It implies that in the steady state the line is sectioned into two independent halves.

If $E_m = E$, the power transmission characteristic is given by

$$P = \frac{2E^2}{X_l} \sin \frac{\delta}{2}. \tag{136}$$

This is shown as the upper curve in Figure 31. The maximum transmissible power is $2E^2/X_l$, twice the steady-state limit of the uncompensated line. It is reached when $\delta/2 = \pi/2$, that is, with a transmission angle of 90° across each half of the line, and a total transmission angle δ of 180° across the whole line.

(a) Illustration of Dynamic Working. The power transmission characteristic expressed by Equation 90 can be interpreted as a sinusoid whose amplitude P'_{max} varies as s varies. From Equation 135a, P'_{max} varies also

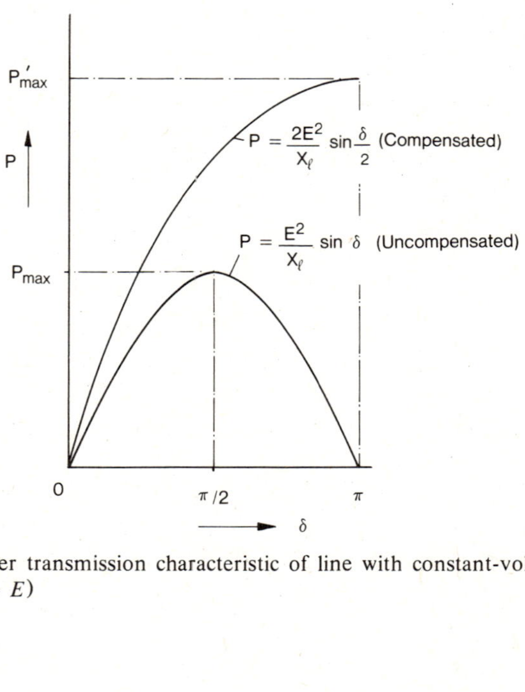

FIGURE 31. Power transmission characteristic of line with constant-voltage compensator at midpoint. ($E_m = E$)

with δ itself and, therefore, also with P (through Equation 135b). This can be written

$$P = P'_{max} \sin \delta , \tag{137}$$

where

$$P'_{max} = \frac{E_m E}{X_l \cos (\delta/2)} . \tag{138}$$

(δ can always be determined from Equation 135b, because E, E_m and X_l are constants.) This gives rise to the concept of a series of sinusoids with varying amplitudes P'_{max} as shown in Figure 32 (dotted curves). Each corresponds to a fixed value of the compensating susceptance B_γ. For example, suppose P is equal to the steady-state limit of the uncompensated line, $P_{max} = E^2/X_l$. From Equation 136 with $E_m = E$, $\delta = 2 \sin^{-1} (½) = 60°$. Operation is at the point A in Figure 32. From Equation 138, $P'_{max} = P_{max}/\cos(30°) = 1.155\ P_{max}$; that is, the "current" static power transmission characteristic has a maximum transmissible power of 1.155 P_{max}.

If the transmitted power were increased to 1.2 P_{max}, δ would increase to $2 \sin^{-1} (1.2/2) = 73.74°$, and P'_{max} would increase to $P_{max}/\cos (36.87°) = 1.250\ P_{max}$. Operation is at the point B in Figure 32. The

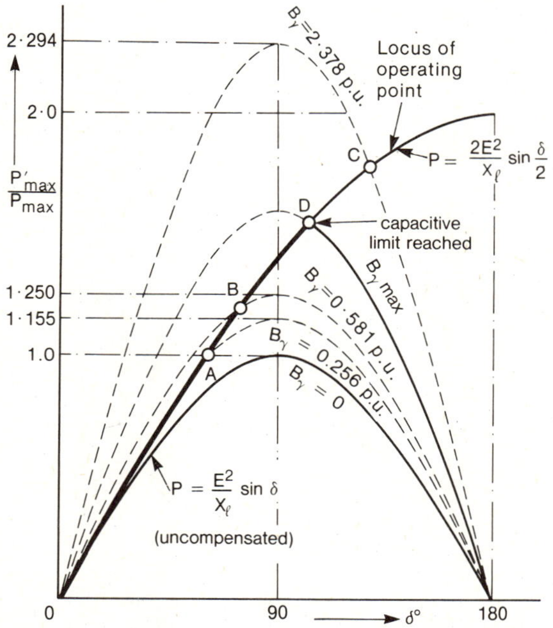

FIGURE 32. Dynamic working of midpoint shunt compensator.

controlling effect of the midpoint compensator constrains the operating point to move along the locus given by Equation 136, passing smoothly from one constant-B_γ sinusoid to another as P varies.

For $P > \sqrt{2}\, P_{max}$, δ is greater than 90°. The system is stable as long as $dP/d\delta > 0$ (i.e., as long as an increase in transmission angle is accompanied by an increase in transmitted power; see Section 2.2.6). However, the operating point is now on the unstable side of the "current" constant-B_γ sinewave. For example, if $P = 1.8\, P_{max}$, $\delta = 2 \sin^{-1}(1.8/2) = 128.32°$, and $P'_{max} = 2.294\, P_{max}$. Operation is at point C in Figure 32. The system owes its stability to the fact that if the transmission angle δ increases slightly, the compensator responds by changing B_γ immediately in such a way as to keep V_m constant and so to increase P'_{max} to the value which satisfies Equation 138, so that the operating point moves along the stable characteristic, $P = 2P_{max} \sin(\delta/2)$.

Typically, in transmission systems which are compensated by sectioning and which are operating at high power levels, the effective compensating susceptance B_γ is capacitive. In practice there is an economic limit to the capacitive reactive power of the compensator. With many types of compensator, when the limit is reached the compensator ceases to maintain constant voltage at its terminals and behaves instead like a fixed susceptance. The operating point then leaves the dynamically stable characteristic $P = 2\, P_{max} \sin(\delta/2)$, and moves on to the constant-B_γ sinusoid corresponding to the maximum value of B_γ, that is, $B_{\gamma max}$.† This departure is shown at point D in Figure 32. The higher the value of $B_{\gamma max}$, the steeper is the (negative) slope of the constant-B_γ characteristic at the point of departure. If $B_{\gamma max}$ were smaller, the point of departure could occur on the stable side of one of the dotted sinusoids in Figure 32, such as at point A or B. However, the power transmitted P and the transmission angle δ would then have to be kept to much lower values. In other words, the capacitive current rating of the compensator is one of the main factors limiting the achievable increase in the maximum transmissible power.

(b) Reactive Power Requirements at the Terminals. From Equations 95, 133, and 134 it can be shown that the reactive power requirement at the sending end is given by

$$Q_s = \frac{2E^2}{X_l}\left[\left(1 - \frac{X_l B_c}{8}\right) - \frac{E_m}{E}\cos\frac{\delta}{2}\right] \qquad (139)$$

with $E_s = E_r = E$. By symmetry, $Q_r = -Q_s$.

† For a synchronous condenser or saturated-reactor compensator $B_{\gamma max}$ is the maximum capacitive current divided by E_m.

(c) Compensator Control and Reactive Power Requirements. A practical control system for varying B_γ would not be based on Equation 134, because there is no information as to the value of δ available at the intermediate compensator station. Instead, a feedback control system would be used to keep $V_m = E_m$. In the saturated-reactor compensator an inherent regulating process achieves the same end.

With constant $V_m = E_m$, the compensator reactive power $Q_\gamma = V_m^2 B_\gamma$ is a function of δ also:

$$Q_\gamma = V_m^2 B_\gamma = -E_m^2 \left\{ \frac{4}{X_l} \left[1 - \frac{E}{E_m} \cos\frac{\delta}{2} \right] - \frac{B_c}{2} \right\} \tag{140}$$

Through the parameter δ, Q_γ is related to the transmitted power P by Equations 135 and 140. Note that $X_l B_c = \omega^2 a^2 lc = \theta^2$, θ being the electrical length of the line in radians.

2.6.3. Example of Line Compensated by Sectioning†

As an example of compensation by sectioning, or "dynamic shunt compensation," consider a 400-mi line with a midpoint constant-voltage compensator. As in Section 2.5.3, $X_l = 0.8108\ Z_0$ and $B_c = 0.8108/Z_0$. The power transmission characteristic is given by Equation 132 as

$$P = \frac{2 \times 1^2}{0.8108} \sin\frac{\delta}{2} P_0 = 2.4667\ P_0 \sin\frac{\delta}{2} \tag{141}$$

with $E_s = E_r = E = 1.0$ pu. P'_{max} is quite close to the value $2.6084\ P_0$ achieved with 50% series compensation (with shunt reactors connected). The reactive power requirement of the compensator is given by Equation 140 as

$$Q_\gamma = -\left[4.9334 \left(1 - \cos\frac{\delta}{2} \right) - 0.4054 \right] P_0 \tag{142}$$

and the terminal reactive powers are given by Equation 139 as

$$Q_s = -Q_r = 2.4667 \left(0.9178 - \cos\frac{\delta}{2} \right) P_0 . \tag{143}$$

† The two halves of the line are again represented by equivalent-π circuits.

At no-load $Q_s = -0.2027\ P_0$, corresponding to the line-charging reactive power of the 100 mi of line nearest to the sending end. (Note that this differs from the value $-0.2055\ P_0$ calculated from Equation 30. This is due to the approximation involved in the π-equivalent circuit.) At no-load, $Q_\gamma = 0.4054\ P_0$, corresponding to the line-charging reactive power of the central 200 mi of the line.

The variation of the main parameters is shown in Table 7 as the transmitted power varies. It can be seen that the reactive power absorbed or supplied by each terminal is just half that of the compensator. The two vary in concert, the terminal synchronous machines compensating the extreme 100-mi portions of the line, while the compensator compensates the central 200-mi portion.

Although the overall transmission angle δ exceeds 90°, transmission remains stable up to 180°, as long as the midpoint voltage remains constant. It can be seen that the power transmission up to 1.25 P_0 does not require excessive capacitive reactive power from the compensator.

TABLE 7
Example of a 400-Mile Transmission Line Compensated by a Constant-Voltage Device at its Midpoint[a]

$p = \dfrac{P}{P_0}$	E_m	$\dfrac{Q_s}{P_0} = -\dfrac{Q_r}{P_0}$	$-Q_\gamma/P_0$	δ (°)	δ Without Compensation (°)
0	1.000	−0.2027	−0.4054	0	0
0.25	1.000	−0.1900	−0.3800	11.634	10.440
0.50	1.000	−0.1515	−0.3030	23.390	21.249
0.75	1.000	−0.0859	−0.1718	35.402	32.932
1.00	1.000	0.0091	0.0182	47.832	46.456
1.25	1.000	0.1375	0.2749	60.895	64.966
1.50	1.000	0.3058	0.6116	74.905	unstable
1.75	1.000	0.5256	1.0512	90.380	unstable
2.00	1.000	0.8202	1.6404	108.34	unstable
2.25	1.000	1.2530	2.5060	131.60	unstable
2.4667	1.000	2.2640	4.5280	180.00	unstable

[a] Note the opposite sign conventions for Q_s and Q_γ: Q_s and $-Q_\gamma$ are both negative for absorption; the compensator is then inductive. Q_s and $-Q_\gamma$ are both positive for generation; the compensator is then capacitive.

Whether the system has transient stability for major faults at this level of transmission is another question, which will be dealt with in the next chapter.

REFERENCES

1. A. Boyajian, "The Physics of Long Transmission Lines," *Gen. Electr. Rev.* 15–22, July 1949.
2. F. Iliceto and E. Cinieri, "Comparative Analysis of Series and Shunt Compensation Schemes for AC Transmission Systems," *IEEE Trans. Power Appar. Syst.* **96**(6), 1819–1830 (1977).
3. F. G. Baum, "Voltage Regulation and Insulation for Large Power Long Distance Transmission Systems," *J. AIEE* **40**, 1017–1032 (1921).
4. C. L. Fortescue and C. F. Wagner, "Some Theoretical Considerations of Power Transmission," *J. AIEE* **43**, 106–113 (1924).
5. E. W. Kimbark, "How to Improve System Stability Without Risking Subsynchronous Resonance," *IEEE Trans. Power Appar. Syst.* **96** (5), 1608–1613 (1977).
6. G. L. Wilson and P. Zarakas, "Anatomy of a Blackout," *IEEE Spectrum,* **15**(2), 38–46 (February 1978).
7. S. B. Crary, *Power System Stability,* Wiley, New York, 1945, 1947.
8. E. W. Kimbark, *Power System Stability,* Wiley, New York, 1948.
9. R. T. Byerly and E. W. Kimbark, *Stability of Large Electric Power Systems,* IEEE Press, 1974.
10. C. A. Gross, *Power System Analysis,* Wiley, New York, 1979.
11. G. Jancke, N. Fahlén, and O. Nerf, "Series Capacitors in Power Systems," *IEEE Trans., Power Appar. Syst.* **94**, 915–925 (1975).
12. S. A. Miske, "A New Technology for Series Capacitor Protection," *Electr. Forum,* **5**(1), 18–20 (1979).
13. G. D. Breuer, H. M. Rustebakke, R. A. Gibley, and H. O. Simmons Jr., "The Use of Series Capacitors to Obtain Maximum EHV Transmission Capability," *IEEE Trans., Power Appar. Syst.* **83**, 1090–1102 (1964).
14. L. O. Barthold et al., "Static Shunt Devices for Reactive Power Control," *CIGRE Paper* 31–08, (1974).
15. E. Friedlander and K. M. Jones, "Saturated Reactors for Long Distance Bulk Power Lines," *Electr. Rev.,* 940–943 (June 1969).
16. R. Elsliger et al., "Optimization of Hydro-Québec's 735-kV Dynamic-Shunt-Compensated System Using Static Compensators on a Large Scale," *IEEE PES Winter Power Meeting, Paper A*78 107-5, New York, 1978.
17. D. A. Woodford and M. Z. Tarnawecky, "Compensation of Long Distance AC Transmission Lines by Shunt Connected Reactance Controllers," *IEEE Trans., Power Appar. Syst.* **94**, 655–664 (1975).
18. E. Friedlander, "Transient Reactance Effects in Static Shunt Reactive Compensators for Long AC Lines," *IEEE Trans., Power Appar. Syst.* **95**, 1669–1680 (1976).
19. J. D. Ainsworth et al., "Long Distance AC Transmission Using Static Voltage Stabilizers and Switched Linear Reactors," *CIGRE Paper 31-01, (1974).*
20. J. D. Ainsworth et al., "Recent Developments towards Long Distance AC Transmission Using Saturated Reactors," *IEE Conf. Publ.* 107, 242–247 (1973).
21. M. Boidin and G. Drouin, "Performance Dynamiques des Compensateurs Statiques à Thyristors et Principes de Regulation," *Rev. Gen. Electr.,* **88**(1), 58–73 (1979). (In French)

Chapter 3

REACTIVE POWER COMPENSATION AND THE DYNAMIC PERFORMANCE OF TRANSMISSION SYSTEMS

R. L. HAUTH

3.1. INTRODUCTION

3.1.1. The Dynamics of an Electric Power System

An electric power system is never in equilibrium for very long. Frequent changes disturb the equilibrium so that the system is almost always in transition between equilibrium or steady-state conditions. The theory in Chapter 2 is valid in the steady-state and when changes are taking place slowly.

This chapter deals with the dynamic behavior of the electric power system during the transition from one equilibrium condition to another. Figure 1 depicts the full range of dynamic phenomena that characterize power system performance. The steady-state conditions treated in Chapter 2 are shown at the far right. Here we are more concerned with dynamic transitions which occupy a time span near the center of Figure 1. In this regime there is a need to control voltage excursions and redistributions of momentum among the synchronous machines which result from faults, switching operations, and load changes. The control must be rapid and accurate; otherwise, the stability of the system may be lost, either locally or throughout the system.

The emphasis in this chapter is on the controlling and stabilizing influence of compensating devices such as shunt and series capacitors, shunt reactors, synchronous condensers, and especially static reactive

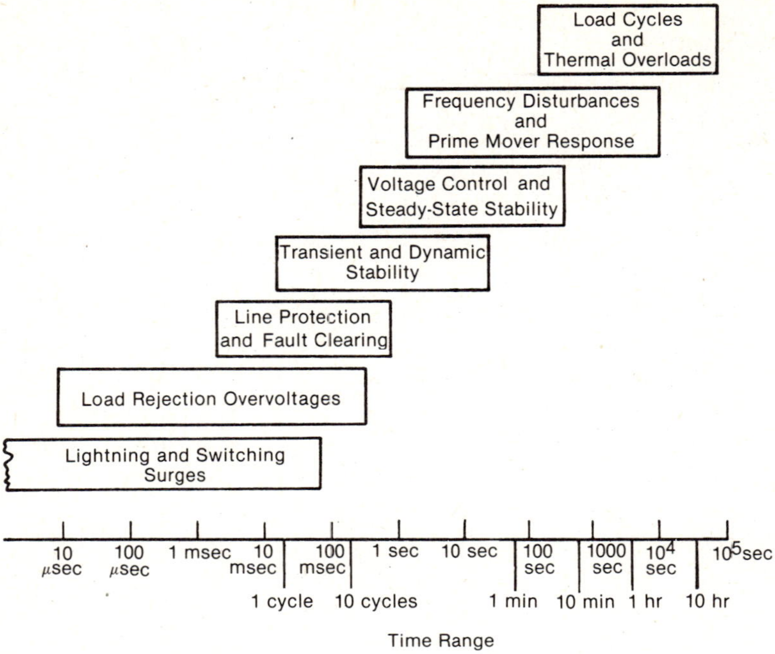

FIGURE 1. Time ranges of transitions in the state of a power system.

power compensators. Although these devices are not the only ones capable of controlling voltage and power swings, they are becoming more important because of the high cost of transmission facilities and the need to transmit as much power as possible through the smallest number of lines.

3.1.2. The Need for Adjustable Reactive Compensation

The need for adjustable reactive power compensation can be divided into three basic classes:

1. The need to maintain the stability of synchronous machines. We shall see that voltage control by reactive power compensation can have a positive stabilizing influence on the system during disturbances which cause the rotor angles of synchronous machines to change rapidly. Both the transient stability and the dynamic stability of a system can be enhanced. It is even possible with controlled compensators to drive voltages deliberately out of their normal steady-state bounds for several seconds following a fault or other major disturbance to enhance the stabilizing influence still further.

2. The need to control voltage within acceptable bounds about the desired steady-state value to provide quality service to consumer loads. Following certain abrupt changes in the load, or in the network configuration as a result of switching actions, it may be necessary to make a voltage correction in as short a time as a few cycles of the power frequency. For other voltage disturbances, a correction within a few seconds will suffice. Uncorrected voltage deviations, even if temporary, may lead to an outage or damage to utility or consumer-owned equipment. Even small variations, particularly those that cause flicker, are often objectionable. (The flicker problem is given special attention in Chapter 9 and is not discussed in this chapter.)

3. The need to regulate voltage profiles in the network to prevent unnecessary flows of reactive power on transmission lines. To this end, reactive power compensation can be used to maintain transmission losses at a practical minimum. While the reactive compensation must be adjusted or changed periodically to maintain minimum losses, the adjustments can be made quite infrequently with several minutes to effect the desired change.

For convenience in describing the influence of the various compensation methods on system dynamic performance, the transition period between equilibria is divided into distinct time periods which are defined in Section 3.2. Because of the role that compensation plays in maintaining stability, the various forms of instability that can occur are also reviewed in Section 3.2.

3.2. FOUR CHARACTERISTIC TIME PERIODS

In general, any disturbance that causes material variations in voltage can be viewed in terms of four periods or stages, beginning at the start of the disturbance and ending when a new quasi-steady-state is reached. The periods are defined with the aid of Figure 2 shown for a typical line fault initiated disturbance. They are characterized not by their duration, which can be variable within wide limits, but by the events and processes taking place within them.

The first several cycles following a disturbance such as a fault make up the *subtransient period.* During this time period there is an initial rapid decay of both ac and dc components of fault current. These components may also include high frequency components which are generally rapidly damped. It is during this period that surge arresters, spark gaps, and nonlinear reactors (including transformer magnetizing reactances) act to prevent extreme voltages which would cause insulation failure.

The *transient period* is second in sequence. As shown in Figure 2*a*, this period lasts for many cycles after the subtransient period. During this time, synchronous machines are sometimes approximately represented by an emf behind a constant reactance, the *transient reactance* X'_d. In this chapter we restrict the term *transient period* to mean a period when it can be assumed that no appreciable changes in the rotor angles of synchronous machines take place.

Many disturbances in a power system are so small that they do not cause appreciable change in the speeds or rotor angles of synchronous machines. Changes in rotor angles of synchronous machines may be negligible even for large disturbances if the disturbance is very distant (electrically) from the generators. When such is the case, this chapter refers to the time periods in Figure 2*a*. When rotor angle swings are noticeable at the point of interest, the subdivision in Figure 2*b* is used. Figure 2*b* shows the total transition period made up of subtransient, first swing, and oscillatory periods.

The *first-swing period* refers to the time for the first swing (half oscillation) of the rotor angle(s) or synchronizing power swings following a large disturbance, such as a fault. This period typically lasts for about 0.5–1 sec. In this period also, synchronous machines are sometimes approximately characterized by constant flux-linkages (constant internal source voltage) behind the machine's transient reactance. It is often the critical period during which transient stability (Chapter 2, Section 2.1.2) is maintained or lost.

The *oscillatory period* follows the first-swing period. During this period significant cyclic variations in voltages, currents, and real and reactive power take place. Synchronizing power swings caused by synchronous

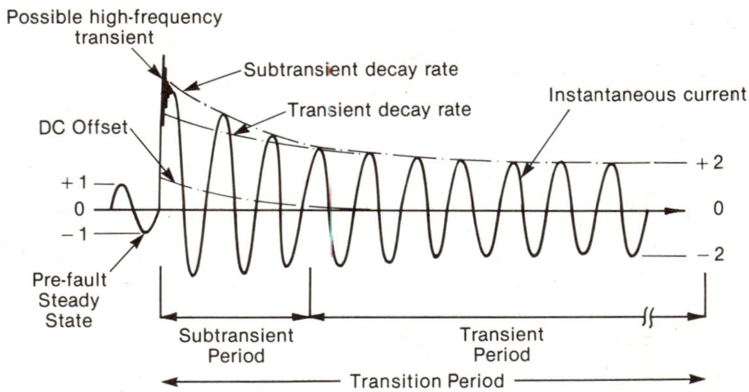

(a)

FIGURE 2. Characteristic time periods: subdivisions of total transition period. (*a*) Subtransient and transient periods, used when transmission angle swings are insignificant.

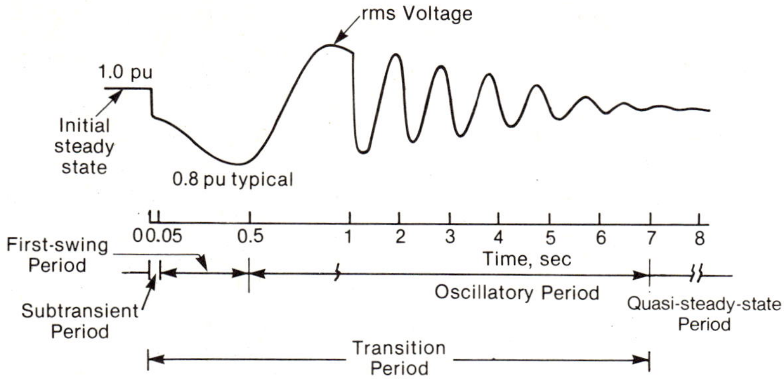

(*b*)

(*b*) First-swing and oscillatory periods, used when transmission angle swings are significant.

machine rotor angle swings may last for 3–20 sec after a severe fault. The precise duration depends on the damping contributed by excitation controls and amortisseur currents in synchronous machines, speed governor control on turbines, and the loads.

The *quasi-steady-state period* is the final period which is reached when the synchronizing power and rotor angle swings have died out. The theory discussed in Chapter 2 describes the system's performance in this regime.

Many disturbances on high voltage power systems are either too mild or occur too far from generators to cause significant rotor angle variations. In these cases, the subtransient and transient periods constitute the entire transition period, and there is no need to discuss the first-swing or oscillatory periods.

The response mechanisms of the compensated power system are visualized in terms of voltage and current phasors in most of this chapter. The phasor visualization is not rigorously correct during the subtransient and transient periods. The reason is that, immediately following a disturbance, the voltages and currents are not pure sine waves. They include dc offset currents, unbalances between phases, and harmonics when nonlinear circuit elements are involved. For this reason, several examples of the response of compensated systems are given, which have been obtained from field tests, computer simulations, and modeling on the transient network analyzer (TNA).

In most of the cases studied in this chapter the subtransient period is skipped and the transient period is assumed to begin instantly following the disturbance. This assumption is justified by the fact that the reactive compensation methods which are of central interest are intended to be effective in the transient and subsequent periods. Some compensators are helpful in the subtransient period in limiting overvoltages, but in general, this period is too short for reactive compensation to be effective, leaving only the questions of the protection of the compensator and its coordination with surge arresters and other high-speed protective devices.

3.2.1. The Transient Period

To illustrate the electrical behavior of a system during this period, consider one phase of the positive-sequence equivalent circuit of a two-machine system in Figure 3*a*. The system is assumed to be lossless (no resistance in the generators or network elements) and the line capacitance is ignored. The theory would apply with circuit resistances and charging included, and these assumptions are made for ease of illustration only.

The reactances are chosen so the system is electrically symmetric around the midpoint bus on which the voltage is shown to be $\mathbf{V}_m$. The two source voltages $\mathbf{E}_1'$ and $\mathbf{E}_r$ are assumed to be equal in magnitude. The only load in the system is in the receiving-end system which is represented by a fixed voltage source $\mathbf{E}_r$. The sending-end generator is delivering power to the receiving-end system proportional to current I_1 which prevails in the initial steady state. The phasor diagram Figure 3*b* corresponds to these initial conditions.

Suppose breakers *a* and *b* are opened simultaneously thereby disconnecting one of the current-carrying system elements. Immediately, the conditions in phasor diagram Figure 3*c* result.

Let us now examine that phasor diagram. The total angle δ_1 between the synchronous source voltages $\mathbf{E}_1'$ and $\mathbf{E}_r$ remains fixed as they are fixed to inertial rotating phasors which cannot change their rotational speed instantly. Since $\mathbf{E}_1'$ and $\mathbf{E}_r$ are momentarily fixed in magnitude and are displaced by a (momentarily) fixed angle, the sum of the voltage drops across the system remains the same. However, because the reactance between phasors $\mathbf{V}_s$ and $\mathbf{V}_m$ has doubled with the line outage, the current I_2 is reduced from the initial value I_1.

As shown in phasor diagram Figure 3*c* all the bus voltages $\mathbf{V}_s$, $\mathbf{V}_m$, and $\mathbf{V}_r$ are instantly shifted in phase relative to each other and the source voltages $\mathbf{E}_1'$ and $\mathbf{E}_r$. The magnitudes of the bus voltages increase momentarily during the transient period. If both source voltage phasors were to remain fixed in relative phase and fixed in magnitude, the transition period would be complete and phasor diagram (Figure 3*c*) would represent the final equilibrium state.

Let us assume the receiving-end system is so large in MVA capacity that $\mathbf{E}_r$ is essentially fixed in magnitude and rotates continuously at a constant synchronous speed (377 rad/sec in a 60-Hz system). The phasor $\mathbf{E}_1'$, however, is associated with a generating unit of finite MVA rating. If we assume the power delivered to the sending end generator by its turbine is constant, the sending end turbine generator unit will tend to accelerate. That occurs because the reduction in current from I_1 to I_2

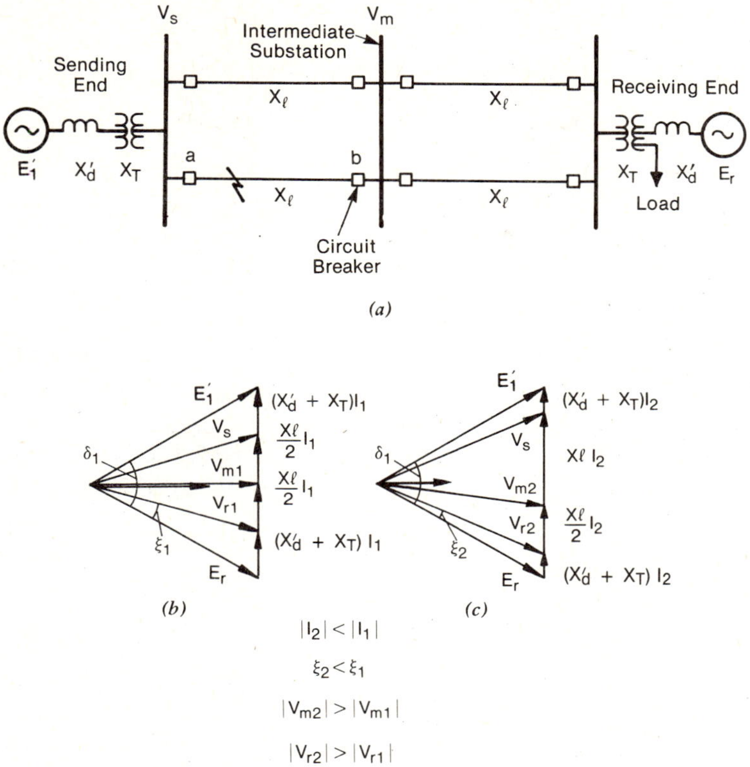

FIGURE 3. (*a*) One-line diagram of a two-machine power system. (*b*) Phasor diagram showing initial conditions. (*c*) Phasor diagram immediately after tripping circuit breakers *a* and *b*.

appears as a reduction in electrical power output from the generator. The momentary mismatch between input power (from turbine) and output power causes the unit to accelerate during which time the angle δ_1 will become larger.

When a disturbance is large enough to cause the unit to change speed (such as a major line trip in this example) even if momentarily, the transient period behavior is followed by "first-swing" and "oscillatory" period behavior which are discussed in subsequent sections.

Finishing this example, however, the voltage phasor $\mathbf{E}_1'$ will rotate so as to increase angle δ_1 and subsequently settle at a larger angle consistent with the initial active power delivered by the sending-end generator. The magnitudes of voltages $\mathbf{V}_s$, $\mathbf{V}_m$, and $\mathbf{V}_r$ will be changed back *toward* their initial values by the unit's voltage regulator action. Adjustment of regulator setpoints, switchable passive reactive compensation and/or controllable compensators can provide additional "control" to restore the voltages in the system to their nominal or desired magnitudes.

3.2.2. The First-Swing Period and Transient Stability

In the line switching example discussed in Figure 3, the post disturbance power input to the sending-end generator (from its prime mover) was assumed to remain constant at the prefault level, while the power output of the generator decreased. This caused the generator to accelerate. If a short-circuit fault had preceded the line outage and caused the breakers to operate under relay control, the unit's acceleration would be greater than that caused by merely opening the breakers without the fault. The transmission angle δ is still a direct measure of the mechanical phase angle between the rotors of the sending-end and receiving-end synchronous machines. The postfault acceleration of the sending-end generator results in a much greater increase in δ and an associated transient redistribution of angular momentum, which must be limited; otherwise, synchronism will be lost. In many systems the corrective action necessary to prevent this loss of "transient stability" must be taken within a fraction of a second.

These dynamic variations are summarized in Figure 4 for the fault discussed above. Figure 3*b* represents the initial conditions for this example. Figure 4*a* shows the phasor diagram for the time $t = t_p$ when $\delta = \delta_{max}$. Figure 4*b* maps the transition of electrical power P versus the angle δ on the transient power-angle diagram. Finally, Figure 4*c* shows the machine's load angle δ, the electrical power P, and the midsystem bus voltage V_m, for the entire scenario, including the fault period and the first-swing period.

The need for dynamic reactive compensation is evident in the V_m trace in Figure 4*c* and the phasor diagram in Figure 4*a*. As the power P increases, the voltage V_m drops to a minimum value at time t_p. Reactive compensation at bus m could minimize the otherwise large voltage variations during this period. The nature of this improvement is illustrated in Figure 5*a*. The compensator can be a synchronous condenser or a static compensator, and the performance of different compensators is described more fully in Sections 3.4 and 3.5. The voltage support provided at intermediate points tends to reduce the transmission angle variations, as indicated in Figure 5*b*. In some cases it can maintain transient stability in a system that would otherwise become unstable.

The electromechanical behavior of synchronous machines during the first-swing period differs from that described under the transient period. During the first-swing period the internal voltage E' may increase as a result of rapidly increasing field current forced by the exciter. From Equation 38, Chapter 2, this tends to increase the power transfer capability of the generator and transmission system at any given δ. This results in a reduction of first-swing angular (δ) excursion during the post-disturbance synchronizing power swings.

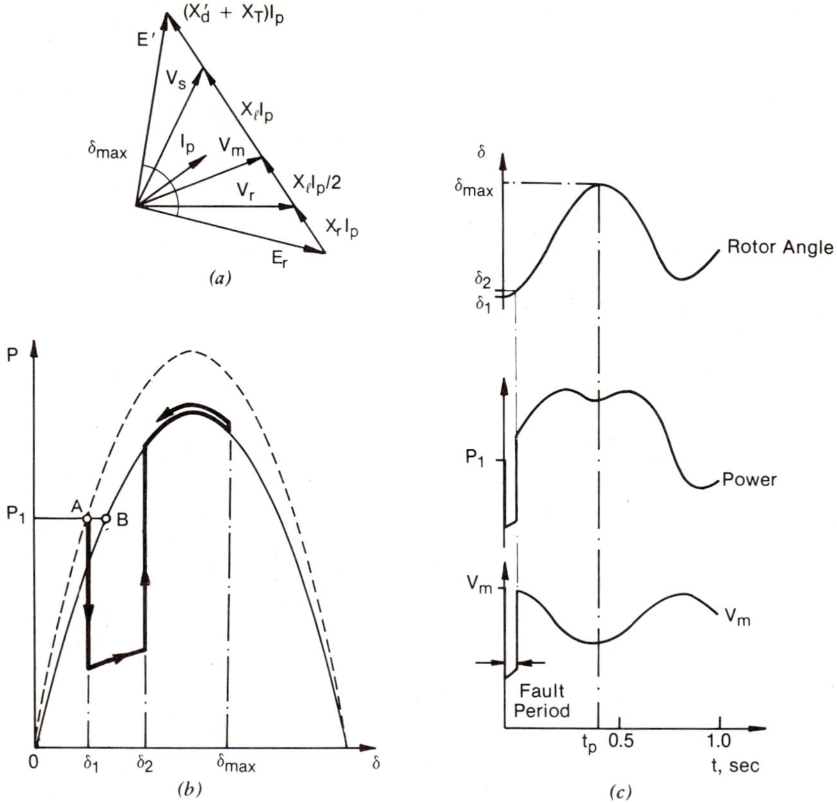

FIGURE 4. (*a*) Phasor diagram for system of Figure 3*a* at maximum transmission angle; $t = t_p$. (*b*) Power-versus-angle curves: ---- = before disturbance; —— = after line section is disconnected; A = initial operating point; B = final operating point. (*c*) Transient response of system of Figure 3*a*.

3.2.3. The Oscillatory Period

The third characteristic time period following a disturbance is defined in Figure 2 as the oscillatory period, that is, the period between the first swing in synchronizing power (and machine rotor angle) and the time at which a quasi-steady state is reached. Depending on the characteristics of the power system, the oscillations will damp out (if they damp at all) in anywhere from 2 to 20 sec. Sometimes, so-called negative damping influences can prevail, causing the postdisturbance oscillation to grow until one or more generators loses synchronism. This form of instability is sometimes called "dynamic instability." Net negative damping in post-disturbance oscillations is most likely to be found in systems in which high power levels are transmitted over long (electrical) distances. To maintain transient stability of the generators in these systems, rapid-

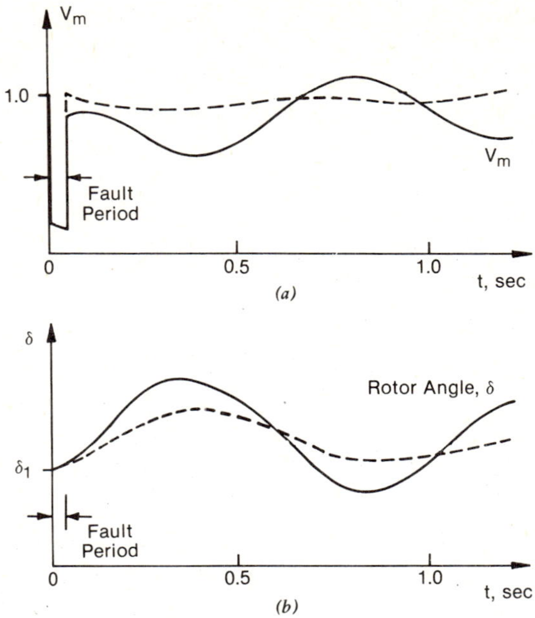

FIGURE 5. Effect of dynamic reactive compensation on voltage and power angle swings. (*a*) Voltage at intermediate substation. (*b*) Transmission angle: —— = uncompensated; ---- = with large dynamic compensator at bus *m*.

response, high-gain excitation systems are employed. Working through the generators' relatively large field time constants, these can themselves have a tendency toward oscillatory instability, and supplementary stabilizing circuits are often used in conjunction with their exciter control systems. These stabilizers sense the oscillation in rotor speed, deviations in bus frequency, or power, and modulate the voltage regulator reference signal in the excitation controller so as to damp out the oscillation. High-voltage dc transmission links can also be utilized to damp out the oscillatory synchronizing power swings in the adjoining ac system. Special damping controls act on the dc power flow in response to ac bus frequency deviations, thereby causing a damping influence on ac power swings. We shall also see later in this chapter that by modulating the reactive power flows in the system, compensators can exert a significant positive damping influence.

The steady-state reached at the end of the oscillatory period does not persist without change for very long. Even if no more major disturbances occur, the system load continually changes at a varying rate. A typical load cycle spanning a 24-hour period for a large interconnected power system is shown in Figure 6. Although the rates of change appear to be slow relative to those seen during and after a fault, the morning load rise

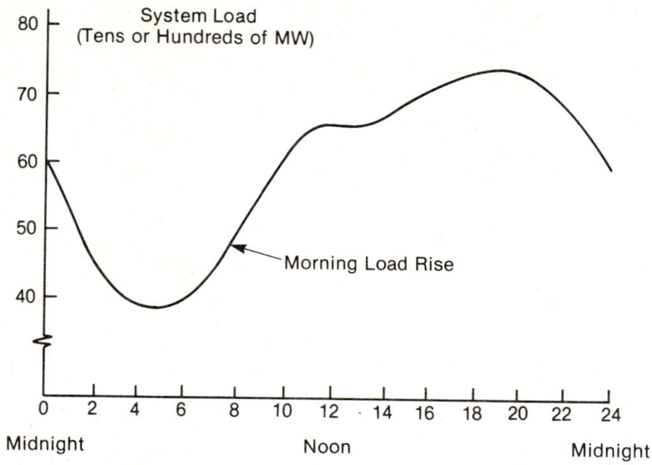

FIGURE 6. Daily load cycle in a typical interconnected power system. Winter and summer profiles differ in overall height, and there are significant differences between one system and another.

can be of the order of 100 MW per minute or more. Figure 7 shows the effect these load changes would have on the system voltage, with and without the benefit of voltage regulators on the generators. The residual voltage error is corrected in general by slow-response methods, such as transformer tap-changing and the switching of lines, capacitors, and sometimes reactors. Rapid-response compensators are not usually required for this function per se. However, if installed for other purposes (such as transient stability improvement), they can assume a useful role in the slowly changing regime also, as described in Chapter 2.

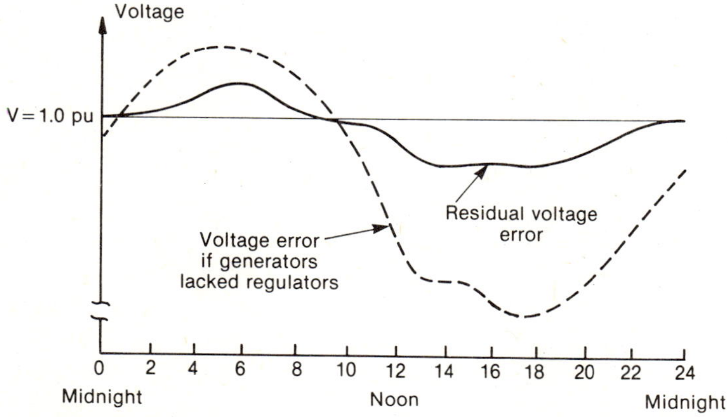

FIGURE 7. Effect of the daily load cycle on overall system voltage.

3.2.4. Compensation and System Dynamics

All forms of compensation, whether passive or controlled, shunt connected or series connected, have an impact on the dynamic behavior of systems in transition. Their effects are discussed in detail in the next four sections. Fixed capacitors and reactors are discussed in Section 3.3, static compensators in Section 3.4, synchronous condensers in Section 3.5, and series capacitors in Section 3.6. The discussions follow a common format. Each compensating method's influence on the system behavior during the four characteristic time periods is discussed in turn.

3.3. PASSIVE SHUNT COMPENSATION

Shunt connected capacitors and/or reactors could be located at any of the buses in Figure 3. For example, shunt capacitors may be connected to prevent low voltage during peak load conditions. During light load conditions shunt reactors may be connected to cancel part of the capacitive reactive power of the line and prevent unduly high voltages. The presence of capacitors and reactors also influences the dynamics of the system following a disturbance. Since they are usually not switched on or off during disturbances, their effects follow from the fact that they modify the network parameters, in particular, its surge impedance, its electrical length, and the driving-point impedances at the system buses.

3.3.1. Transient Period

The concept of the system's reactive load line, introduced in Chapter 1 to describe the quasi-steady state, can be extended to visualize the effect of fixed capacitors and reactors on the positive-sequence component of voltage during other transition periods, including the transient period. For instance, we can view the system in Figure 3*a* from bus *m*, in terms of the Thévenin equivalent circuit shown in Figure 8*a*, where node *m* is the bus *m* in Figure 3*a*. The reactive load line is shown in Figure 8*b*. Figures 8*c* and 8*d* illustrate the use of this load line concept when a capacitor is added to bus *m*. Figures 8*e* and 8*f* apply the same visualization to operation with a shunt reactor at bus *m*.

Without a shunt reactive device at bus *m*, the line outage disturbance discussed with respect to Figure 3 would cause the load line transition in Figure 9*a*. This shows the open-circuit voltage V_m decreasing from E_1 to E_2 and the slope of the load line increasing from X_{s1} to X_{s2}. Figure 9*b* shows that, in the presence of a shunt capacitor with reactance $X_{C\gamma}$, the voltage change ($V_{C1} \rightarrow V_{C2}$) is less than ($E_1 \rightarrow E_2$) without the capacitor. Figure 9*c* illustrates that with an inductor of reactance X_L at bus *m*, the voltage drop from ($V_{L1} \rightarrow V_{L2}$) is greater than ($E_1 \rightarrow E_2$) without the

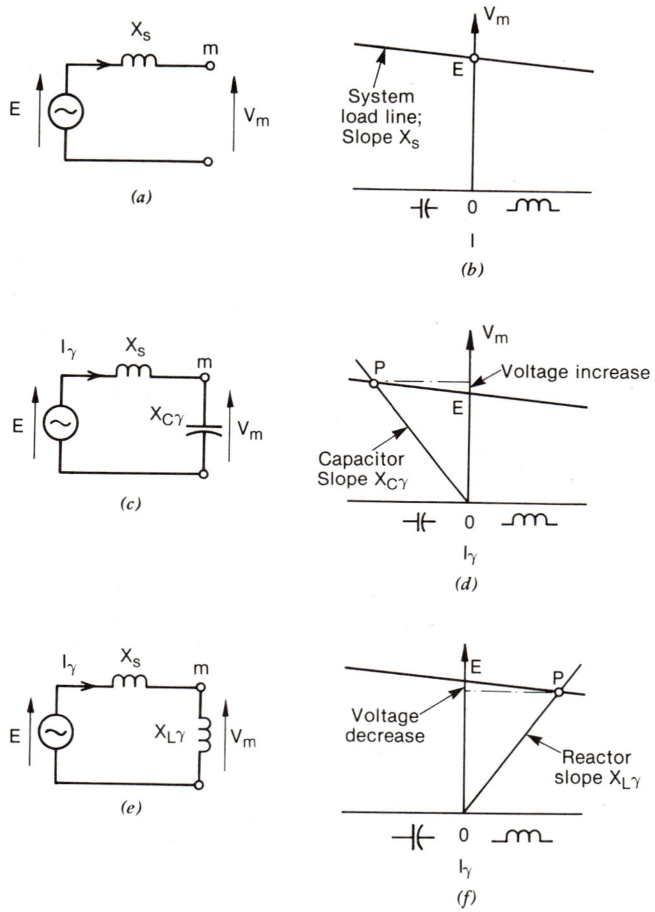

FIGURE 8. Equivalent circuits, load lines, and operating points. (*a*) Thévenin equivalent circuit of system of Figure 3*a*, viewed from bus *m*. (*b*) System load line viewed from bus *m*. (*c*) Shunt capacitor at bus *m*. (*d*) Operating point P with shunt capacitor. (*e*) Shunt reactor at bus *m*. (*f*) Operating point P with shunt reactor.

inductor. In the general case, both capacitors and reactors can influence the magnitude of the voltage change and its direction (increase or decrease). For example, in Figure 9*b* if the capacitor rating were doubled, $(X_{C\gamma}/2)$, a voltage rise would take place instead of the decrease indicated. That is, the voltage would rise from V'_{C1} to V'_{C2}.

Fixed reactors tend to reduce the steady-state voltage, particularly during peak power flow periods. They therefore reduce the steady-state power transfer capacity of the line. The temptation to switch them off during peak power flow periods is usually avoided, because with conventional controls and switchgear they cannot be reconnected fast enough to

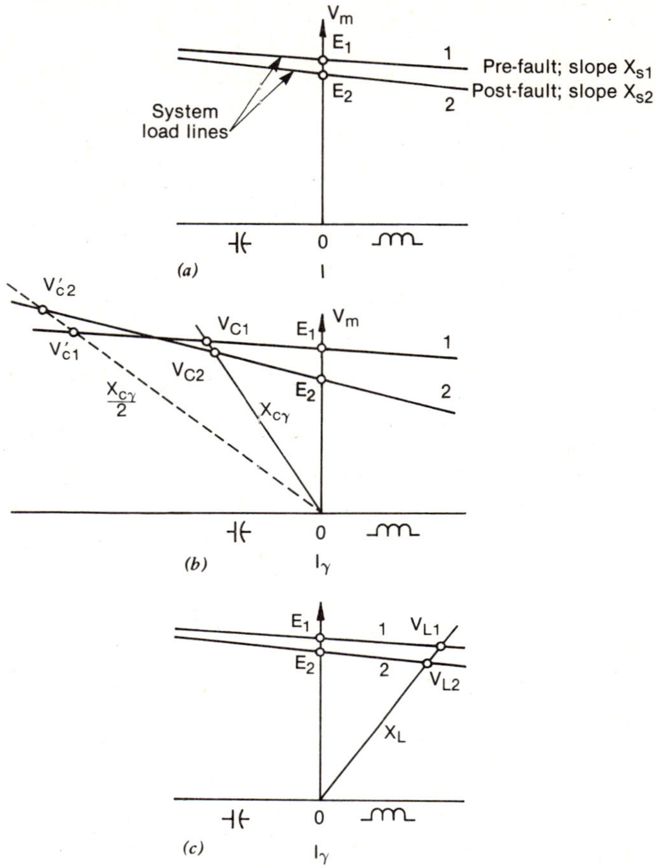

FIGURE 9. Effect of shunt reactor or capacitor on operating point. (*a*) Transition of system load line at bus *m* resulting from disconnection of line section in Figure 3*a*. (*b*) Effect of capacitor at bus *m*. (*c*) Effect of reactor at bus *m*.

suppress overvoltages in the event of a sudden load rejection. High-speed mechanical switches capable of reconnecting the reactor in about 2–3 cycles are available but they may not be suitable for repeated operations (two or more times per day for operator-directed steady-state voltage control), because of excessive wear on moving parts and contacts.

Fixed shunt capacitors raise different questions. Because their reactive power contribution diminishes with the square of the voltage during voltage depressions, it can be uneconomic to rely too heavily on fixed capacitive voltage support to improve transient stability. With large banks of capacitors there is also the possibility of overvoltages following a load rejection. The size of the bank is thus limited by the feasibility of switching them off very quickly under these conditions, which may require expensive and sophisticated switchgear and controls.

3.3.2. First-Swing Period

A typical system response to a disturbance that causes rotor angle, power, and voltage swings was shown in Figure 5 and is reproduced in Figure 10. The first-swing period lasts for about half a second in this example. Unless they are switched on and off at the required moments, shunt capacitors and reactors have only a limited effect on the total change in voltage during the first swing. To be truly effective in reducing the large voltage dip during the first-swing period, shunt capacitors not already energized would have to be switched on line during or immediately following the fault. If sufficient capacitance is switched on at precisely the time the fault is removed, the dashed voltage curve in Figure 10*a* could result. The minimum voltage experienced during the first swing is higher, tending to increase the postfault power and, therefore, to decelerate the generators which accelerated during the fault, reducing the load angle excursions. (This is not shown in Figure 10, but see Figures 25 and 26.)

There are several reasons why conventionally switched shunt capacitors (or reactors) are rarely used in the manner just described. Referring to Figure 10 again, it was assumed that capacitors were switched in at the

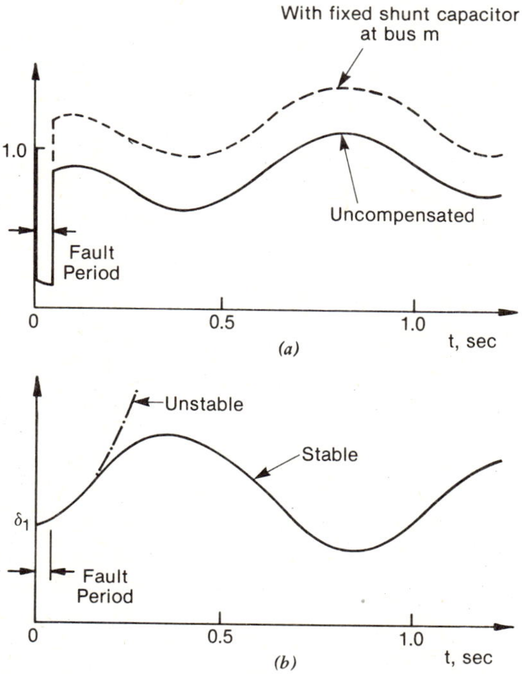

FIGURE 10. Effect of fixed shunt capacitor on postfault response of system of Figure 3*a*. (*a*) Voltage at intermediate bus *m*. (*b*) Transmission angle.

precise time the fault was removed. Since modern circuit breaker and relay operating times are in the 3- to 7-cycle range the capacitor switches would be required to operate in this time. Conventional capacitor switching devices generally operate in 6–30 cycles. The slower switches would not be fast enough to impact the first swing voltage dip. Ideally, the capacitor bank might have to be switched off again immediately after the first swing to prevent sustained high voltages during subsequent oscillations.

Here again, high-speed switches capable of opening or closing in 2–3 cycles could be used but repeated operation may result in prohibitive wear of the switching mechanism and contactors.

3.3.3. Oscillatory Period

As in the case of the first-swing period, fixed shunt compensation has only a limited influence on voltage, power, and machine angle swings during the oscillatory period. To be effective in damping the oscillations in voltage, power, and machine angles, shunt capacitors and reactors would have to be repeatedly switched on and off at precise times to effectively increase and decrease the transfer reactance between synchronous machines. This switching can be more reliably achieved with thyristor switches than with conventional mechanical switchgear (see Section 3.4).

3.3.4. Summary – Passive Shunt Compensation

This section has shown that shunt capacitors and reactors do influence the postdisturbance voltage variations, and, to a limited extent, the stability of generators in the system. Fixed shunt capacitors and reactors merely bias the average value of the voltage up or down during the postdisturbance transition period.

If they are to correct for momentary (half-second) overvoltages or voltage dips, they must be switched on or off rapidly, and in some cases in a repeated manner. This is not generally practical with conventional switches.

3.4. STATIC COMPENSATORS

The static compensator can be thought of as an adjustable shunt susceptance. Its capabilities go far beyond those of fixed shunt reactors and capacitors. We have already seen (Chapter 2, Figure 30) how a continuously adjustable shunt susceptance with a sufficiently rapid response can behave as a constant-voltage device. Such compensators can be controlled in such a way as to improve the power system response during all four of the characteristic time periods defined in Figure 2.

In practice there are many different types of static compensators. Their properties vary quite widely. Chapter 4 describes most of the main operating principles and control characteristics. In this chapter we concentrate on the commonest types, the thyristor-controlled reactor (TCR), the thyristor-switched capacitor (TSC), and the saturated reactor (SR). The TCR and SR compensators are both usually used in parallel with capacitor banks which may be switched by a variety of means. We use the TCR compensator as the main vehicle for describing the generic dynamic properties of static compensators. We continue to use the symbol $\mathbf{V}_m$ for the phasor voltage on the compensator bus.

3.4.1. Transient Period

Thyristor Controlled Reactor with Unswitched (Fixed) Capacitor (TCR-FC). As applied on transmission systems, the TCR includes a closed-loop voltage regulator and thyristor gating-angle control system. These controls possess a slightly delayed time response in correcting for a change in V_m resulting from disturbances on the transmission system. The time lag is statistical in nature, but one can assume a pure delay ranging from 1/6 cycle to 1/2 cycle before the TCR reacts.

The transient response of a TCR-compensated system can be visualized in terms of Figure 11. In the most general case the voltage/current characteristic of the compensator extends into the lagging and leading current regions, the TCR being biased by a shunt capacitor as described in Chapters 1 and 4. Consider a sudden disturbance in the power system that causes the load line at the compensator bus to change from 1 to 2 (Figure 11). Before the disturbance, operation was at the steady-state in-

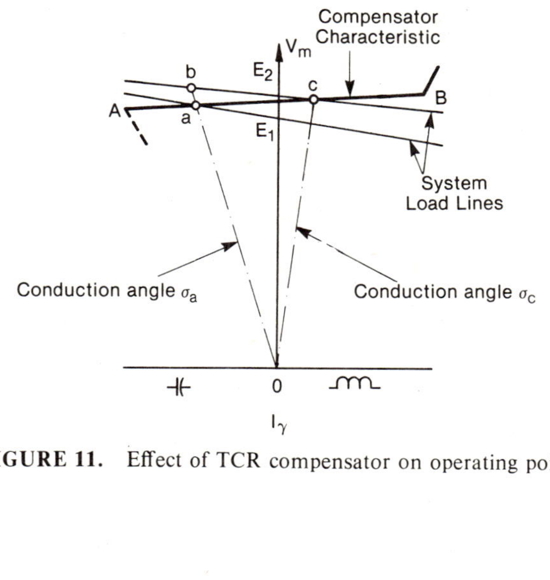

FIGURE 11. Effect of TCR compensator on operating point.

tersection, point *a*. If the response of the TCR control system were slow, the operating point would move straight to point *b*. The thyristor conduction angle would initially be the same as at *a*. Subsequently, the voltage-regulating control system would increase the conduction in the TCR to bring the operating point to *c*. The highest voltage experienced on the compensator bus during this sequence is the voltage at point *b*, which in this example overshoots the voltage E_2 which would be experienced and sustained on this bus in the absence of the compensator. In practice, the rapid response of the TCR causes point *c* to be reached within about 1.5 cycles. With such rapid response the voltage rise on any phase is of very short duration, and may never even reach point *b*.

(a) Simulator Study. To illustrate the benefits of this fast response, a simulation was performed using the actual control system of a TCR wired into a system model.[6] The system model included a three-phase load connected to the compensator bus in a configuration similar to Figure 3*a*. An oscillogram of the load current in one phase is shown in Figure 12 along with the voltage v_{12} between phases 1 and 2 on the compensator bus. The load current was reduced abruptly causing the phase-to-phase

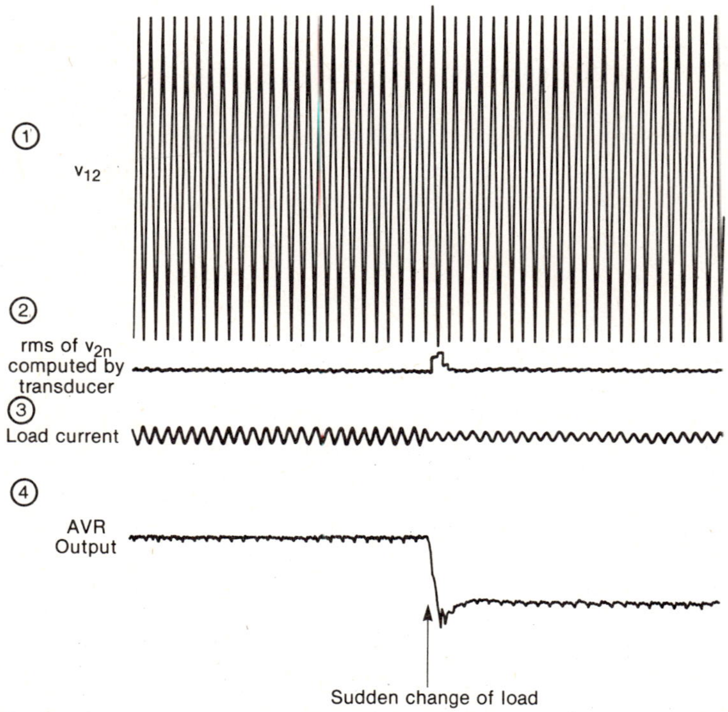

FIGURE 12. Oscillogram of TCR compensator response to step change in load. (AVR = Automatic Voltage Regulator)

voltage to increase. The settling time is most clearly evident in the second trace in Figure 12, which is a measure of the phase 2-to-neutral voltage computed by the compensator's voltage-measuring transducer. The transducer continuously measures the phase-to-neutral voltage and computes an effective rms value every cycle of the voltage waveform. A discrete output is plotted for each half cycle and it is clear from the oscillogram that the voltage was corrected back to the desired set point (the initial value) in about 1.5 cycles after the disturbance.

The thyristor current can be phased from full-on to full-off in one half cycle in each phase in response to a step change in gating angle demand for that phase (Chapter 5). For such a step input given simultaneously to all three phases, the effective positive sequence current in the three phase inductor can be changed in about one cycle. The voltage took 1.5 cycles to change in Figure 12 due to the small time constant in the voltage regulator which converts the voltage error into a gating angle demand signal.

(b) Large Disturbances. Very large system disturbances may cause the compensator current to be driven outside its normal control range AB, as illustrated in Figure 13. Consider first a balanced three-phase fault somewhere on the transmission system which causes the system load line to change from 1 to 2 and remain there for several cycles (Figure 13*a*). At first the voltage on the compensator bus drops from point *a* to point *b*. The voltage at point *b* is not much greater than the voltage E_2 which would be experienced without the compensator, because the compensator current I_γ decreases linearly with voltage. After the small (1/6–1/2 cycle) delay mentioned earlier, the TCR controls correct the voltage to point c_1 or c_2, depending on the capacitive rating of the compensator. Point c_1 results if the capacitive rating is too small to restore normal system voltage. With a much larger capacitive rating (I_{Cmax2} instead of I_{Cmax1} in Figure 13*a*) the compensator could restore approximately normal voltage (c_2) even during the fault. Because faults are generally cleared within a few cycles, such a large capacitive rating is rarely justified. The capacitive rating is more often chosen to allow recovery to near-nominal voltage during the *post*fault period (as discussed in Section 3.4.2).

A second example, Figure 13*b*, shows the effect of a large load rejection in which the compensator current is driven outside its control range in the lagging region. With a modest TCR rating the operating point follows the trajectory abc_1 within 1–2 cycles, but once into "full conduction" ($\sigma = 180°$, see Chapter 4) the TCR acts as a plain linear reactor and cannot hold the voltage on the compensator bus below point c_1. With a larger TCR rating (I_{Lmax2} in Figure 13*b*) the voltage is held down to point c_2. For an extreme example, in the absence of the compensator, the voltage might rise to the level $E_2 \cong 1.4$ pu. The provision of a sufficiently large TCR to hold nominal voltage indefinitely following

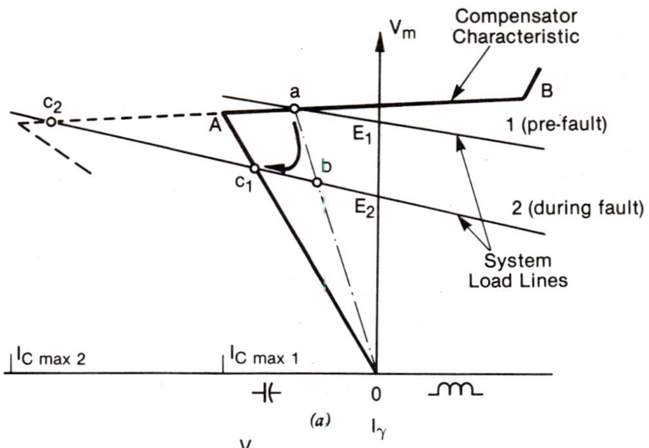

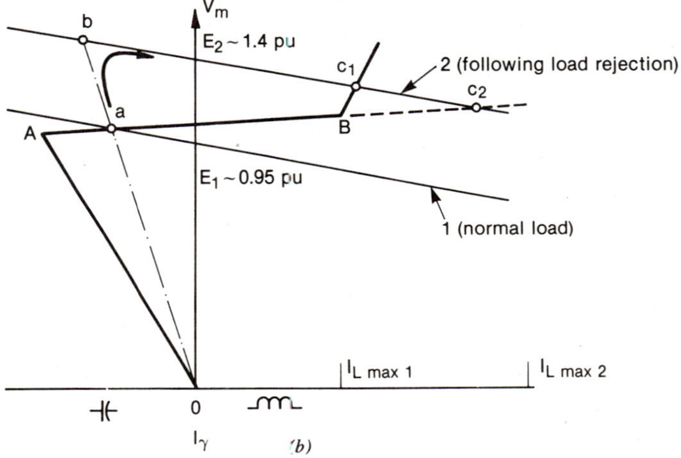

FIGURE 13. Over-ranging the TCR compensator. (*a*) During a voltage depression. (*b*) During a voltage rise.

worst-case load rejections can be expensive. However, if the shunt capacitors can be rapidly switched off, it is possible to design the TCR with only a short-time overload rating such as that corresponding to I_{Lmax2}. When the capacitors are switched off, the TCR is relieved of their biasing current, which results in an effective increase in the inductive I_{Lmax1} for the same reactor current.

(c) Field Test Involving a Large Disturbance. The oscillograph traces shown in Figure 14 were measured during field tests performed on a compensator installed for transmission system voltage control. The compensator consists of a 30-MVAr unswitched capacitor bank and a 40-MVAr TCR connected to the 13.8-kV tertiary winding of a 230/115-kV autotransformer. It regulates the 115-kV bus voltage to minimize voltage

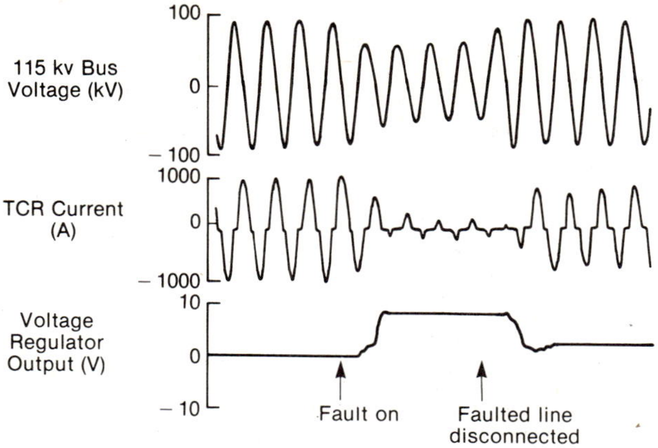

FIGURE 14. Field test results showing response of a TCR compensator to a single line-ground fault.

variations and to enhance the power transfer capacity of the 115-kV network. The test[8] involved a staged single line-to-ground fault applied about 15 mi from the compensator on a 115-kV circuit. The fault was cleared through primary relay and breaker action in about 5 cycles.

The effect of the fault is clearly evident in the 115-kV voltage trace. Note that the TCR current was effectively "turned off" within a cycle of fault inception. (The current in a TCR is never controlled to exactly zero lest gating control be lost.) As soon as the inductor virtually ceases to absorb reactive power, the transmission system benefits immediately from all the reactive power available from the capacitor bank at the prevailing supply voltage.

Upon removal of the faulted line by breaker action, the voltage recovered to near normal. As noted in the figure, the TCR current recovered to a lower value than it had before the fault, indicating that more capacitive reactive power was required from the compensator to obtain the desired voltage after the disconnection of the faulted line. The regulator output trace shows that the compensator had settled to its new operating point within 1.5 cycles following this disconnection. There were no significant rotor-angle synchronizing swings, so the two transition periods (during the fault and after its removal) are simply transient period behavior, with no subsequent first-swing or oscillatory period behavior.

(d) Transient Response of Compensator Near a HVDC Converter Terminal. The example that follows deals with the high voltage ac bus voltage at the input to a high-voltage dc (HVDC) converter. If the short-circuit capacity of the ac system at an HVDC converter site is low, the ac voltage can be quite sensitive to variations in the active power absorbed

or delivered by the converter. The converter terminals absorb reactive power equal to approximately 60% of the rated dc power. Capacitor banks are often employed on the ac bus to supply the steady-state reactive power demand of the converter. Some of these banks also serve as harmonic filters. Consider the load rejection that occurs when the terminal is operating under load and the converter valves suddenly block. This blocking might be the result of a major fault on the ac system near the terminal at the other end of the dc line which requires a momentary or permanent shutdown of the HVDC power flow to protect the dc converters.

To illustrate the ability of a compensator to control the overvoltage caused by such a load rejection, the system of Figure 15 was studied on a Transient Network Analyzer (TNA).[9] The rectifier terminal was initially operating at its 200 MW rating. The ac filter and capacitor banks were together rated 114 MVAr. The compensator modeled had a net rating as viewed from the 230-kV bus of 0–130 MVAr inductive. Three HVDC shutdown cases were studied.

1. Without compensation.
2. With the compensator controlling the 230-kV bus voltage through its automatic voltage regulator.
3. With the TCR "phased on" fully by a control signal from the HVDC converter. In this case, the dc valve blocking control action and the order to the TCR occurred simultaneously.

The results of these three TNA studies are illustrated in Figures 16*a*, *b*, and *c*, respectively. The phase *a*-to-neutral voltage on the 230-kV bus is shown in all three pictures. The valve-blocking action (load rejection) occurred at the fifth voltage zero from the left.

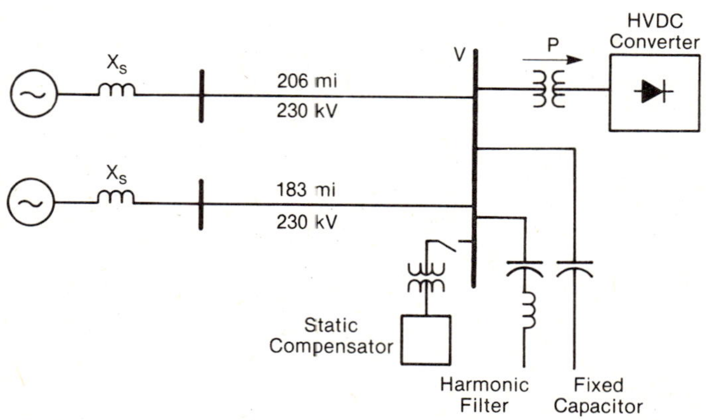

FIGURE 15. System used to study load rejection overvoltages at a HVDC converter terminal. © 1982 IEEE.

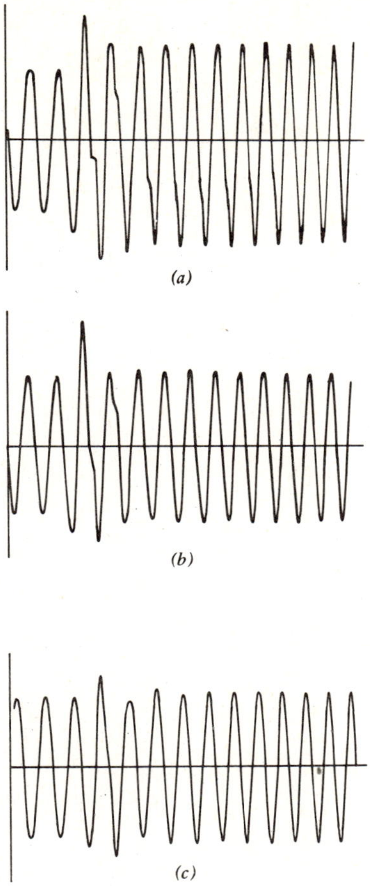

FIGURE 16. Load rejection at the HVDC converter terminal. Traces show HVAC voltage *V*. © 1982 IEEE. (*a*) Without compensation. (*b*) With compensator regulating voltage normally. (*c*) With compensator receiving a signal demanding full inductive current. The signal is derived from the converter as it blocks dc conduction.

The subsequent peak voltages are tabulated in Table 1. When the compensator reacted via its automatic voltage regulator (Case 2), it was unable to reduce the first (−) and second (+) peaks. When given an advance signal from the dc controls (Case 3), the compensator was able to reduce those peaks. In both cases when the compensator was in service, the overvoltage was reduced significantly. A compensator with a greater inductive rating could have limited the overvoltage even further. The one in the example was more than fully utilized and temporarily operated above its normal voltage control range.

Saturated-Reactor Compensator with Fixed Capacitor (SR-FC). Saturated-reactor compensators generally do not employ an electronic control system. Nevertheless, there is a small delay in their response, so that the settling time following a disturbance is comparable to that of systems with TCR compensators, being of the order of 1–2 cycles, depending on the dynamic properties of the ac system. The response delay is also

TABLE 1
Load Rejection Overvoltage Example on TNA

Time (Cycles)	1 No Compensation V Peak	2 Normal AVR Control V Peak	3 Coordinated Control V Peak
0[a]	± 0.95	± 0.95	± 0.95
0.25	− 1.27	− 1.2	− 1.11
1.75	+ 1.65	+ 1.65	+ 1.2
1.25	− 1.65	− 1.32	− 1.27
1.75	+ 1.27	+ 1.01	+ 0.90
2.25	− 1.52	− 1.08	− 1.01
2.75	+ 1.33	+ 1.01	+ 1.08
Sustained	± 1.33	± 1.02	± 1.02

[a] Time = 0 when dc blocking occurred, at about the fifth voltage zero of phase *a* in the oscillograms.

influenced by the damping circuit in parallel with the series-connected slope-correction capacitor. That damping circuit is necessary to prevent subharmonic instability (see Chapter 4, Section 4.4.2).

As in the case of the TCR compensator, the ability to limit overvoltages is determined by the inductive rating of the saturated reactor, and this is limited by the voltage rating of the slope-correction capacitor which may be protected by a parallel spark gap set to a voltage corresponding to, say, three times normal rated current. Even when the spark gap short-circuits the slope-correction capacitor, the reactor itself retains a fairly flat voltage/current characteristic (typically 8–15% slope) up to a much larger current. The overload capability of the reactor is limited in magnitude by the insulation requirements and the forces on its windings, and in duration by its thermal capacity. For cases with the series-connected slope-correction capacitor installations utilizing spark gaps, the gaps generally need a bypass switch which closes when they operate and reopens when normal voltage is restored.

For small or large disturbances which tend to cause a voltage reduction, the saturated-reactor compensator operates as shown in Figure 13*a*. For small disturbances resulting in a voltage rise it reacts similarly to the TCR in accordance with Figure 11. The saturated reactor is slowed slightly by its natural time constant, just as the TCR is slowed by its small phase-control gating delay.

Thyristor-Switched Capacitor (TSC). As described in Chapter 4, the thyristor-switched capacitor consists of several banks of shunt capacitors, each of which is connected or disconnected, as needed, by thyristor switches. The TSC has a control system that monitors the voltage. When the voltage deviates from the desired value by some preset error (deadband) in either direction, the control switches in (or out) one or more banks until the voltage returns inside the deadband, provided that not all the capacitors have been switched in (out). Figure 17 shows the response locus for a system disturbance that tends to cause a voltage drop. Assuming a TSC with zero, one, or two capacitor banks, operating point *a* would prevail with normal conditions and one bank connected. At the beginning of the disturbance the voltage would drop to point *b* until the TSC control switches in the second bank to bring the voltage up to point *c*.

It is important to note that because of the on/off nature of the TSC control, the compensating current can change only in discrete steps as a result of control action. In high-voltage applications the number of shunt capacitor banks is limited to a small number (say, three or four) because of the expense of the thyristor switches. As a result, the discrete steps in compensating current may be quite large, giving somewhat coarse control. Equally important is the fact that the capacitors can be connected to the line only at instants in the voltage wave when the stored voltage on the capacitor is equal, or nearly equal, to the instantaneous system voltage. This results in an effective delay which can be as much as one cycle (see Chapter 4).

The TSC by itself is incapable of limiting large transient overvoltages except by switching itself off. It has the advantage of then being able to

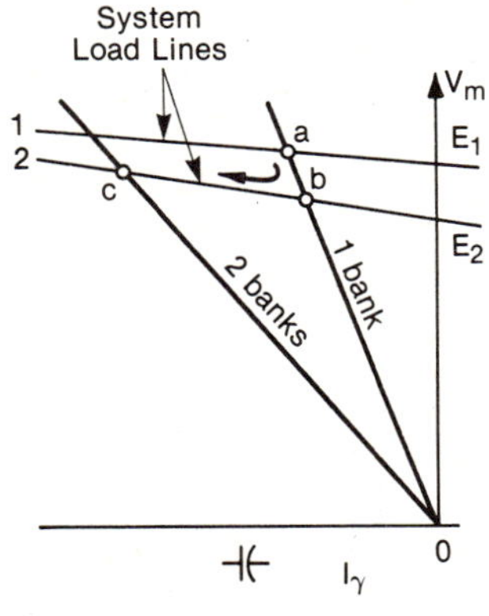

FIGURE 17. Effect of TSC compensator on operating point.

stand by in readiness to help control synchronizing power swings following the disturbance (see Section 3.4.2 of this chapter).

Combined Thyristor-Controlled Reactor and Switched Capacitor. The one-line diagram, Figure 18, depicts a compensator with a thyristor-controlled reactor and four capacitor banks. One of the capacitor banks is always in service with filter reactors to absorb the harmonic currents caused by the phase control action of the TCR. The TCR and the fixed filter capacitor bank (FC) constitute configuration 1 in Figure 18. The voltage/current characteristic for configuration 1 is shown in Figure 19 as 0-1-1′; or 0-1-1″ if short-time overload capacity is provided in the TCR. Three other characteristics are shown which correspond to the three operating states with one, two, or three of the capacitor banks connected. With suitable coordination of the TCR controls and the capacitor switches, the overall voltage/current characteristic is the *continuous* line 0-4-1′ (or 0-4-1″). It is capable of maintaining virtually constant voltage for the range of system load lines shown in Figure 19. The TCR voltage setpoint and control slope (or gain) must be adjusted each time a capacitor bank is switched, in order to obtain the continuous overall characteristic shown.

The voltage control performance of the hybrid compensator during the transient period is shown in Figure 20. Assume initial operation at point *a*, the intersection of the dashed compensator characteristic labeled 0-1-1′ and the first system reactive load line. Suppose a disturbance takes place that alters the system load line as shown, from 1 to 2. The compensator voltage V_m will begin to fall towards point *b* until the TCR responds to "phase off" entirely, bringing the operation to point *c* within about 1 cycle. So far in the sequence, a capacitor switching has not occurred.

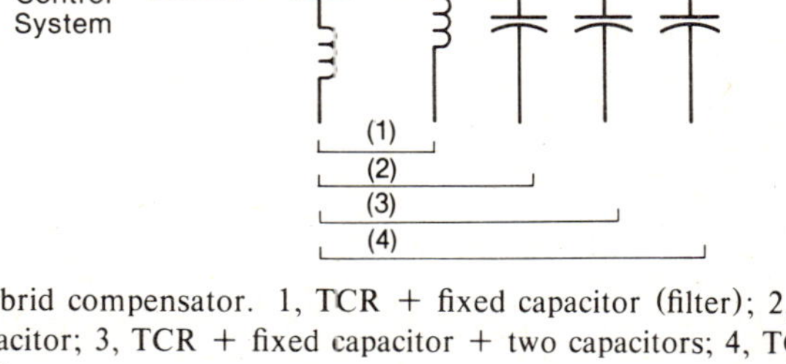

FIGURE 18. Hybrid compensator. 1, TCR + fixed capacitor (filter); 2, TCR + fixed capacitor + one capacitor; 3, TCR + fixed capacitor + two capacitors; 4, TCR + fixed capacitor + three capacitors.

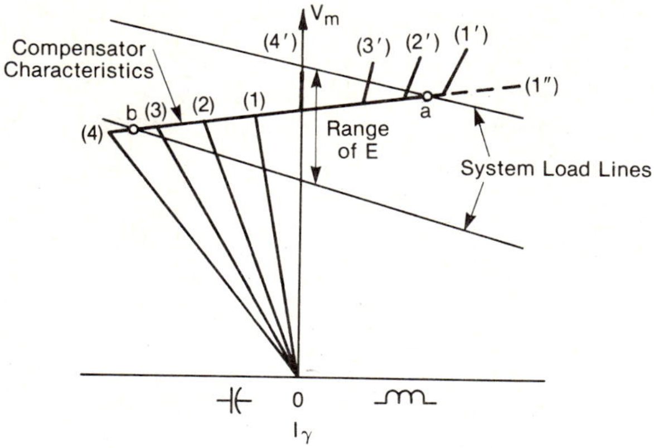

FIGURE 19. Characteristics of hybrid compensator.

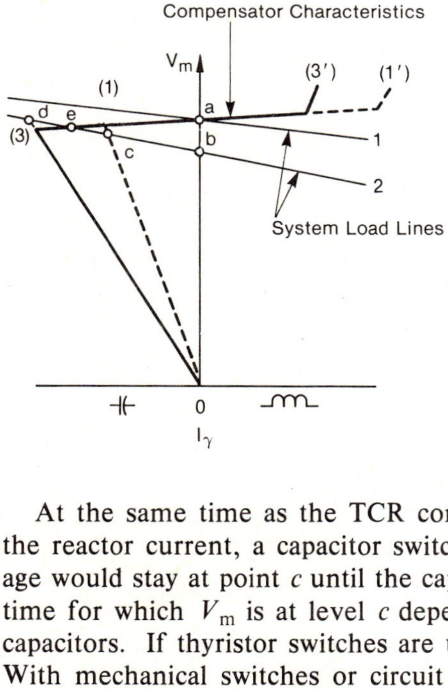

FIGURE 20. Effect of hybrid compensator on operating point.

At the same time as the TCR controls order the reduction to zero of the reactor current, a capacitor switching order can be issued. The voltage would stay at point *c* until the capacitors are energized. The period of time for which V_m is at level *c* depends on the method of switching the capacitors. If thyristor switches are used, it can be as short as one cycle. With mechanical switches or circuit breakers it would be several cycles. Upon energization of the capacitor, the voltage would jump to point *d* until the TCR corrects the voltage back to the final point *e*. Employing the fastest available mechanical switches (2 cycles) or thyristor switches (1 cycle) to switch in the capacitors, control action could cause the sequence *a-b-c-d-e* to be completed in less than 2 cycles, with pauses at point *b* and *d* equal to no more than half a cycle. In principle, the sequence could be reduced to *a-b-e,* with little or no overshoot toward point *d,* by fully utilizing the timing precision of thyristor switches for both the TCR and the capacitors.

Combined Saturated Reactor (SR) and Switched Capacitor. The sequence discussed with the aid of Figure 20 is achieved with this type of static compensator also. The only difference is that the capacitor switching is under "forced control" while the reactor is self-adjusting or "inherently controlled." The capacitor switching decisions can be made in response to a combination of signals representing the voltage at the compensator bus and the current in the reactor.

Summary – Behavior of Static Shunt Compensation During Transient Period. Although static compensators generally have neither the capacity nor the speed of response to limit extremely fast voltage transients, such as those produced by lightning or switching surges, their response is certainly fast enough to stabilize the transmission voltage within the "transient" period. Different compensators achieve more or less the same results in different ways (but with significant differences in overall performance when cost, losses, and harmonic performance are considered). The TCR and especially the saturated-reactor compensators both have apparently large overvoltage-limiting capability, but each application should be studied in detail because the real system response depends on many system parameters including the initial operating point.

3.4.2. First-Swing Period

So far, we have considered disturbances that can be represented by step changes in the system reactive load line at the compensator bus. This is generally valid only for a few cycles after the disturbance, and only if the magnitude of the initiating disturbance is very small and/or the source voltages are generated by synchronous machines of very large rating relative to the compensating means considered. If, however, the disturbance also upsets the distribution of momentum among the synchronous machines, the load line at the compensator bus will change (rise and fall) continually for a longer time, and perhaps by a large amount during the first-swing period.

The performance of a typical shunt compensator during the first-swing period is very similar to that described for the transient period. A compensator characteristic and a sequence of system reactive load lines are shown in Figure 21*a*. The sequence of reactive load lines 1, 2, 3 arises as follows.

1. This represents normal prefault conditions in a system of the type shown in Figure 3.

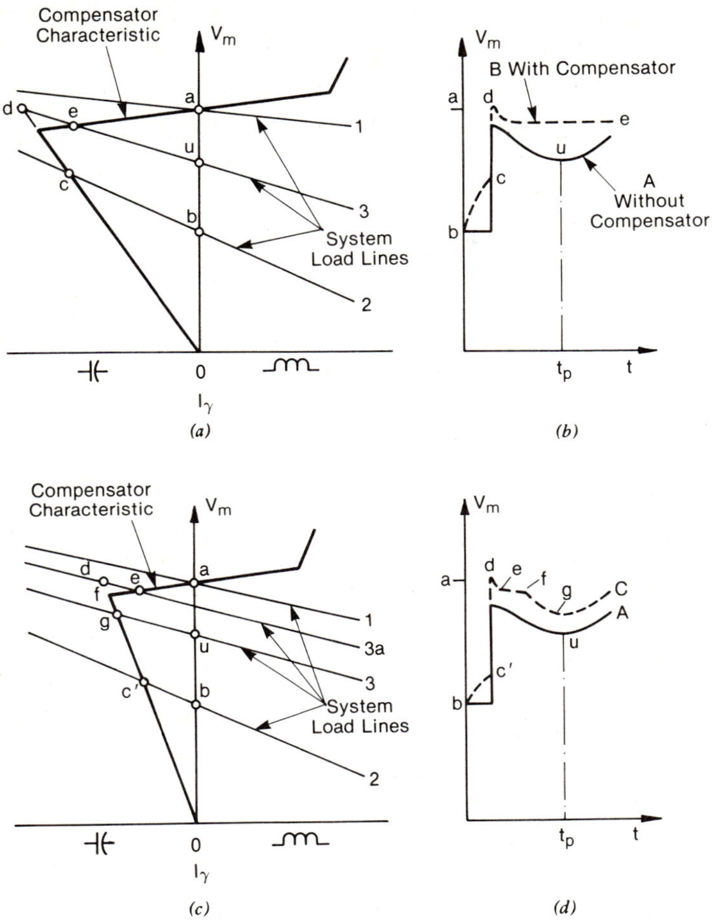

FIGURE 21. Response of static compensator in first-swing period. (*a*) Large compensator. (*b*) Transient response of V_m with large compensator. (*c*) Small compensator. (*d*) Transient response of V_m with small compensator.

2. A fault occurs to the right of circuit breaker *a* in Figure 3, depressing the load line by a large amount. Load line 2 is assumed to remain fixed for the duration of the fault.
3. After the faulted line section is removed from the system in Figure 3, by opening breakers *a* and *b*, the load line jumps initially above position 3, and drops during the first-swing period to its lowest (minimum voltage) position 3 as shown in Figure 21*a*.

The disturbance causes an oscillatory transient in the power, in the voltage V_m, and in the transmission angle δ. The first half of the V_m oscillation is shown in Figure 4*c*, and reproduced as trace A in Figure 21*b*.

Load line 3 moves up and down the plot as time passes. Correlating the voltage trace A with the sequence of load lines, the voltage drops instantly from point *a* to point *b* along the ordinate in Figure 21*b*. The voltage remains at point *b* during the fault and recovers to some value between points *u* and *a* when the faulted line section is disconnected. As δ increases, the voltage drops to a minimum value (point *u*) as shown in Figure 21*b* at time t_p.

If a relatively large compensator were in service on bus *m* of Figure 3, the dashed voltage trace B in Figure 21*b* would result. It is assumed that the compensator was operating at point *a* before the fault occurred, with zero compensating current. As soon as the fault occurs, the voltage V_m drops to point *b* even with the compensator in service. Within about a half cycle the TCR is "phased-off" and the voltage rises to point *c* on the fixed-capacitor line in Figure 21*a*. When the faulted line is disconnected, the voltage jumps to point *d*. Following a slight delay, less than or equal to 1/2 cycle, the TCR "phases on" part way and stabilizes the voltage at point *e*. The dashed trace B in Figure 21*b* shows the best voltage control performance achievable with a relatively large compensator at bus *m* in the system considered.

Figures 21*c* and *d* show the expected performance with a smaller compensator whose current and voltage are forced outside the normal control range during the postfault power swings. With a smaller capacitive rating than the one considered in Figure 21*a*, the slope *0f* is attained. The solid trace A in Figure 21*d* again shows the uncompensated response. The dashed voltage trace C shows the effect of the smaller compensator.

The voltage trace with the smaller compensator follows the same sequence during the fault: *a* to *b*, then to *c'*. Voltage level *c'* is slightly lower than *c* in Figures 21*a* and 21*b* owing to the smaller capacitive rating. Neither compensator does much to correct the voltage during the fault period.

When the fault is cleared and the postfault load line recovers to line 3a (located between load lines 3 and 1), the voltage V_m recovers also. The voltage trace C shows that the compensator was able to operate within its voltage control range, holding the voltage nearly constant for a short time between points *e* and *f*, until the power swing caused the operating point to drop over the left-most end of the compensator characteristic at point *f* and reach a minimum position *g* on load line 3. Even though the compensator is over-ranged, it limits the minimum voltage to point *g* instead of point *u* during the first swing.

In this section we have tacitly assumed that the variation in the system load line is unaffected by the action of the compensator. In general, the compensator will tend to reduce the variation in the transmission angle δ, which in turn affects the load line variation. We next consider the influence of the compensator on the transmission angle and the transient stability of the system.

3.4.3. The Effect of Static Shunt Compensation on Transient Stability

Section 2.6 of Chapter 2 introduced the concept of *compensation by sectioning,* in which a transmission system is sectioned into two or more parts by shunt compensators that hold the voltage essentially constant at intermediate points. In particular, it was shown that the *steady-state* power transfer capacity of a symmetrical line could be increased by adding a dynamic shunt compensator at the electrical midpont. In the limit, with perfect voltage control at the midpoint bus, the steady-state power transfer capacity was doubled (Figure 32, Chapter 2).

For the symmetrical system shown in Figure 22, this doubling is shown by the increase in the P/δ curve from a to b in Figure 23. Without the compensator, curve a is given by Equation 38 of Chapter 2, where E is now the "emf behind transient reactance" (E') of the two generators and X_ℓ is the total series reactance, equal to the sum of the line and transformer reactances, $X_\ell + 2X_T$ and the *transient* reactances X_d' of the generators (which are assumed identical). The effect of shunt capacitance is ignored. With an ideal compensator that holds the midpoint voltage constant at the value E', the power-angle curve b is obtained according to the equation

$$P = \frac{2E'^2}{2X_d' + X_T + X_\ell} \sin\frac{\delta}{2}\,. \tag{1}$$

This curve is realizable only if the (transient) compensator voltage/current characteristic is flat, if the compensator responds instantaneously, and if it has sufficient capacitive current capability. In general, none of these conditions is met. The small positive slope in the voltage/current characteristic, and the small response delay, combine to reduce the power-angle curve from b to c, while the limited capacitive current capa-

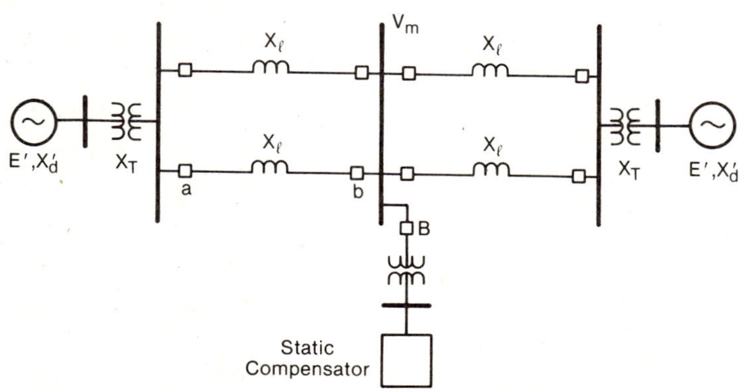

FIGURE 22. Two-machine system with midpoint dynamic compensator.

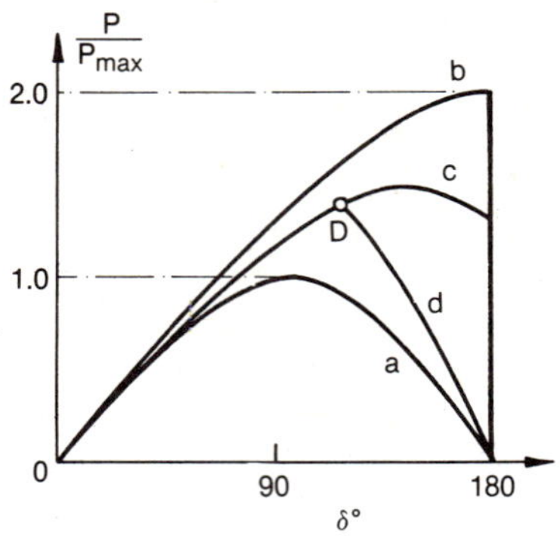

FIGURE 23. Transient power/angle curves with and without dynamic shunt compensation: a, no compensator at bus *m*; b, ideal dynamic compensator at bus *m*; c, compensator with drooping voltage characteristic; and d, compensator with limited rating, reached at D.

bility breaks the curve at point D, the compensator behaving as a fixed capacitor at higher load angles, curve d. (See also Figure 32 of Chapter 2.)

Theory of Transient Stability Improvement. We are now in a position to use the equal-area criterion to compare the relative transient stability with and without a dynamic shunt compensator. First consider a case without compensation.

Imagine a fault that occurs between circuit breakers *a* and *b* in Figure 22 and is cleared by those breakers. Curve 1 in Figure 24*a* illustrates the prefault *transient* power angle curve which has a maximum of

$$P_{max} = \frac{E'^2}{X_\ell + 2(X_T + X'_d)} . \tag{2}$$

Curve 2 prevails during the fault. Curve 3 results after the faulted line section is removed and differs from curve 1 in that X_ℓ is replaced by $3X_\ell/2$. E' is assumed constant through the first-swing period. In the absence of the compensator, the system can, in principle, be preloaded to the transient stability power limit P_1 for the prescribed fault, such that the available decelerating energy A_2 just balances the accelerating energy A_1. In practice, the power level would be somewhat less than this, to provide a stability margin.

The effect of a large, rapid-response midpoint shunt compensator is shown in Figure 24*b*. Curves 1, 2, and 3 of Figure 24*a* are replaced by the higher curves 1, 2, and 3, respectively. For the same prefault power P_1 and the same fault duration, the available decelerating area is now larger and is only partially used up, leaving a margin as indicated in Figure 24*b*. In other words, the transient stability limit is increased, that

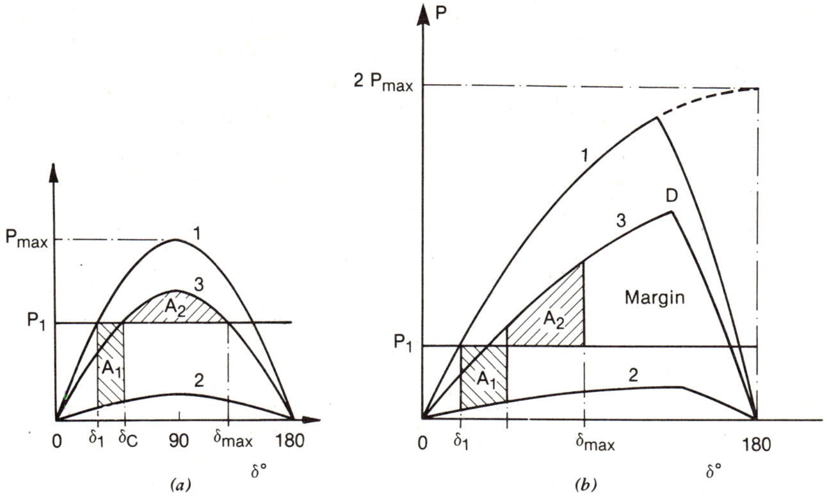

FIGURE 24. Equal-area method illustrating increased transient stability with dynamic shunt compensation. (*a*) No compensator at bus *m*. (*b*) Dynamic shunt compensator at bus *m*, of large rating, maintains linear voltage control during entire first-swing period.

is, the power transfer can be increased up to the point where all of the margin is used up. It is clear that the capacitive current capability of the compensator is a limiting factor in the stability margin, because it reduces the margin area in Figure 24*b* to the right of point D. Nevertheless, a worthwhile improvement in transient stability is still possible with compensators of limited size.

Computer Simulation Examples. The dynamic performance of compensated systems can be studied in greater detail with computer simulations. Three examples are now described which show the performance benefits attainable with a compensator.

Example 1. Increase Transient Stability Margin for a Given Power Transfer

The two-machine model in Figure 22 is now assumed to represent a 1000-MW generator feeding a double-circuit, 500-kV line. The receiving-end machine is assumed to be an infinite bus. Sixty percent of the line-charging current was compensated with fixed shunt reactors. The generator had a high-response excitation system. Figure 25 shows the response of the transmission angle δ, the power, and the midpoint bus voltage following a three-phase fault on one circuit near the generator end. The faulted circuit was tripped and the system remained stable: no pole-slip took place, although large rotor angle swings were experienced.

Figure 26 shows results for the same study system equipped with a compensator consisting of a 300-MVAr TCR and 300 MVAr of shunt ca-

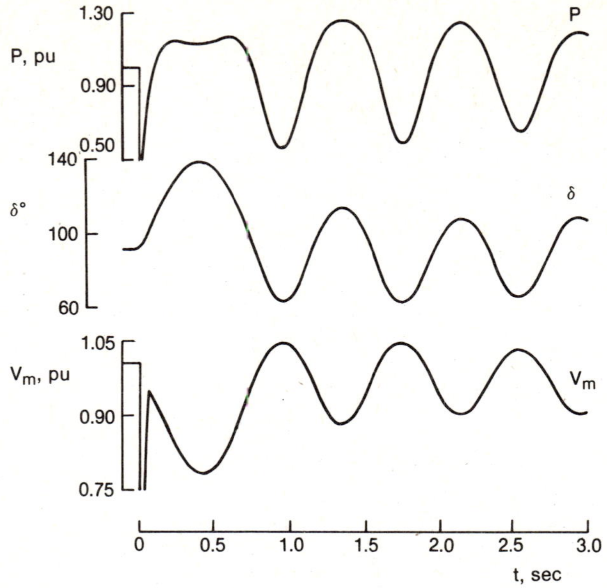

FIGURE 25. Response of system in Figure 22 to a three-phase fault; no compensator. © 1982 IEEE.

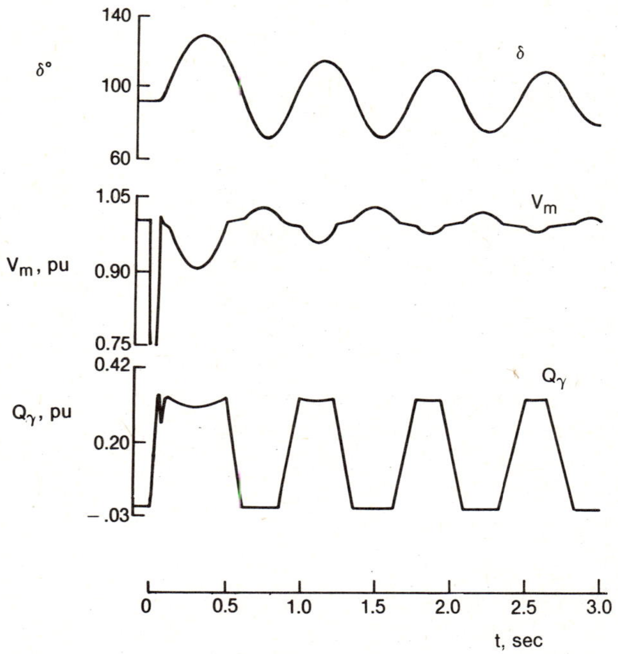

FIGURE 26. Response of system in Figure 22 to a three-phase fault; with static compensator. © 1982 IEEE.

pacitance. The first as well as the subsequent swings in the rotor angle were significantly reduced over those in the uncompensated case, indicating an increase in transient stability. The voltage on the midpoint 500-kV bus is virtually steady only three seconds after the fault. The points at which the compensator becomes over-ranged are clearly seen in the lower two traces. Although these results are based on a simple model, more comprehensive studies have shown that such improvements are obtainable in other realistic power system configurations.

Example 2. Increasing Power Transfer while Maintaining Transient Stability

The system studied in this example was the 500-kV transmission system shown in Figure 27. This system typefies a transmission tie between a remote generating plant and a major power grid system. The transmission circuit consisted of two lines, 300 mi long, with an intermediate switching station, bus 2. There was assumed to be 500 MW of load at the remote generation site. For the transient stability investigations, the system disturbance was a zero-impedance, three-phase fault at point X, the faulted line being tripped in 5 cycles.

The objective of the study was to demonstrate the utilization of static shunt compensators to increase the power that could be transferred over a given (fixed) set of transmission lines without risking transient instability for a specific fault. This would be analogous to an actual condition in which approval for additional transmission lines is difficult to obtain or in which more economical alternatives are desired. In contrast to the fixed transmission it was presumed that the generation capability (and associated transformers) would be increased, consistent with the increased power transfer capability made possible by the application of the compensator.

The investigations were made using a digital power system stability program.[12] Compensators were considered to be located at bus 1 and bus 2.

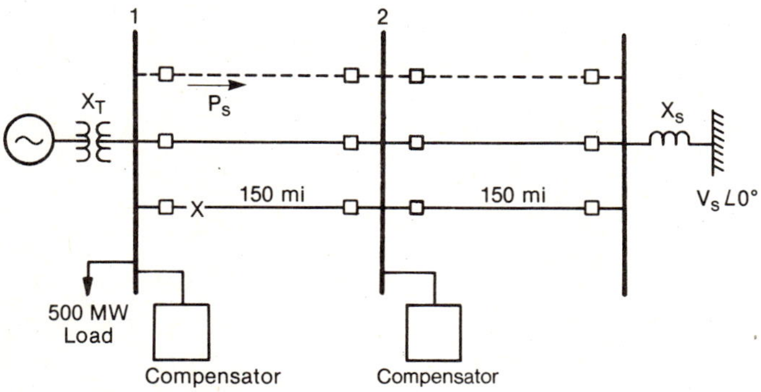

FIGURE 27. 500-kV system. © 1982 IEEE.

For a number of power levels P_S, and a number of capacitive MVAr ratings for the compensator at bus 1, multiple program runs were made to determine the minimum capacitive MVAr rating of the compensator at bus 2 required to maintain transient stability for the stated fault. For each P_S the generator and transformer per-unit reactances were kept constant while the MVA base was scaled up. The *ohmic* value of the line impedance was kept constant. This gave the effect of trying to transmit the output of larger and larger generating stations through the same transmission line.

The results are shown in Figure 28 and indicate for each P_S the combination of capacitive ratings for the two compensators that will maintain transient stability. For example, for a P_S of 1100 MW, ratings of 600 MVAr at bus 1 and 230 MVAr at bus 2 are required. The power levels investigated ranged from 720 MW, which is the uncompensated transient stability limit, to 1250 MW, which is the limit with a third line (shown dashed in Figure 27) but no compensators. In all cases, the MVA rating of the generators and associated transformers was made equal to the prefault power flow. The control settings for the compensators were modeled such that compensator output was zero under initial (prefault) conditions.

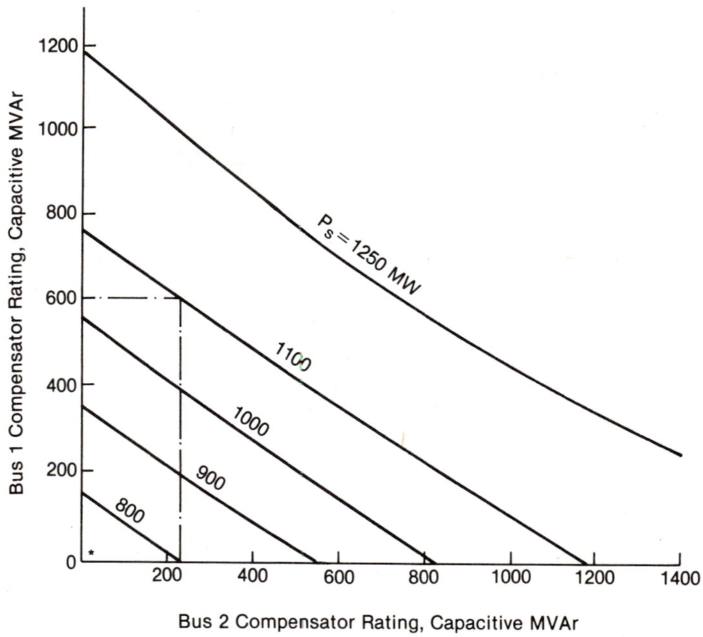

FIGURE 28. Minimum combination of compensator ratings required for transient stability at various prefault power levels P_S. © 1982 IEEE.

For the circuit investigated, the results show that the compensators increase the transient stability limit. The increase for this circuit configuration is approximately equal to 100 MW for each 200 MVAr of added compensation, and can be used to achieve from two transmission lines the power level associated with three lines but no compensators.

The results also indicate the preferred location for the compensators, that is, that which provides the greatest increase in power transfer capability for a given combined compensator rating. In the application of compensators to an actual system, a number of fault conditions and contingencies must be considered, and the final choice of location(s) may be a compromise. However, for the fault studied in this example, the preferred location for a single compensator is at bus 1. This is due to the fact that bus 1 is closest to the electrical midpoint of the postfault circuit. Without compensators, this location experiences the greatest voltage dip during the generator angle swings that follow the fault clearing.

An approximate transient power-angle curve for the compensated system is illustrated by curve *b* in Figure 29. This curve is contrasted to the corresponding curves *a* and *c* for the two- and three-line circuits respectively, without compensators. Since the effects on transmitted power (P_s) are of most interest, the calculations for the curves were based on the assumptions of no local load and a 1000-MVA fixed generation capacity. The curves were calculated for the postfault circuit condition, with the generators modeled as a constant voltage behind transient reactance. The discontinuity on curve *b* is the point at which the compensators reach the

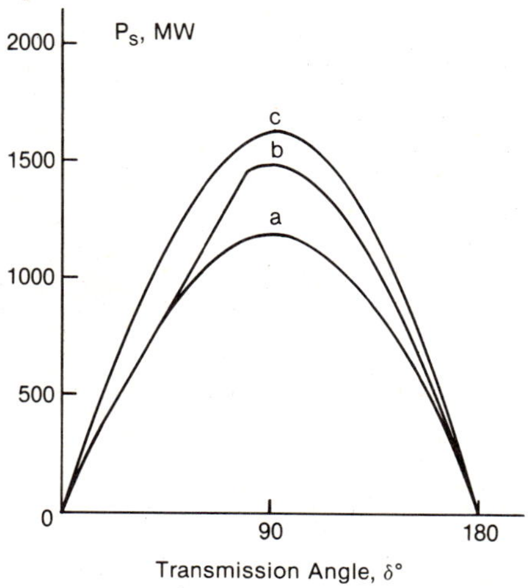

FIGURE 29. Postfault power/angle curves for the system of Figure 27: a, two lines with no compensator; b, two lines with compensator; c, three lines with no compensator. © 1982 IEEE.

end of their control range and become essentially fixed capacitors for the lower voltages associated with larger transmission angles.

Additional dynamic simulations were also performed to investigate the effect of other compensator parameters on the compensator rating required to achieve transient stability for specific power levels. In all previously discussed simulations, the compensators were modeled with a control gain of 100 (voltage droop of 1% over the compensator's total control range) and a single lag time constant of 33 msec. (see Chapter 4). Two variations from this were investigated and the results are shown in Figure 30. One variation was a reduction in control gain to 20 (total control range voltage droop of 5%). This reduction caused a slight increase in the required capacitive ratings.

The second variation considered was a delay in the availability of reactive power output from the compensator after the fault was cleared. It is during the first major angular swing and associated voltage depression that the compensator output is most important in preserving first-swing system stability. The compensator capacitors were switched on following a three cycle (50 msec) delay after fault clearing. Such a delay might be obtained if the capacitors are switched with mechanical switches. As is shown in Figure 30, the switching delay required a 20% increase in the capacitive rating of the compensators to maintain transient stability for the specific power levels examined in the simulation.

An additional consideration in the required compensator ratings is the size of the line-connected shunt reactors. The circuit of Figure 27, as modeled, included 81-MVAr shunt reactors permanently connected to

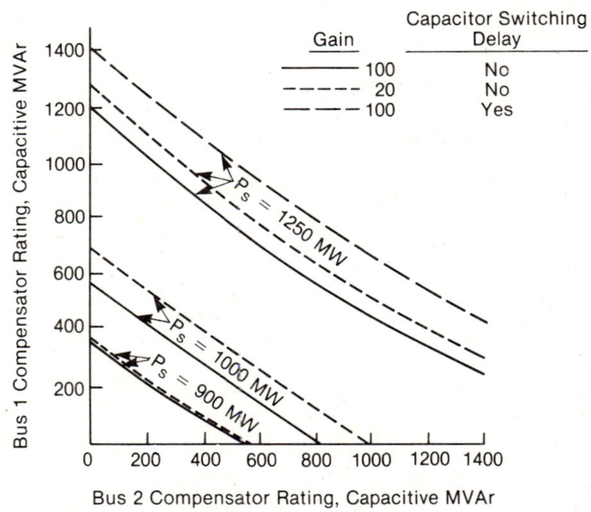

FIGURE 30. Effect of gain and delay on compensator ratings required to maintain transient stability at various prefault power levels P_S.

each end of each line section. This achieved 60% compensation of the line capacitance, as is typical for such a circuit. These fixed reactors were modeled in all cases, whether or not compensators were present. In practice, the location, rating, and switching strategy of shunt reactors must be coordinated with the application of compensators. For the circuit studied, such coordination might have resulted in a reduction of the required capacitive rating for the compensators.

Example 3. Increase Critical Fault Clearing Time for a Given Power Transfer

Another investigation of transient stability was performed on the 345-kV system shown in Figure 31. In this example the first swing stability enhancement of the system achieved by a compensator at bus 2 was quantified in terms of an increase in the critical clearing time. Zero impedance, three-phase faults at either locations A or B were the system disturbances. These two cases bracket the extremes of practical postfault system conditions. The permanent fault at point B results in a line outage. The fault at point A results in the loss of a stub line carrying no initial load. A fault at point B with instantaneous reclosing would yield the same result as fault A. In all cases, the ratings of the generators and transformers, and the prefault power (1200 MW), were the same.

The effect of the compensator capacitive rating on the critical clearing time for a fault at A is shown in Figure 32 for various lengths of the total transmission circuit. The curves show a definite increase in the critical fault clearing time with increased compensator capacitive rating. Fixed high-voltage shunt reactors, normally used to compensate for line charging, were neglected in these studies.

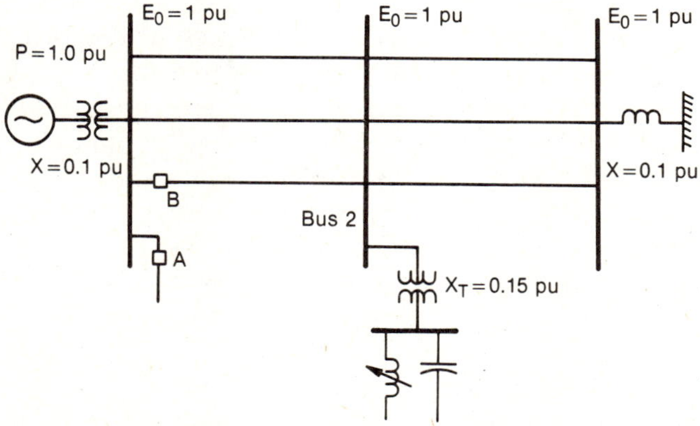

FIGURE 31. 345-kV system. Base MVA = 1200 except for X_T of compensator. (X_T was 0.15 pu on varied static compensator capacitive rating) © 1982 IEEE.

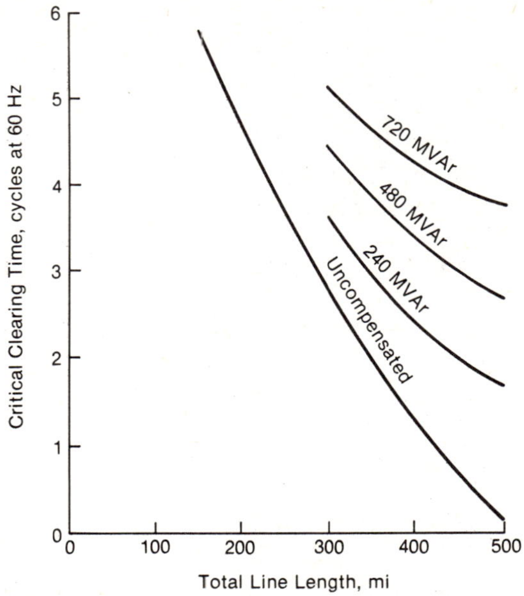

FIGURE 32. Effect of compensator capacitive rating on critical clearing time for various total line lengths (fault at A in Figure 31) © 1982 IEEE.

The results for a fault at B are shown in Figure 33. Here again, a significant increase in critical clearing time is obtained through the stabilizing action of the compensator.

3.4.4. Oscillatory Period

Just as in any control system, the effective gain and response delay in a static compensator influence the modes (eigenvalues) of the power system. In discussing the damping of power swings in Figures 25 and 26, the compensator was described as having a high voltage-regulating gain and a rapid response. These dynamic properties of electronically controlled compensators can be modified or optimized to improve the damping of power swings beyond what is possible with a plain "constant voltage" reactive compensator. (Supplementary stabilizers are used on the excitation systems of synchronous generators to the same end.) The use of supplementary damping controls on a compensator regulator is best illustrated with a simulation example. The system used in this example was structurally the same as shown in Figure 22.

Consider the damping controls illustrated in Figure 34 which are designed to modulate the compensator reactive power in response to both

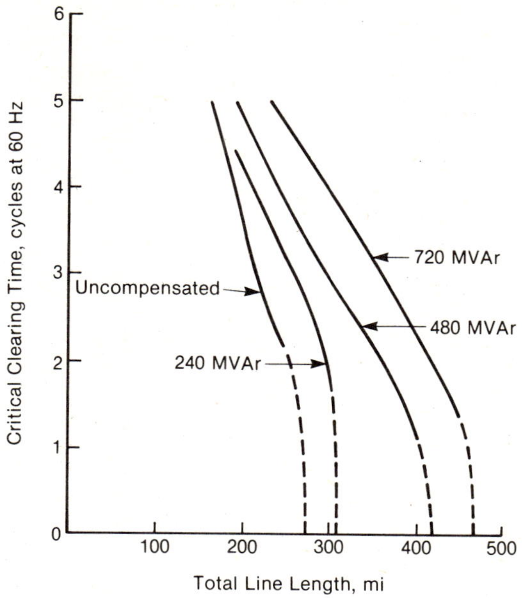

FIGURE 33. Effect of compensator capacitive rating on critical clearing time for various total line lengths (fault at B in Figure 31).

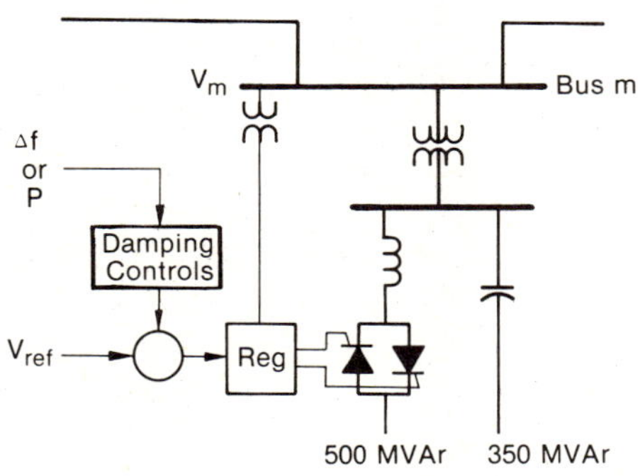

FIGURE 34. Supplementary damping control for providing power-angle swing damping.

voltage error and deviation in bus frequency or the oscillations in ac power flow.

The comparison of simulated performance in Figure 35 shows that damping of the angular swings was improved by feeding a properly conditioned signal derived from power flow on the line to the voltage regulator. The use of bus frequency has been shown to provide better damping in general. Notice that the voltage control capability was compromised slightly in order to add damping to the power-angle swings. This compromise is necessary if stability enhancement is of primary importance.

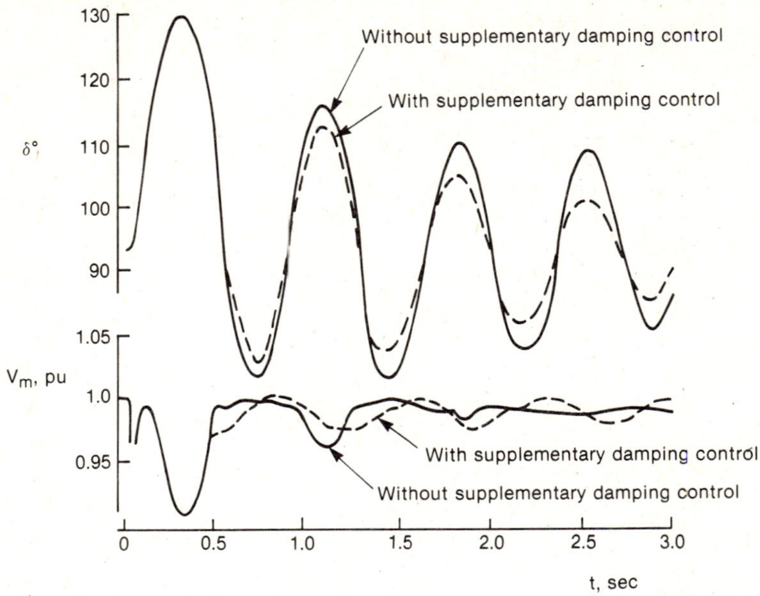

FIGURE 35. Damping of rotor angle oscillations using supplementary damping controls.

3.4.5. Preventing Voltage Instability with Static Compensation

A voltage instability can occur in the steady-state[11] if the reactive demand of a load increases as the supply voltage decreases, causing a further decrease in voltage. This voltage-collapse sequence can be triggered by a sudden disturbance. It can be particularly disruptive in the system exemplified later, in which continuous-process industrial plants are stalled, causing extreme penalties in lost production and loss of revenues to the utility company.

The ability of a static shunt compensator to prevent this form of instability is best shown by the following simulation. A system that would exhibit the voltage collapse phenomenon in the absence of dynamic voltage support is shown in Figure 36. The system represents a mixed residential and industrial load area with local generation available to serve about half of the MW demand. The load area is thus dependent on power imported over a transmission link. For the example, the disturbance chosen was to trip one of the 60-MW generators by opening breaker *a*. At the time of this disturbance, the only transmission link in service was assumed to be a single 138-kV line.

Proper modeling of the load proved to be important in the simulation of the voltage collapse phenomenon. The 220-MW load selected consisted of 160 MW of induction motors, with the remainder evenly split

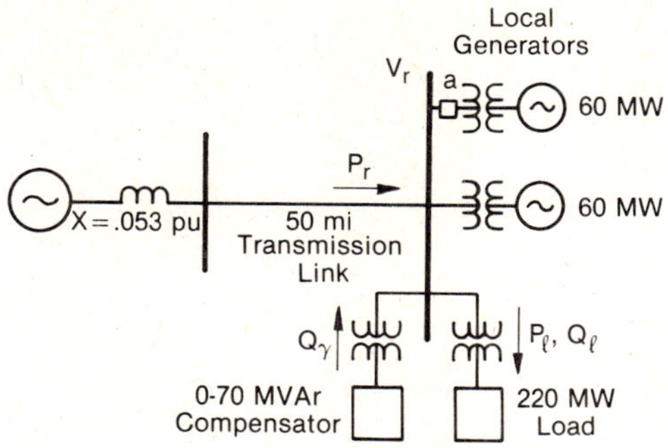

FIGURE 36. 138-kV system. Voltage instability (base MVA = 100). © 1982 IEEE.

between constant-current and fixed-impedance load types. The induction motor load was modeled in detail, including its inertia. In the initial steady state the reactive power requirement of the motor load was 77 MVAr, which was compensated to unity power-factor by a fixed shunt capacitor (not part of the compensator). The other load types were assumed to have no reactive power requirements. Before the disturbance, the local generators were operating at rated MVA and were supplying reactive power to the transmission line in order to maintain 1 pu voltage at the receiving-end bus. The local generators were modeled with rapid-response exciters. The compensator was assumed to be operating initially at zero current. It was modeled with a gain of 100 (1% voltage slope) and a single time constant of 33 msec (see Chapter 4).

Figure 37 shows the effect of the generator trip with and without the compensator. Without the compensator, the loss of one local generator precipitates a sudden voltage collapse and causes the induction motors to stall. However, with the compensator this collapse is avoided.

The reasons for the voltage collapse can be understood in the context of the quasi-steady-state receiving-end voltage-versus-power characteristic[11] of the transmission link, shown in Figure 38. (See Chapter 2, Figure 8.) Each curve is for a specific level of reactive power at the receiving end. The curves are plotted for fixed shunt capacitors or reactors in increments of 25 MVAr (based on 1 pu voltage). The reactive power (or power factor) at the receiving end greatly influences the maximum transmissible power. Superimposed on the family of curves of Figure 38 are the trajectories of the system response, starting at point 1, for the first 3.0 seconds following the generator trip, with and without the compensator. It is obvious that the compensator (together with the remaining generator) provides the extra reactive power necessary to maintain stability while the imported power is increased.

3.4.6. Summary – Compensator Dynamic Performance

The dynamic response of compensators to typical disturbances on the power system has been described in terms of fundamental-frequency, bal-

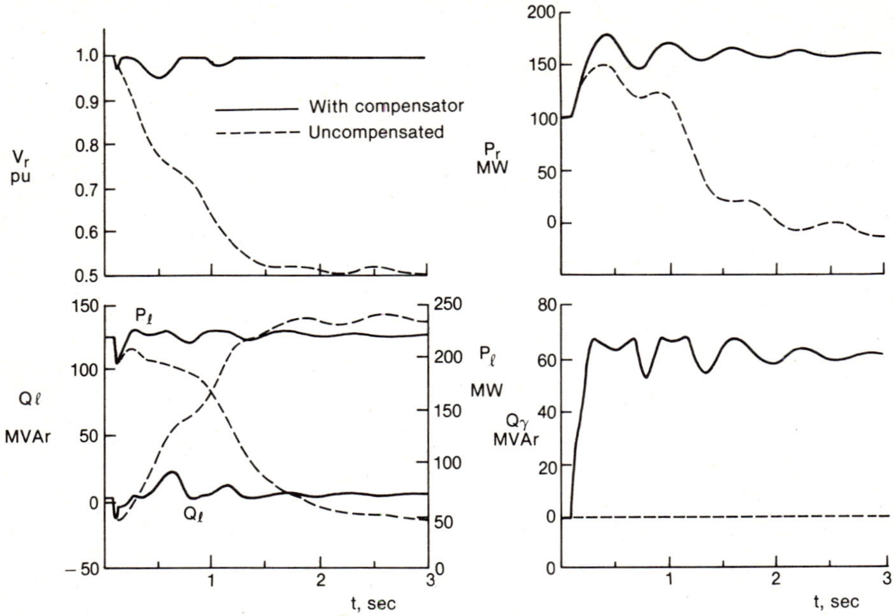

FIGURE 37. Response of system of Figure 36 to local generator trip, with and without compensator. © 1982 IEEE.

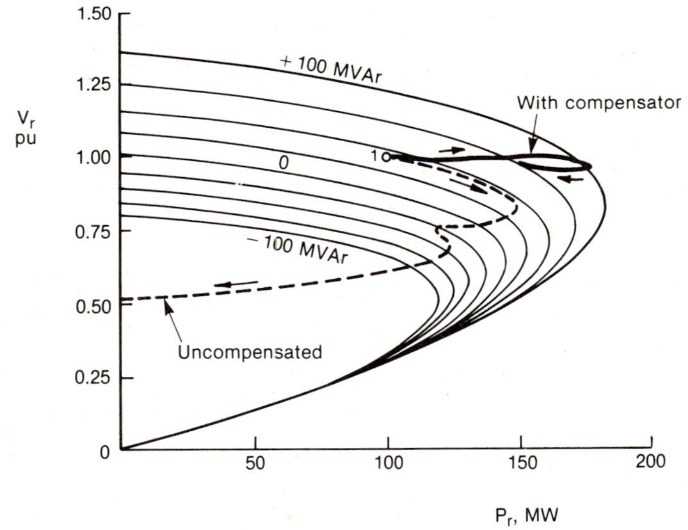

FIGURE 38. Receiving-end voltage/power characteristics. © 1982 IEEE.

anced three-phase effects. Extending the well-known theories of synchronous machines and of power system stability, the dynamic transitions were explained in terms of a sequence of quasi-steady-state conditions, using voltage/current characteristics and related concepts based on phasor analysis.

The graphical approach of illustrating dynamic compensator performance was reinforced and extended by oscillograms from actual tests and comprehensive model tests obtained on a transient network analyzer, as well as by digital computer simulations.

The ability of the static shunt compensator to aid power system stability through rapid adjustment of its reactive power was demonstrated by several simulation examples. The speed of response and the capacitive and inductive ratings are the key parameters which determine its benefit to power system dynamic performance. The use of one or more compensators can compete with the alternative of additional transmission lines in meeting the need for greater power transfer capability.

The controllability of TCR-based compensators permits the use of supplementary damping signals to accelerate the settling of the power system following disturbances.

3.5. SYNCHRONOUS CONDENSERS

Synchronous condensers, or synchronous compensators as they are sometimes called, are synchronous machines designed for shunt reactive power compensation. They fall into the class of active shunt compensators described in Chapter 2. The synchronous generator also falls in this category because reactive power is exchanged with the transmission system through the same electromagnetic principles. The application of synchronous condensers is described in Chapter 8. Here we consider its basic operation as an active shunt compensator.

In the steady state the synchronous condenser with fixed field current can be represented approximately by a generated emf $\mathbf{E}_0$ in series with the synchronous reactance X_d (Figure 39*a*). This equivalent circuit has voltage/current characteristics shown by the series of steep lines in Figure 40. Each line represents a fixed value of field current or equivalent open-circuit voltage E_0. The slope of these solid lines is proportional to

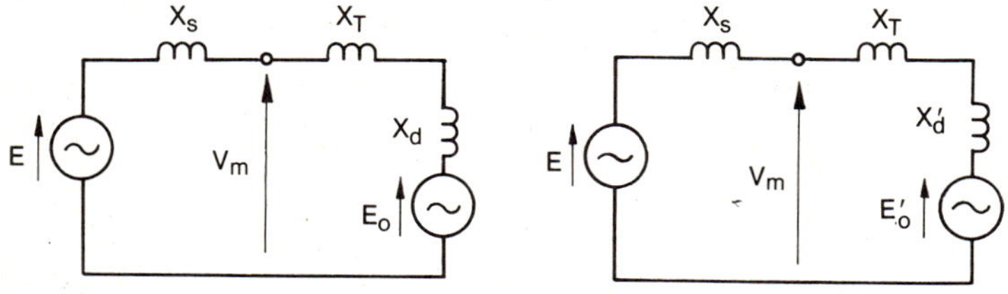

FIGURE 39. Synchronous condenser equivalent circuit. (*a*) Steady-state conditions. (*b*) Transient conditions.

$(X_d + X_T)$. These characteristics are normally applicable only when field current is fixed, such as under "maximum excitation limit" used to limit condenser overloading.

Under transient conditions the condenser behaves as though its synchronous reactance were reduced to the transient reactance X'_d, which is much smaller than X_d. Even without changing the field current, therefore, the condenser tends to have a flatter *transient* voltage/current characteristic than the constant field current lines of Figure 40. The slope of the dashed transient characteristic is proportional to $(X'_d + X_T)$. With a fast-acting voltage regulator controlling V_m, the machine can be made to operate *continuously* very near to the dashed transient characteristic. The effective equivalent circuit for this period is as shown in Figure 39*b*.

3.5.1. Transient Period

The equivalent circuit shown in Figure 39*b* has a fundamental positive-sequence voltage/current characteristic similar to that of a static compensator. However, there is one fundamental difference, which we now discuss briefly.

The internal voltage E'_0 is a true emf generated by the air gap flux linkages which tend to remain fixed in the short term (such as during the transient period). Because of this, it is impossible for transient voltage excursions to deviate very far from the transient voltage/current characteristic shown in Figure 40. Indeed, the faster the voltage tends to change, the flatter the slope of this characteristic becomes, as a result of the induced (*sub*transient) currents in the amortisseur windings. This is the opposite of the behavior of the TCR compensator, whose voltage/current characteristic tends to revert to the much steeper slope of the uncontrolled reactors during very rapid changes, and which must be corrected by control action that cannot be instantaneous. Even so, when expressed in per-unit on the appropriate system base, the slope of the synchronous condenser's voltage/current characteristic is limited to a cer-

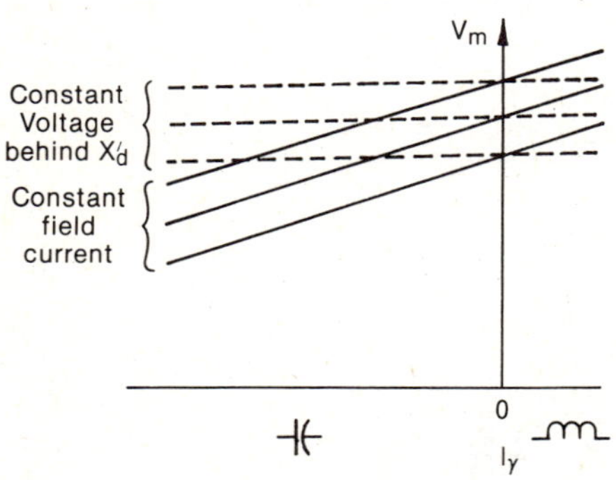

FIGURE 40. Steady-state and transient voltage/current characteristics of synchronous condenser.

tain minimum value by economics, in much the same way as in the saturated-reactor compensator.

Figure 41 illustrates a condenser operating initially at zero current, point *a*. Consider a system disturbance that causes the load line to change suddenly from 1 to 2. In the absence of the condenser, the voltage at the compensator bus would decrease to E_2. With the condenser, the operating point follows the trajectory $a-b$. This trajectory tends to be a little flatter than the transient characteristic because of the subtransient currents induced in the amortisseur windings, the effect of which dies out in the few cycles required to transition from point *a* to point *b*.

The voltage at point *b* is higher than E_2, thus illustrating the ability of the condenser to instantly supply additional reactive power. At *b* the condenser is operating overexcited, but at less than its original terminal voltage. The voltage regulator will automatically increase the field current to restore condenser terminal voltage to operating point *c*. If the condenser's rating is sufficient, point *c* is the final steady state. However, this operating point can often be beyond the condenser's normal rating in which case operation at point *c* would be limited to a minute or more depending on the condenser's short-time overload capability. If operating point *c* is beyond the condenser continuous rating, then field current will be reduced (normally by automatic adjustment of the voltage regulator set point) to a new operating point such as *d*.

As in the case of the static compensator, we have kept the discussion confined to balanced three-phase effects. Depending on the nature of the system disturbance, the voltage at the compensator bus may be higher in the first one or two half cycles, and it may be unbalanced during this time also. If the condenser is the only means for limiting these voltage excur-

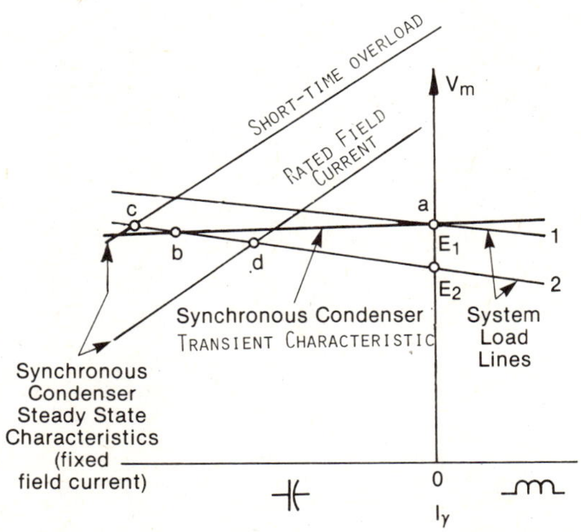

FIGURE 41. Response of synchronous condenser to voltage drop condition.

sions, it does so very quickly, in a period commensurate with its short-circuit armature time-constant.

3.5.2. First-Swing and Oscillatory Periods

Like other active shunt compensators, the synchronous condenser can reduce power and voltage swings following a disturbance, and its effectiveness in both the first-swing and oscillatory periods is shown by examples in Chapter 8. Like the electronically controlled TCR compensator, its voltage regulator (which is similar to that of the TCR) can be supplied with supplementary damping signals for improving the general response of the system.

3.6. SERIES CAPACITOR COMPENSATION

Series capacitors were discussed in Chapter 2 as a means of reducing the effective inductive reactance of transmission lines. The benefits to steady-state operation included better voltage (profile) control, reactive power management, reduced losses and improved power transfer capability. Chapter 7 discusses the series capacitor application in more detail. This chapter extends the theory covered in Chapter 2 to deal with the dynamic behavior and transient stability of series-compensated systems. In keeping with earlier sections, the transient, first-swing, and oscillatory periods will be studied in order.

3.6.1. Transient Period

By reducing the equivalent series reactance of a transmission line, the series capacitor makes the voltage less sensitive to most disturbances. An example of this is given in detail in Section 2.5.3 of Chapter 2 for steady-state conditions. Provided the capacitor bypass (protective) equipment does not operate, the performance in the transient period is much the same as if a range of powers in Table 5, Chapter 2, were swept through in a short period.

During large disturbances (such as those caused by faults) the capacitor protective equipment can have a very significant effect. This will be discussed further in the next section and in Chapter 7.

3.6.2. First-Swing Period and Transient Stability

The notion of line length compensation with series capacitors was introduced in Chapter 2. By reducing the effective inductive reactance of those lines in which the capacitors are installed, they can increase the relative stability of the system generators by reducing the transmission angle at a given level of power transfer (Equation 107, Chapter 2). An increase in the peak synchronizing power is also achieved.

This latter effect can be seen with reference to the radial system in Figure 42. Assume that the total reactance X_t were reduced by 10% using series capacitors in the two parallel lines. The peak synchronizing power capability (P_{max} in Figure 24*a*) would be increased by 10%. The generators would have more stability margin for the same initial operating power P_1.

With a 10% increase in peak synchronizing capability, the available decelerating energy (area A_2) in Figure 24*a* would also increase. For the same initial power, fault, and fault clearing time, the system would be more stable on first swing with series capacitors than without. With a corresponding increase in P_{max}, curve 2, the accelerating energy (area A_1) would also decrease in the direction toward greater first-swing stability, provided that the capacitor is not bypassed by its protective equipment during the fault period.

Series capacitors may be automatically bypassed (short-circuited) during faults to protect them from damaging fault currents. The bypass usually operates at a predetermined level of capacitor voltage, this being proportional to the current. For example, a voltage-sensitive spark gap can be employed, triggering a bypass switch. Alternatively, zinc-oxide nonlinear resistors can be used (see Chapter 7). Since the bypass only operates when the voltage across the capacitors exceeds the predetermined threshold, not all series capacitors in a given system will necessarily be bypassed, but only those close enough to the fault to have high voltages. For worst case faults, enough capacitors may be bypassed to make the postfault system too weak to avoid transient instability unless some or all of the capacitors are reinserted rapidly (by opening the bypass circuit) after the fault is cleared.

Figure 43 illustrates the effect of series capacitor reinsertion on machine stability. Figure 43*a* shows a worst-case fault condition with the series capacitor bypass remaining in effect on the unfaulted lines long after the faulted line is removed. This case is clearly unstable because the accelerating energy A_1 is greater than the available decelerating energy A_2.

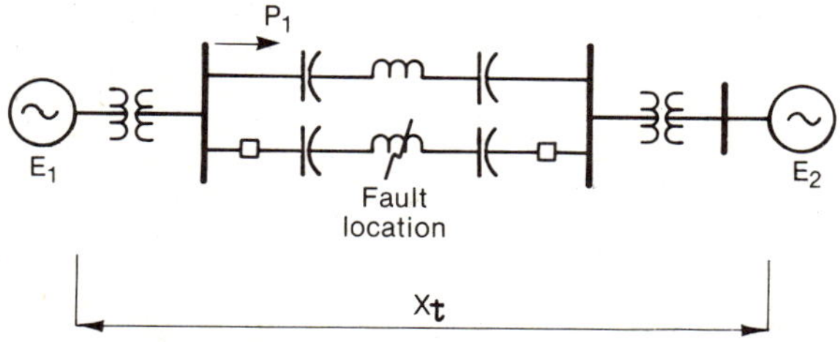

FIGURE 42. Series-compensated system.

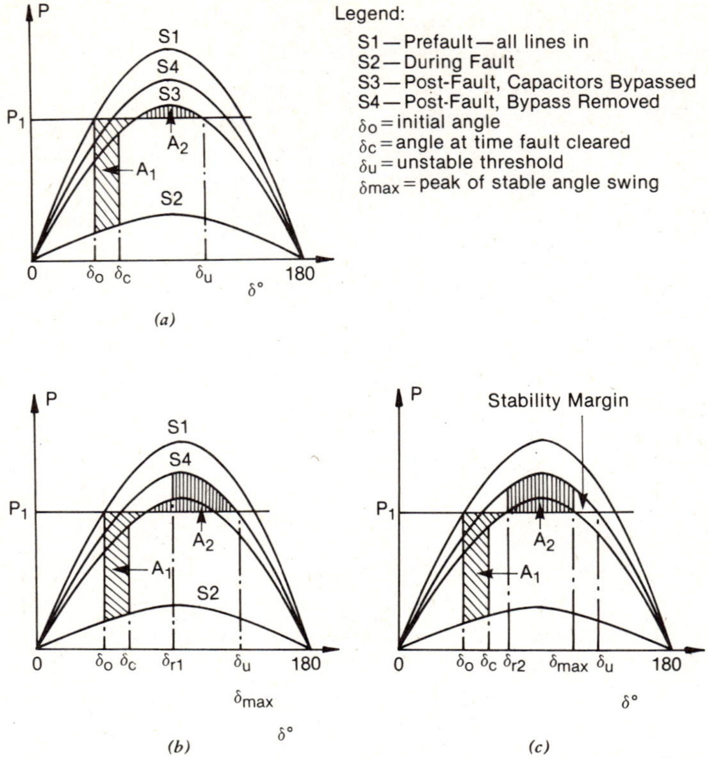

FIGURE 43. Effect of series capacitor bypass and reinsertion. *(a)* $A_2 < A_1$, unstable; bypass not removed. *(b)* $A_2 = A_1$, critically stable, $\delta_{max} \cong \delta_u$. *(c)* $A_2 = A_1$, stable with a margin, $\delta_{max} < \delta_u$.

Figure 43*b* shows the same fault, but now the capacitors in the unfaulted lines are reinserted when the transmission angle reaches δ_{r1}. This example is chosen such that the system is just stable: the peak angular swing δ_{max} is equal to the critical value δ_u. There is no "stability margin" because the total available decelerating (synchronizing) energy is fully utilized; that is, the total available A_2 just equals the accelerating energy A_1.

Figure 43*c* is the same, except that the capacitors in the unfaulted lines are reinserted even earlier. The angle at the time of reinsertion, δ_{r2}, is less than δ_{r1} in Figure 43*b*. In this final case, δ_{max} is less than the critical value δ_u and while $A_2 = A_1$, A_2 is less than the total available decelerating energy. The unused portion of available synchronizing energy constitutes a "margin" of stability as noted in Figure 43*c*.

If the series capacitor protective bypass scheme employs a zinc-oxide device, the reinsertion is essentially instantaneous. Moreover, the capacitors are only partially bypassed during the fault itself (see Chapter 7, Figure 10*b*). The series capacitors in the unfaulted lines are reinserted at the time of fault clearing ($\delta = \delta_c$). In such a case, an even greater stability margin than that shown in Figure 43*c* is achievable. In practice, this extra margin would be partially used up by operating at a higher initial power transfer, thus utilizing the transmission system more effectively.

3.6.3. Oscillatory Period

By the time this period commences, following a single fault and its initial swings, most capacitor bypass circuits will have been reopened in the remaining compensated line sections. The remaining series capacitors tend to reduce the power and angle swings compared with the oscillations that would take place without them.

Series capacitor compensation aids the system's generators in developing synchronizing torque. This can be seen by comparing transient power angle curves S3 and S4 in Figure 43. Curve S3 characterizes the postfault system with the capacitor bypassed. Curve S4 results after reinsertion, and it has a greater peak synchronizing power than curve S3. For any given oscillation in power the angle oscillations are smaller in magnitude for curve S4 on curve S3.

3.7. SUMMARY

This chapter has presented a description of dynamic system behavior following any disturbance while the system is in transition from one equilibrium state to another. For convenience, the transition period was broken up into basic periods: the transient, first-swing, and oscillatory periods. The final equilibrium, when discussed, was labeled as the quasi-steady-state period.

Various forms of passive and controllable reactive power compensation devices were characterized by how they perform and effect system dynamic performance during the basic transition periods. Included in this discussion were shunt capacitors, shunt reactors, series capacitors, static compensators, and synchronous condensers.

Both passive and active (controllable) compensators play an important role in ac power system steady-state performance. Switched and/or controlled compensators can serve to enhance the dynamic behavior of the power system during transitions between equilibrium states. The magnitude of the swings in active power, angle, and voltage during these transition periods can be reduced and damped by suitably controlled reactive compensation.

Field test oscillograms and computer simulations were used to illustrate the effects of compensation on system dynamic behavior. Later chapters covering static compensators, synchronous condensers, and series capacitors contain additional evidence that dynamic performance of a power system can be improved with proper application of reactive compensation equipment.

REFERENCES

1. *Test Procedures for Synchronous Machines,* IEEE Pub. 115.
2. IEEE Task Force On Terms and Definitions, "Proposed Terms and Definitions for Power System Stability," *IEEE paper* 81WM 082-7, Atlanta, Georgia (Feb. 1981).
3. E. W. Kimbark, *Power System Stability,* Wiley, New York, 1948 (Vol. I) and 1956 (Vol. II).
4. S. B. Crary, *Power System Stability,* Wiley, New York, 1945 (Vol. I) and 1947 (Vol. II).
5. R. T. Byerly and E. W. Kimbark, *Stability of Large Electric Power Systems,* IEEE Press, 1974.
6. R. L. Hauth and R. J. Moran, *The Performance of Thyristor Controlled Static Var Systems in HVAC Applications,* Part I. IEEE Tutorial Course Text 78EH0135-4-PWR, 56–64 (July 1978).
7. R. L. Hauth, R. J. Newell, and T. Humann, "Application of a Static Var System To Regulate System Voltage in Western Nebraska," *IEEE Trans. Power Appar. Syst.,* **97** (5), 1955–1964 (Sept./Oct. 1978).
8. R. J. Newell, M. D. Risan, D. D. Raatz, and R. L. Hauth, "Staged Field Testing of The Victory Hill Static Var Control," *IEEE Trans. Power Appar. Syst.* **99** (2), 426–433 (March/April 1980).
9. R. L. Hauth et al., *Static Reactive Power Compensators for High Voltage Power Systems,* Report for U.S. Department of Energy and Los Alamos National Laboratory, April 1981, DOE/NBM-1010
10. E. Kimbark, "How to Improve System Stability Without Risking Subsynchronous Resonance," *IEEE Trans. Power Appar. Syst.* **96** (5), 1608–1619 (Sept./Oct. 1977).
11. B. M. Weedy, *Electric Power Systems,* 2nd Ed., 161–164 Wiley, New York, 1979.
12. E. V. Larsen and W. W. Price, *MANSTAB/POSSIM Power System Dynamic Analysis Programs — A New Approach Combining Nonlinear Simulation and Linearized State-Space/Frequency Domain Capabilities,* PICA Conference Proceedings, May 1977.

Chapter 4

PRINCIPLES OF STATIC COMPENSATORS

T. J. E. MILLER and R. W. LYE

4.1. COMPENSATOR APPLICATIONS

The desirability of reactive compensation in electric power systems has been explained in Chapters 1 through 3. In this and subsequent chapters we examine some of the principles by means of which compensators are realized in practice.

This chapter is concerned with *static* shunt compensators. These compensating devices fall into the class of active compensators (Chapter 2, Section 2.3.2). *Static* means that, unlike the synchronous condenser, they have no moving parts. They are used for surge-impedance compensation and for compensation by sectioning in long-distance, high-voltage transmission systems. In addition they have a variety of load-compensating applications. Their applications are listed in greater practical detail in Table 1. The main headings in Table 1 will be recognized as the fundamental requirements for operating an ac power system, as discussed in Chapters 1–3. Other applications not listed in Table 1, but which may nevertheless be very beneficial, include the control of ac voltage near HVDC converter terminals, the minimization of transmission losses resulting from local generation or absorption of reactive power (Chapter 11), and the suppression of subsynchronous resonance. Some types of compensator can also be designed to assist in the limitation of dynamic overvoltages (Chapter 3).

4.1.1. Properties of Static Compensators

In Chapter 2 it was explained how a constant ac voltage could be maintained at the terminals of a controlled susceptance. Figure 30 in

TABLE 1
Practical Applications of Static Compensators in Electric Power Systems

Maintain voltage at or near a constant level

Under slowly varying conditions due to load changes

To correct voltage changes caused by unexpected events (e.g., load rejections, generator and line outages)

To reduce voltage flicker caused by rapidly fluctuating loads (e.g., arc furnaces)

Improve power system stability

By supporting the voltage at key points (e.g., the midpoint of a long line)

By helping to improve swing damping

Improve Power Factor

Correct Phase Unbalance

Chapter 2, showed this principle in terms of the voltage/current characteristic or *control* characteristic. Figure 1 represents an idealized static reactive power compensator. The ideal compensator is a device capable of continuous adjustment of its reactive power, with no response delay, over an unlimited range (lagging and leading).

The most important property of the static compensator is its ability to maintain a substantially constant voltage at its terminals by continuous adjustment of the reactive power it exchanges with the power system. This constant-voltage property is the first requirement in dynamic shunt

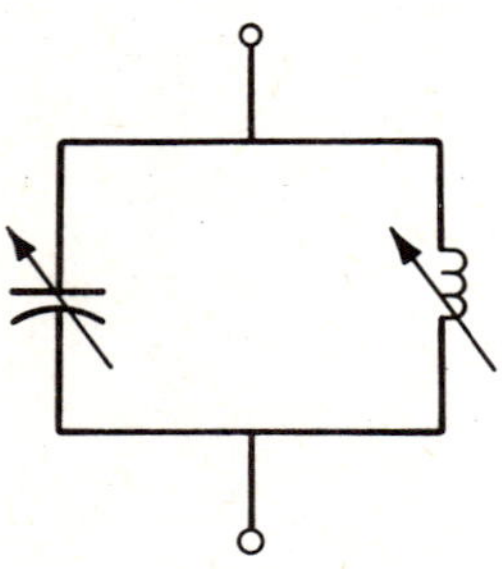

FIGURE 1. Idealized static reactive power compensator.

compensation or compensation by sectioning (Chapter 2), and is equally important in reducing flicker and other voltage fluctuations caused by variable loads.

A second important property is the speed of response. The reactive power of the compensator must change sufficiently quickly in response to a small change in terminal voltage. It is hard to generalize about what constitutes a sufficiently rapid response. In transmission systems the time constants governing the re-establishment of a steady state following a disturbance (i.e., the system modes or eigenvalues) depend as much on the external power system as on the compensator, and they can vary with the system configuration. Although a fast response is generally desirable, it can happen that other factors limit the stability of the system in such a way that there is no point in specifying a compensator with the fastest response that is theoretically possible. In load compensation, the reduction of flicker is possible *only* with the most rapid types of compensator.

The control characteristic usually has a small positive slope to stabilize the operating point (which is defined as the intersection with the system load line). The reactive current is limited in both the lagging and the leading regimes by factors in the compensator design and its principle of operation. Also, the characteristic can deviate from a straight line, have discontinuities, and change slope according to the rates of change of current and voltage.

Several principles have been used to design static compensators with control characteristics similar to Figure 30 in Chapter 2. We concentrate on three main types: the thyristor-controlled reactor (TCR), the thyristor-switched capacitor (TSC), and the saturated reactor (SR). These and their variants account for the majority of static compensator applications in both transmission and distribution systems. The synchronous condenser, which is the oldest type of controllable shunt compensator, is described in Chapter 8.

4.1.2. Main Types of Compensator

Figures 2–4 show one-line schematic diagrams of the main types of compensator. Before examining each type in detail, a few general features are worth mentioning. First, it is common to find fixed shunt capacitors in parallel with the controlled susceptance. The fixed shunt capacitors are very often tuned with small reactors to harmonic frequencies which may be of integer or noninteger order. The reason for the tuning is to absorb harmonics generated by the controlled susceptance (TCR or SR), or to avoid troublesome resonances (see Chapter 10). The fixed capacitors bias the reactive output of the compensator into the leading (generating) regime. A second common feature of the compensators in Figures 2–4 is the step-down transformer. This is not always present, but when it is it

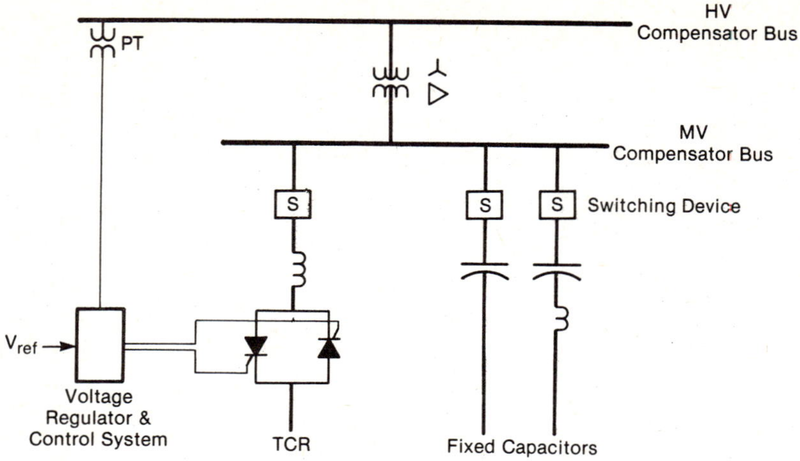

FIGURE 2. One-line diagram of TCR compensator with fixed shunt capacitors.

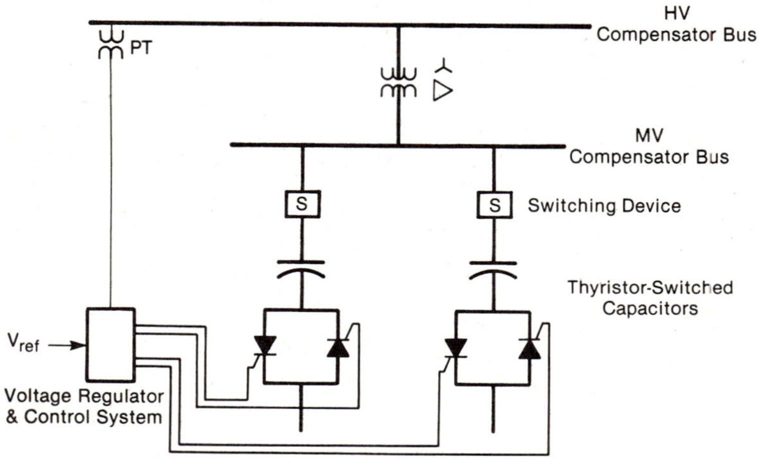

FIGURE 3. One-line diagram of TSC compensator.

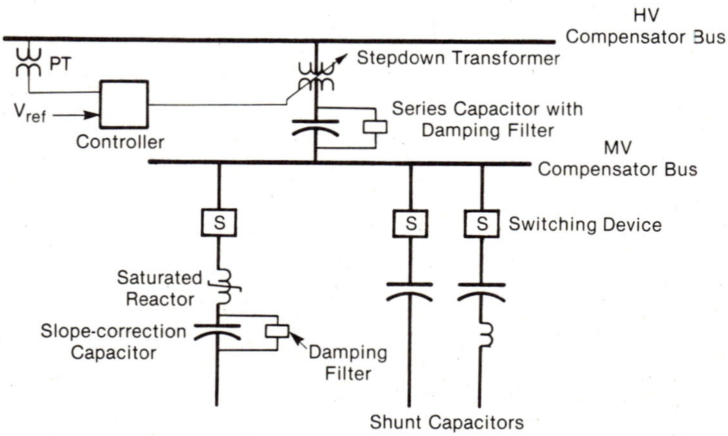

FIGURE 4. One-line diagram of saturated-reactor compensator with shunt and slope-correcting capacitors.

can significantly affect the performance of the compensator, particularly with respect to harmonics, losses, and overvoltage performance. The fixed shunt capacitors are sometimes connected on the high-voltage side of this transformer, but more commonly they share the medium-voltage compensator bus with the controlled element. Sometimes the compensator is connected on the tertiary winding of an existing transformer in the power system. With TCR compensators, connecting the shunt capacitors on the high-voltage side may require a larger step-down transformer and this can adversely affect the losses (discussed later). The same is true with the thyristor-controlled transformer (TCT), a derivative of the TCR. Finally, the similarity in the one-line schematic diagrams in Figures 2–4 is largely conceptual. The substation layouts and appearances of the different compensators can vary quite widely. For example, the saturated reactor is of transformer-type construction, whereas the thyristor controllers of the TSC and TCR compensators are physically separated from their capacitors and reactors, and are often housed in a simple building for weather protection.

4.2. THE THYRISTOR-CONTROLLED REACTOR (TCR) AND RELATED TYPES OF COMPENSATOR

4.2.1. Principles of Operation

The basis of the thyristor-controlled reactor (TCR) is shown in Figure 5. The controlling element is the thyristor controller, shown here as two oppositely poled thyristors which conduct on alternate half-cycles of the supply frequency. If the thyristors are gated into conduction precisely at the peaks of the supply voltage, full conduction results in the reactor, and the current is the same as though the thyristor controller were short-circuited. The current is essentially reactive, lagging the voltage by nearly 90°. It contains a small in-phase component due to the power losses in the reactor, which may be of the order of 0.5–2% of the reactive power. Full conduction is shown by the current waveform in Figure 6*a*.

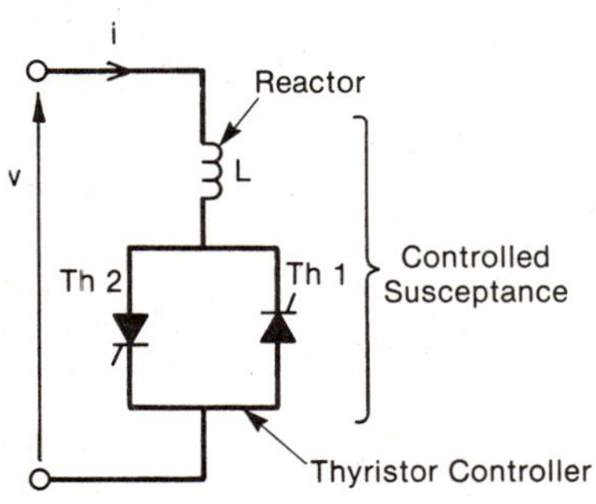

FIGURE 5. Elementary thyristor-controlled reactor.

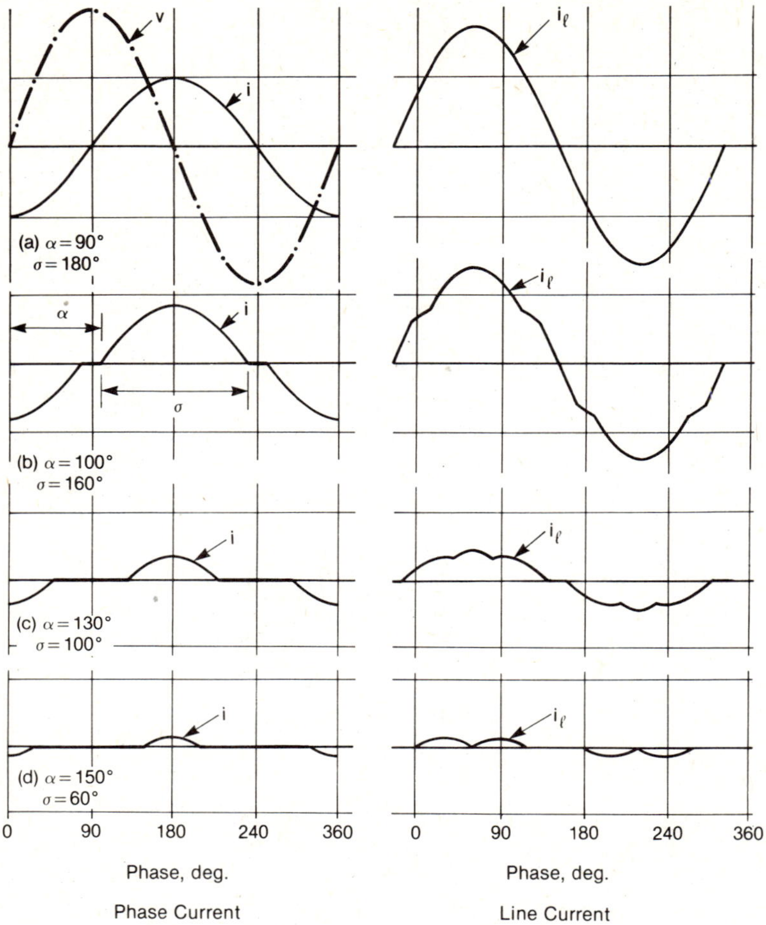

FIGURE 6. Phase and line current waveforms in delta-connected TCR.

If the gating is delayed by equal amounts on both thyristors, a series of current waveforms is obtained, such as those in Figures 6*a* through *d*. Each of these corresponds to a particular value of the gating angle α, which is measured from a zero-crossing of the voltage. Full conduction is obtained with a gating angle of 90°. Partial conduction is obtained with gating angles between 90° and 180°. The effect of increasing the gating angle is to reduce the fundamental harmonic component of the current. This is equivalent to an increase in the inductance of the reactor, reducing its reactive power as well as its current. So far as the fundamental component of current is concerned, the thyristor-controlled reactor is a controllable susceptance, and can therefore be applied as a static compensator.

The instantaneous current i is given by

$$i = \begin{cases} \dfrac{\sqrt{2}V}{X_L}(\cos\alpha - \cos\omega t), & \alpha < \omega t < \alpha + \sigma \\ 0, & \alpha + \sigma < \omega t < \alpha + \pi \end{cases} \tag{1}$$

where V is the rms voltage; $X_L = \omega L$ is the fundamental-frequency reactance of the reactor (in Ohms); $\omega = 2\pi f$; and α is the gating delay angle. The time origin is chosen to coincide with a positive-going zero-crossing of the voltage. The fundamental component is found by Fourier analysis and is given by

$$I_1 = \frac{\sigma - \sin\sigma}{\pi X_L} V \qquad \text{A rms.} \tag{2}$$

σ is the *conduction angle*, related to α by the equation

$$\alpha + \sigma/2 = \pi\,. \tag{3}$$

We can write Equation 2 as

$$I_1 = B_L(\sigma)V\,, \tag{4}$$

where $B_L(\sigma)$ is an adjustable fundamental-frequency susceptance controlled by the conduction angle according to the law,

$$B_L(\sigma) = \frac{\sigma - \sin\sigma}{\pi X_L}\,. \tag{5}$$

This control law is shown in Figure 7. The maximum value of B_L is $1/X_L$, obtained with $\sigma = \pi$ or 180°, that is, full conduction in the thyristor controller. The minimum value is zero, obtained with $\sigma = 0$ ($\alpha = 180°$). This control principle is called *phase control*.

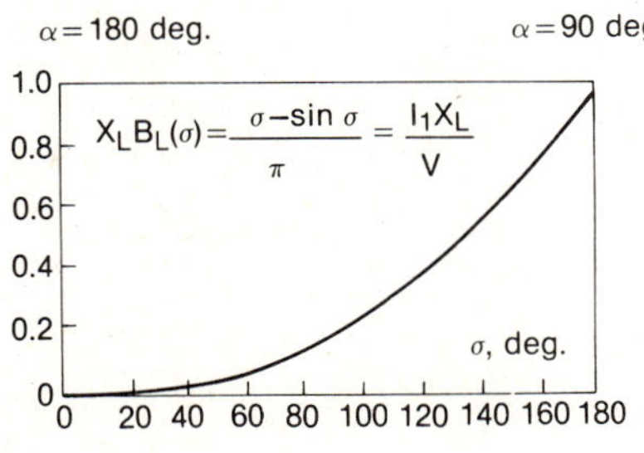

FIGURE 7. Control law of elementary TCR.

4.2.2. Fundamental Voltage/Current Characteristic

The TCR has to have a control system that determines the gating instants (and therefore σ), and that issues the gating pulses to the thyristors. In some designs the control system responds to a signal that directly represents the desired susceptance B_L. In others the control algorithm processes various measured parameters of the compensated system (e.g., the voltage) and generates the gating pulses directly without using an explicit signal for B_L (discussed later). In either case the result is a voltage/current characteristic of the form shown in Figure 8. Steady-state operation is shown at the point of intersection with the system load line. In the example, the conduction angle is shown as 130°, giving a voltage slightly above 1.0 pu, but this is only one of an infinite number of possible combinations, depending on the system load line, the control settings, and the compensator rating. The control characteristic in Figure 8 can be described by the equation

$$\mathbf{V} = \mathbf{V}_k + jX_s\mathbf{I}_1 , \quad 0 < I_1 < I_{max} \tag{6}$$

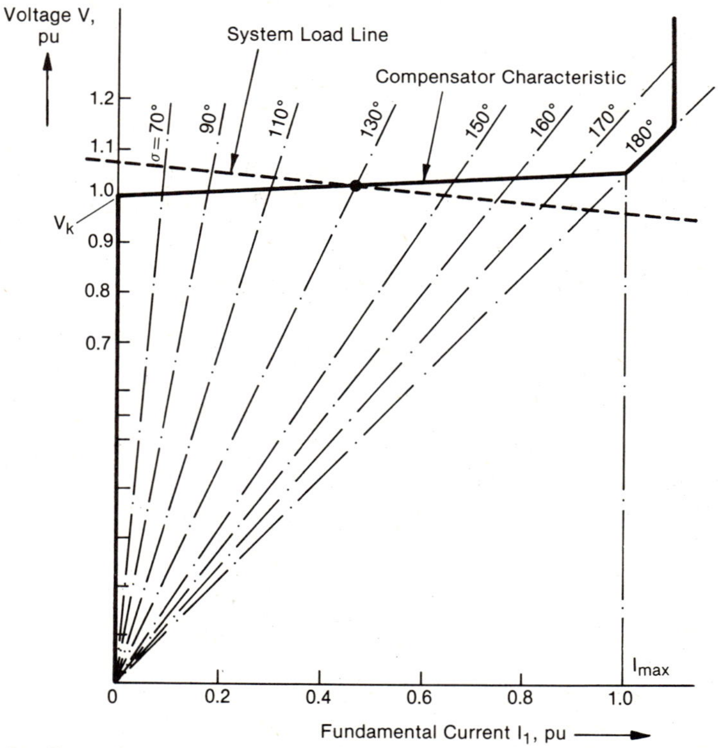

FIGURE 8. Formation of fundamental voltage/current characteristic in the TCR compensator.

which has been used in Chapters 1 and 3. In Figure 8, I_{max} is normally the rated current of the reactors, shown here as 1.0 pu.

4.2.3. Harmonics

Increasing the gating angle (reducing the conduction angle) has two other important effects. First, the power losses decrease in both the thyristor controller and the reactor. Secondly, the current waveform becomes less sinusoidal; in other words, the TCR generates harmonic currents. If the gating angles are balanced (i.e., equal for both thyristors), all odd orders are generated, and the rms value of the nth harmonic component is given by

$$I_n = \frac{4}{\pi}\frac{V}{X_L}\left[\frac{\sin(n+1)\alpha}{2(n+1)} + \frac{\sin(n-1)\alpha}{2(n-1)} - \cos\alpha\frac{\sin n\alpha}{n}\right], \quad (7)$$

$$n=3,5,7,\ldots .$$

Figure 9a shows the variation of the amplitudes of some of the major (lower-order) harmonics with the conduction angles, and Figure 9b the variation of the total harmonic content. Table 2 gives the maximum amplitudes of the harmonics down to the 37th. (Note that the maxima do not all occur at the same conduction angle.)

The TCR we have described so far is only a single-phase device. For three-phase systems the preferred arrangement is shown in Figure 10; that is, three single-phase TCR's connected in delta. When the system is balanced, all the triplen harmonics circulate in the closed delta and are absent from the line currents. All the other harmonics are present in the line currents and their amplitudes are in the same proportions as shown in Figure 9 and Table 2. However, the waveforms differ from those of the phase currents, and examples are given for particular conduction angles in Figure 6.

It is important in the TCR to ensure that the conduction angles of the two oppositely poled thyristor switches are equal. Unequal conduction angles would produce even harmonic components in the current, including dc. They would also cause unequal thermal stresses in the oppositely poled thyristors. The requirement for equal conduction also limits σ to a maximum of 180°. However, if the reactor in Figure 5 is divided into two separate reactors (Figure 11), the conduction angle in each leg can be increased to as much as 360°. This arrangement has lower harmonics than that of Figure 5, but the power losses are increased because of currents circulating between the two halves.

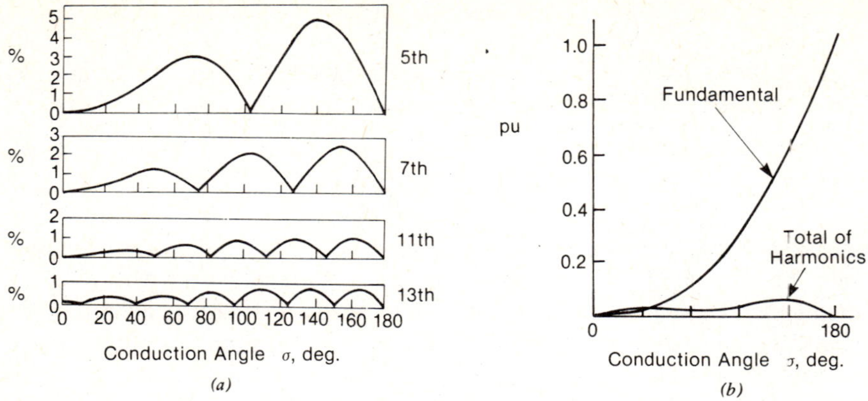

FIGURE 9. TCR Harmonics. (*a*) Major harmonic current components of TCR. Each is shown as a percentage of the fundamental component at full conduction. The percentages are the same for both phase and line currents. (*b*) Total harmonic content of TCR current, as a fraction of the fundamental component at full conduction. The percentage is the same for both phase and line currents.

TABLE 2
Maximum Amplitudes of Harmonic Currents in TCR[a]

Harmonic Order	Percentage
1	100
3	(13.78)
5	5.05
7	2.59
9	(1.57)
11	1.05
13	0.75
15	(0.57)
17	0.44
19	0.35
21	(0.29)
23	0.24
25	0.20
27	(0.17)
29	0.15
31	0.13
33	(0.12)
35	0.10
37	0.09

[a] Values are expressed as a percentage of the amplitude of the fundamental component at full conduction. The values apply to both phase and line currents, except that triplen harmonics do not appear in the line currents. Balanced conditions are assumed.

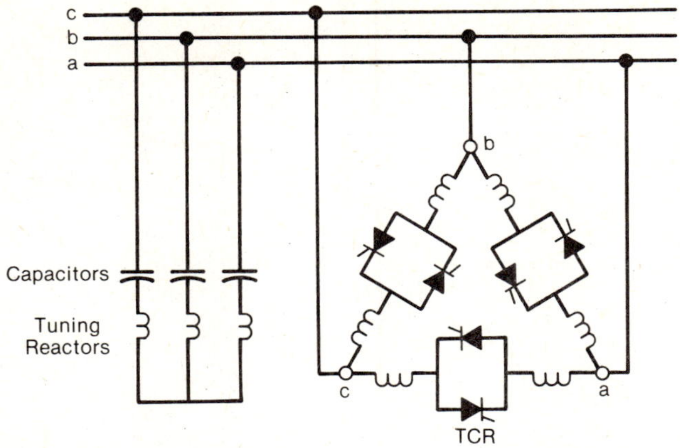

FIGURE 10. Three-phase TCR with shunt capacitors. The split arrangement of the reactors in each phase provides extra protection to the thyristor controller in the event of a reactor fault.

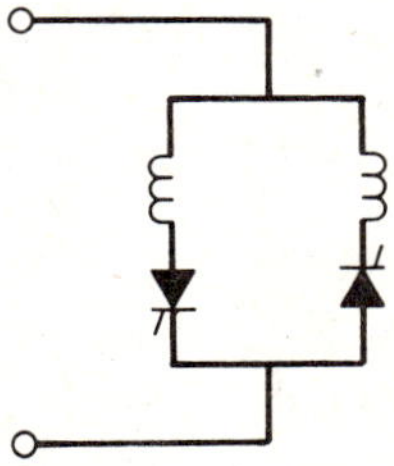

FIGURE 11. TCR with more than 180° of conduction in each leg to reduce harmonic currents.

As already noted, TCR harmonic currents are sometimes removed by filters (Figure 10). An alternative means for eliminating the 5th and 7th harmonics is to split the TCR into two parts fed from two secondaries on the step-down transformer, one being in wye and the other in delta, as shown in Figure 12. This produces a 30° phase shift between the voltages and currents of the two TCR's and virtually eliminates the 5th and 7th harmonics from the primary-side line current. It is known as a 12-pulse arrangement because there are 12 thyristor gatings every period. The same phase-multiplication technique is used in HVDC rectifier transformers for harmonic cancellation, and has affinities with the polyphase saturated reactor compensator of the frequency-multiplier type to be described later. With the 12-pulse scheme, the lowest-order characteristic harmonics are the 11th and 13th. It can be used without filters for the 5th and 7th harmonics, which is an advantage when system resonances occur near these frequencies. For higher-order harmonics a plain capacitor is often sufficient, connected on the low-voltage side of the step-down transformer. Otherwise a high-pass filter may be used. The generation of

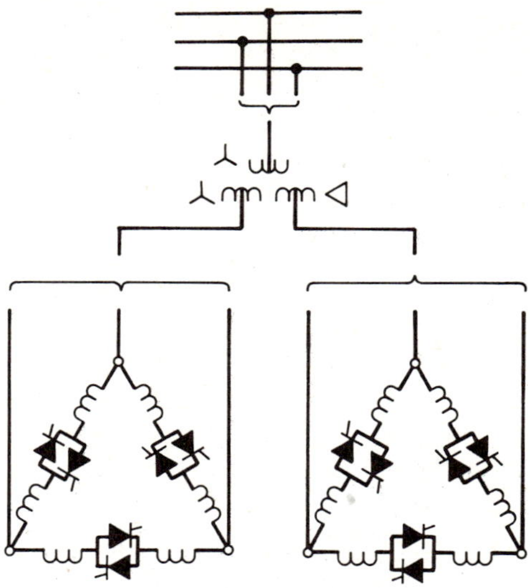

FIGURE 12. Arrangement of 12-pulse TCR with double-secondary transformer.

third-harmonic currents under unbalanced conditions is similar to that in the 6-pulse arrangement (Figure 10).

With both 6-pulse and 12-pulse TCR compensators, the need for filters and their frequency responses must be evaluated with due regard to the possibility of unbalanced operation. The influence of other capacitor banks and sources of harmonic currents in the electrical neighborhood of the compensator must also be taken into account. Harmonic propagation computer programs are used for this purpose, and sometimes it is necessary to cover quite a large portion of the interconnected power system.

The 12-pulse connection has the further advantage that if one half is faulted the other may be able to continue to operate normally. The control system must take into account the 30° phase shift between the two TCR's, and must be designed to ensure accurate harmonic cancellation. A variant of the 12-pulse TCR uses two separate transformers instead of one with two secondaries.

4.2.4. The Thyristor-Controlled Transformer

Another variant of the TCR is the thyristor-controlled transformer (TCT, Figure 13). Instead of using a separate step-down transformer and linear reactors, the transformer is designed with very high leakage reactance, and the secondary windings are merely short-circuited through the thyristor controllers. A gapped core is necessary to obtain the high leakage reactance, and the transformer can take the form of three single-phase transformers. With the arrangements in Figure 13 there is no secondary bus and any shunt capacitors must be connected at the primary voltage

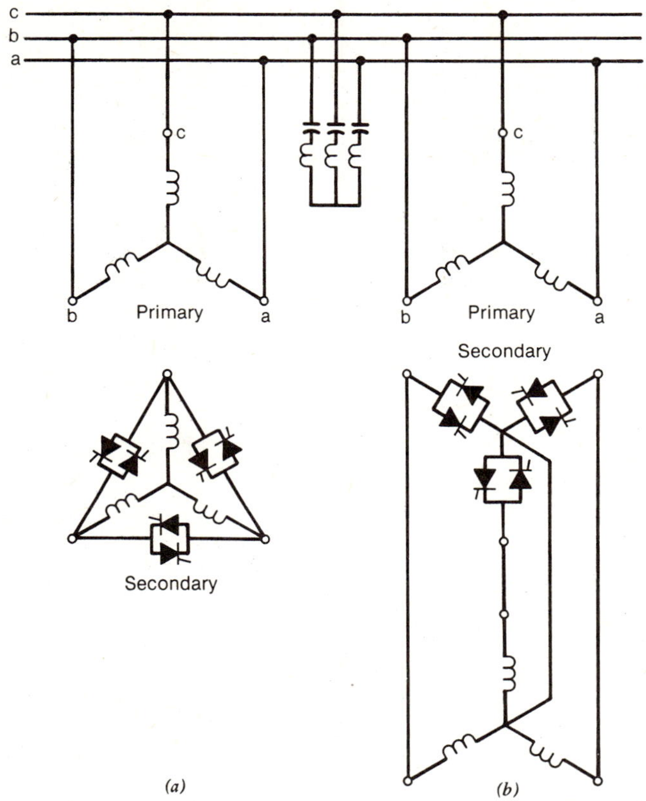

FIGURE 13. Alternative arrangements of thyristor-controlled transformer compensator. (*a*) With wye-connected reactors and delta-connected thyristor controller. (*b*) With wye-connected reactors and thyristor controller (4-wire system).

unless a separate step-down transformer is provided. The high leakage reactance helps protect the transformer against short-circuit forces during secondary faults. Because of its linearity and large thermal mass the TCT can usefully withstand overloads in the lagging (absorbing) regime.

4.2.5. The TCR with Shunt Capacitors

It is important to note that the TCR current (the compensating current) can be varied *continuously,* without steps, between zero and a maximum value corresponding to full conduction. The current is always lagging, so that reactive power can only be absorbed. However, the TCR compensator can be biased by shunt capacitors so that its overall power factor is leading and reactive power is generated into the external system (Chapter 1). The effect of adding the capacitor currents to the TCR currents shown in Figure 8 is to bias the control characteristic into the second quadrant, as shown in Figure 14. In a three-phase system the preferred arrangement is to connect the capacitors in wye, as shown in Figure 10. The current in Figure 14 is, of course, the fundamental positive-sequence component, and if it lies between I_{Cmax} and I_{Lmax} the control characteristic is again represented by Equation 6. However, if the voltage regulator gain is unchanged, the slope reactance X_s will be slightly increased when the capacitors are added.

As is common with shunt capacitor banks, the capacitors may be divided into more than one three-phase group, each group being separately switched by a circuit breaker. The groups can be tuned to particular frequencies by small series reactors in each phase, to filter the harmonic currents generated by the TCR and so prevent them from flowing in the external system. One possible choice is to have groups tuned to the 5th and 7th harmonics, with another arranged as a high-pass filter. The capacitors arranged as filters, and indeed the entire compensator, must be designed with careful attention to their effect on the resonances of the power system at the point of connection (see Chapter 10).

It is common for the compensation requirement to extend into both the lagging and the leading ranges. A TCR with fixed capacitors cannot have a lagging current unless the TCR reactive power rating exceeds that of the capacitors. The net reactive power absorption rating with the capacitors connected equals the difference between the ratings of the TCR and the capacitors. In such cases the required TCR rating can be very large indeed (up to some hundreds of MVAr in transmission system applications). When the net reactive power is small or lagging, large reactive currents circulate between the TCR and the capacitors without performing any useful function in the power system. For this reason the capacitors are sometimes designed to be switched in groups, so that the degree of capacitive bias in the voltage/current characteristic can be adjusted in steps. If this is done, a smaller "interpolating" TCR can be used. An

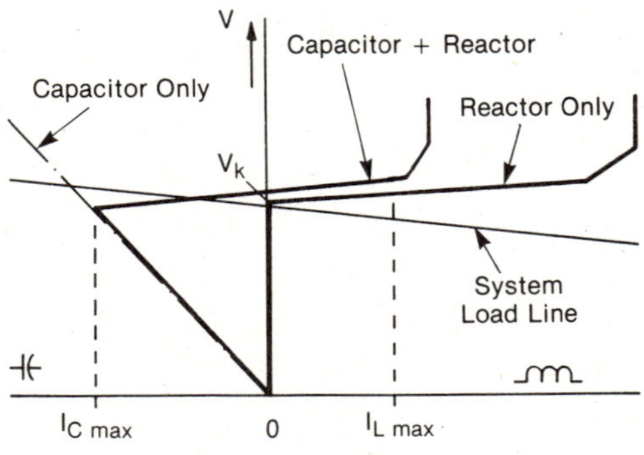

FIGURE 14. Voltage/current characteristic of TCR.

example is shown schematically in Figure 15, having the shunt capacitors divided into three groups. The TCR controller is provided with a signal representing the number of capacitors connected, and is designed to provide a continuous overall voltage/current characteristic. When a capacitor group is switched on or off, the conduction angle is immediately adjusted, along with other reference signals, so that the capacitive reactive power added or subtracted is exactly balanced by an equal change in the inductive reactive power of the TCR. Thereafter the conduction angle will vary continuously according to the system requirements, until the next capacitor switching occurs. An example of the effect of switching the capacitor is given in Chapter 7, Figure 13.

The performance of this hybrid arrangement of a TCR and switched shunt capacitors depends critically on the method of switching the capacitors, and the switching strategy. The least expensive way to switch the capacitors is with conventional circuit breakers. If the operating point is continually ranging up and down the voltage/current characteristic, the rapid accumulation of switching operations may cause a maintenance problem in the circuit breakers. Also, in transmission system applications there may be conflicting requirements as to whether the capacitors should be switched in or out during severe system faults. Under these circumstances repeated switching can place extreme duty on the capacitors and circuit breakers, and in most cases this can only be avoided by inhibiting the compensator from switching the capacitors. Unfortunately this prevents the full potential of the capacitors from being used during a period when they could be extremely beneficial to the stability of the system.

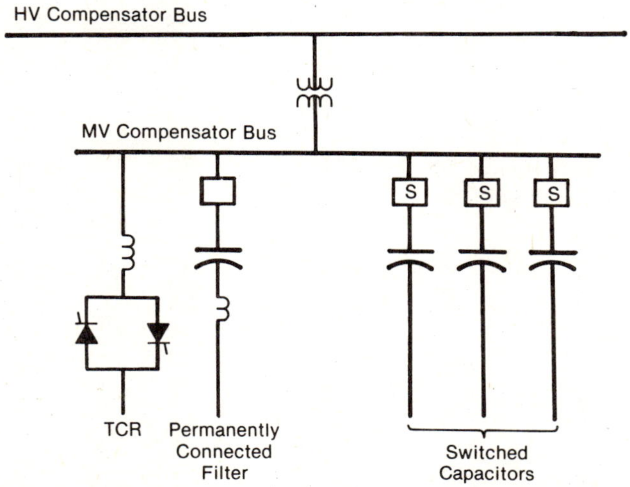

FIGURE 15. Hybrid compensator with switched capacitors and "interpolating" TCR. The switches S may be mechanical circuit breakers or thyristor switches.

In some cases these problems have been met by using thyristor controllers instead of circuit breakers to switch the capacitors, taking advantage of the virtually unlimited switching life of the thyristors. The timing precision of the thyristor switches can be exploited to reduce the severity of the switching duty, but even so, during disturbances this duty can be extreme. The number of separately switched capacitor groups in transmission system compensators is usually less than four.

4.2.6. Control Strategies

For application purposes the primary characteristic of a compensator is its voltage/current characteristic, the properties of which have been discussed in Chapters 1–3. A typical TCR characteristic is shown in Figure 8 (solid line). For all voltages the TCR current is determined by the inductance of the reactors and the conduction angle σ. Each point on the solid line represents a particular value of B_L, according to Equations 4 and 5. It has already been shown (Chapter 1) that the compensator will operate at the point of intersection of the V/I_1 characteristic and the system load line. The control system automatically adjusts the gating instants (and therefore σ) so that this condition is satisfied.

The two basic types of control which can be used are closed-loop and open-loop control. "Closed-loop" implies the classical feedback system, Figure 16*a*. "Open-loop" implies a system without feedback, Figure 16*b*. Here the feedforward transfer function G is "pre-programmed" to produce a certain value of C given a certain value of R. There is no measurement of C made to check the result. In other words, all conditions must be taken into account *in advance* by the function G. If unforeseen circumstances change the characteristics of the external system, then C will not turn out as planned. The main advantage of the closed-loop system is accuracy. The main advantage of the open-loop system is rapid response.

Open-Loop Control. This has been used successfully for flicker reduction applications where rapid response is essential. Several algorithmic

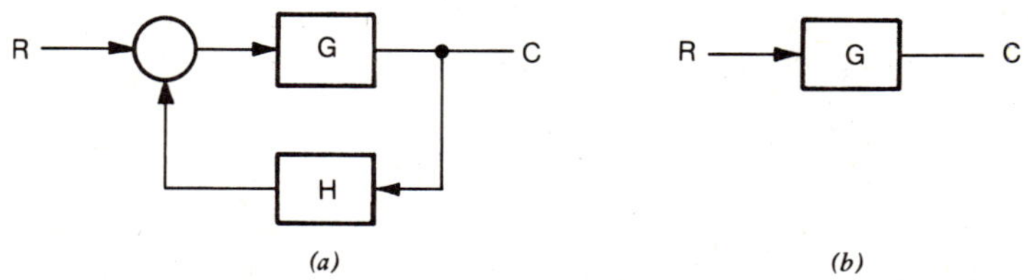

FIGURE 16. Control strategies. (*a*) Closed-loop, $C = RG/(1+GH)$; (*b*) Open-loop, $C = RG$. H = feedback transfer function, G = feedforward transfer function, R = reference, C = controlled variable.

principles have been applied in designing circuitry to generate the thyristor gating pulses, and some specific examples are described in Chapter 9. In general terms, the open-loop arrangement can be described in terms of Figure 17. The susceptance calculators SC compute the desired compensating susceptance from input signals representing the measured voltages and currents of the load and the required compensator characteristics. The desired compensating susceptance may be expressed in a number of different ways (Chapter 1). The signal representing it is processed by a conduction angle calculator (CAC) to produce a signal representing the required conduction angle according to the equation

$$B_\gamma = B_L(\sigma) - B_C, \tag{8}$$

where $B_L(\sigma)$ is given by Equation 5 and Figure 7. In effect, this pair of algebraic equations must be solved for σ. Equation 5 is nonlinear, and one workable method is to construct a circuit with the transfer function

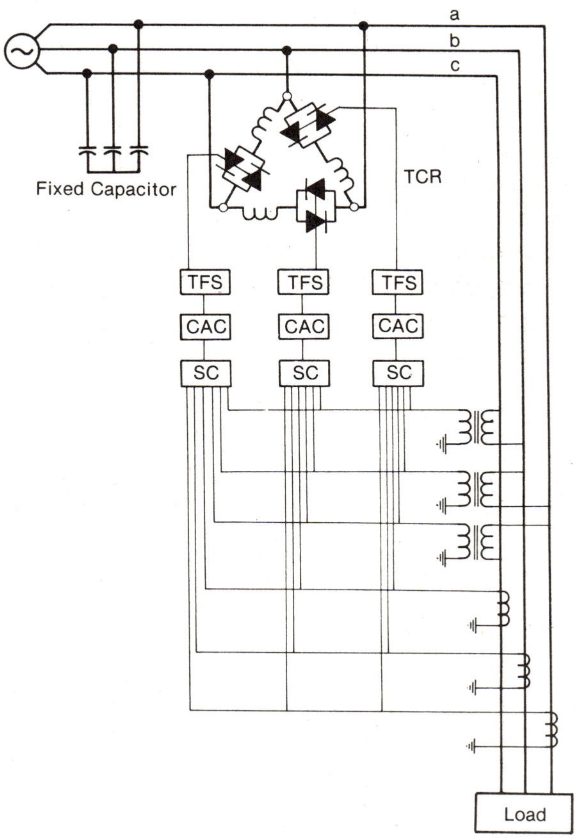

FIGURE 17. Open-loop flicker compensator control. TFS = thyristor firing (gating) system, CAC = conduction angle calculator, SC = susceptance calculator.

$K_2\pi/(\sigma - \sin\sigma)$. When this is put in the control loop ahead of the transfer function of the TCR itself (Equation 5), the result is a linear relationship between the desired susceptance signal and the actual compensating susceptance. This is shown in Figure 18 for one phase. K_1 and K_2 are adjustable gains. It should be pointed out that the explicit generation of a signal representing the desired value of B_L is not performed in all compensator control systems. This step is bypassed in certain alternative approaches, some of which are described in Chapter 9.

The open-loop control strategy described so far is suitable for load compensation where the desired response of the compensator is expressed entirely in terms of admittances computed from the instantaneous voltage and current of the load. There is no explicit voltage-regulating function in this control strategy.

Closed-Loop Control. The open-loop control strategy just described cannot be applied at an intermediate point on a transmission network remote from loads and sources. In these cases the objective is usually to regulate the voltage, and for this a closed-loop control is used. A voltage error (i.e., the difference between the actual system voltage and a reference applied to the compensator) is measured and used to cause the susceptance of the compensator to increase or decrease until the voltage error is reduced to an acceptable level. Figure 19 shows the operation of this type of control.

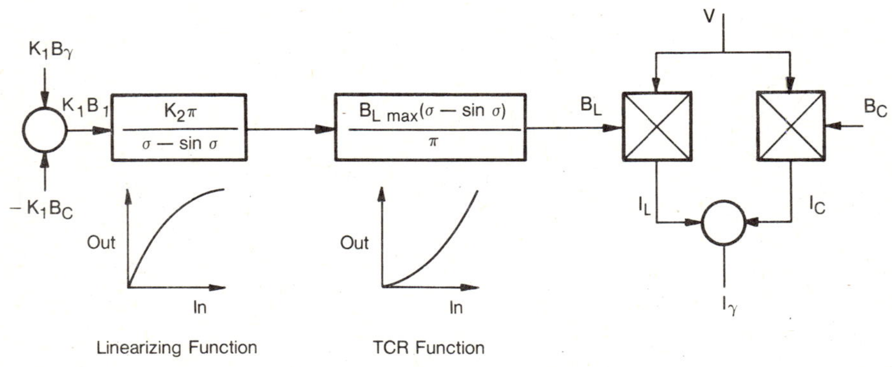

$$B_L = \frac{K_1 B_1 K_2}{(\sigma - \sin\sigma)/\pi} \times \frac{B_{L\,max}(\sigma - \sin\sigma)}{\pi}$$

$$= K_1 K_2 B_1 B_{L\,max}$$

$$= K B_1$$

FIGURE 18. TCR block diagram. $K_1 B_\gamma$ = signal representing required compensating susceptance (calculated from load voltage and current).

The ratio of the compensator current to the voltage error determines the slope of the voltage/current characteristic. The speed of response and stability are determined by the total loop gain and time constants of the regulating system. The loop gain is proportional to the source (or system) impedance. This means that as the system impedance increases the compensator will have a faster response, but will also be less stable. Since the relationship between voltage error and compensator current is not naturally linear, a linearizing network (similar in principle to the susceptance calculator) is required if a straight characteristic is desired. The accuracy of the regulating scheme will depend on the linearity, stability, and accuracy of all the components shown in Figure 19 except the system impedance. This means that the impedances of the capacitors, the reactors, and the transformer must all be known and constant.

The accuracy can be improved by using two control loops (Figure 20). An inner control loop in the form of a current regulator is added to the basic voltage regulator. If the gain of this loop is very high then the current error is negligible. This means that the controlled reactive current will be proportional to the voltage error, independent of the values of X_L, X_C, and X_T, and independent of the gains of the conduction angle calculators, the linearizing network, and the gating pulse generator. The slope of the control characteristic is determined by the gain K_1. The current error amplifier (K_2) may be an integrator so that the current error is essentially zero.

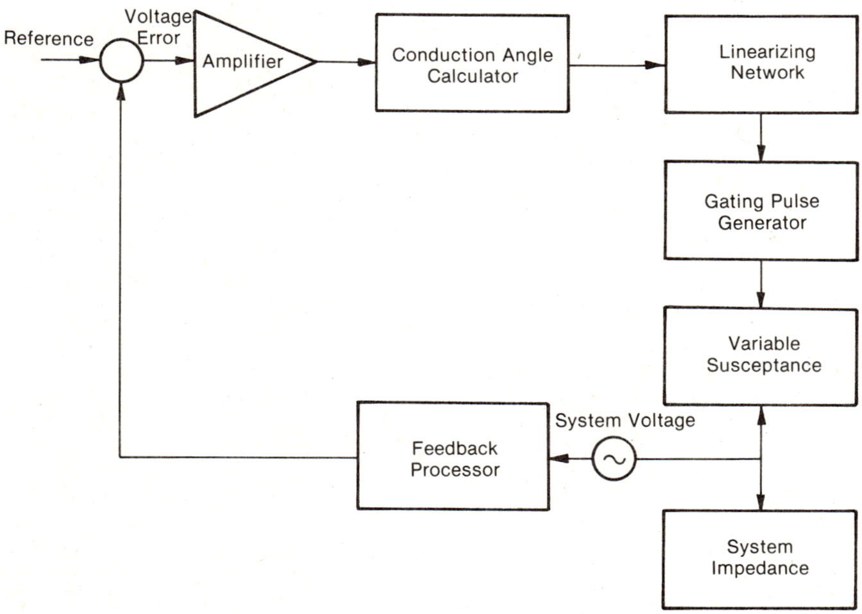

FIGURE 19. Compensator with voltage regulator.

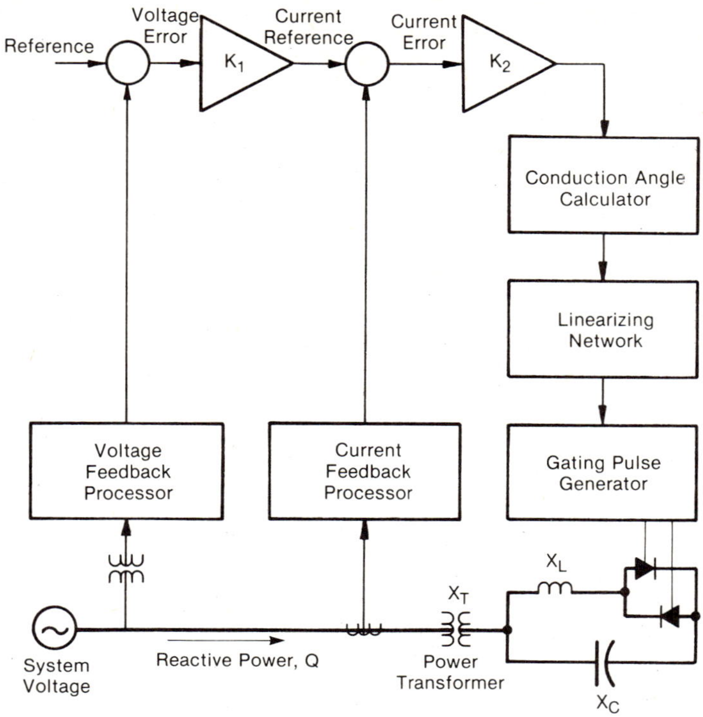

FIGURE 20. Compensator with voltage and current regulator.

Combinations of Open-Loop and Closed-Loop Control. The two basic types of control are often used in combination (Figure 21). For example, in a load compensator (which may have an open-loop control system of the form shown in Figure 17) it is often desirable to add a closed-loop control to maintain some quantity at a prescribed average value. One possibility is to regulate the overall average power factor (of the load plus compensator). Another possibility is to regulate the net compensation to an average of zero MVAr. This strategy means that for a steady load there would be no compensation, but rapid compensation would be applied by the open-loop control for changes in load. The speed of response of the closed-loop power factor or reactive power regulator would be slow compared to the open-loop control. Typically the open-loop control might respond in around 1 cycle whereas the closed-loop control might respond in 1 sec.

4.2.7. Other Performance Characteristics of TCR Compensators

Speed of Response. It is clear from Section 4.2.1 that the conduction angle in any phase of the TCR can change by any amount between successive half-cycles of the supply frequency, provided that σ remains less

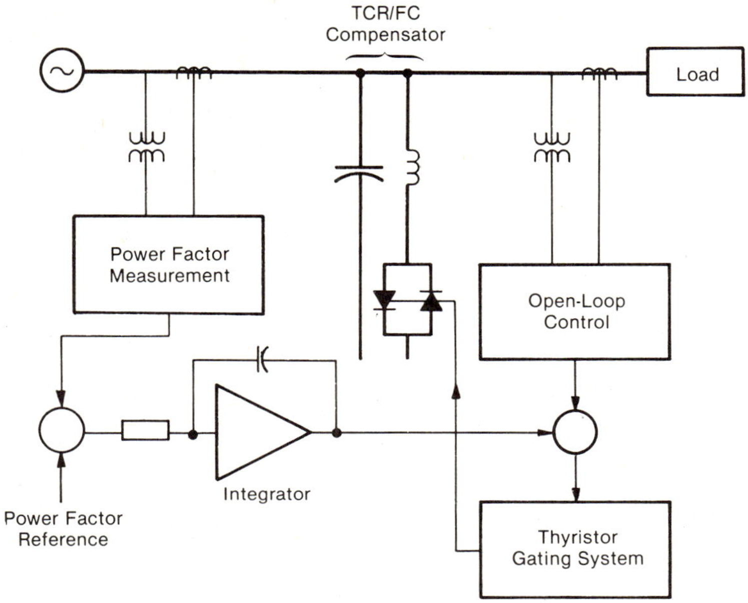

FIGURE 21. Combination of open-loop and closed-loop control.

than 180°. This extremely rapid response is rarely required in high-voltage transmission systems applications, where the stability and accuracy of the control system are usually more important. Even so, the response time of a TCR with fixed capacitor can be as little as 2 cycles of the power frequency (or even less) for the largest voltage disturbances. In arc-furnace applications fast response is of overriding importance, and the nature of the response of both the TCR and the TSC types of compensator is discussed in detail in Chapter 9.

Independent Phase Control. It is inherent in the TCR concept that the three phases can be independently controlled. The TCR can therefore be used for phase balancing, as described in Chapter 1. The plain TSC is also capable of individual phase control, but since only discrete steps of reactive power are possible, the accuracy of phase balancing would not be quite as good. Shunt capacitors used with the TCR do not limit the phase balancing ability: on the contrary, they may enhance it, since the compensating network (Chapter 1) may in general require either inductive or capacitive admittances in any phase. The unbalanced TCR may generate more harmonics than under balanced conditions, increasing the filtering requirement. In particular, triplen harmonics will appear in the line currents.

Response to Overvoltage and Undervoltage. Another important characteristic of all static compensators is their performance under conditions of very high or very low voltages. This question has already been discussed from the systems point of view in Chapter 3. Referring to the steady-state V/I_1 characteristic in Figure 8, if the system voltage rises above the point corresponding to the maximum conduction angle, the controls will be at their limit, and as far as the fundamental current is concerned the TCR will behave as a plain linear reactor. The reactance of the TCR may be too high to limit the rise of system voltage under the most extreme conditions, and it may eventually be necessary to start reducing the conduction angle to protect the thyristors from excessive junction temperatures due to the high currents, particularly if these currents are sustained. Some thyristor controllers have temperature transducers and/or simulation circuits to monitor the junction temperatures (Chapter 5). At high junction temperatures the risk of damage from overvoltage is increased. The current-limiting feature of the control is shown in Figure 8. At still higher voltages than those shown in Figure 8, the step-down transformer will begin to saturate and its magnetizing current will increase rapidly. At the high-voltage bus this will have a beneficial effect in helping to hold down the voltage. The same principle applies in the TCT. The saturation knee-voltage of the transformer must not be set too low, otherwise problems with ferroresonance can arise. At higher voltage levels still, protective measures are necessary, involving the operation of circuit breakers and surge arresters. The provision of an inherently large overload margin on the normal control range of the TCR is a possibility, but this would be expensive.

When the system voltage falls below the control range of the compensator, it behaves as a fixed capacitor, the thyristors being "phased off." If the shunt capacitor is absent or already switched off, the TCR by itself has no effect. There are circumstances in which the voltage can recover after fault clearing to an initially excessive value, and in such cases the TCR will be of benefit in limiting the overvoltage; indeed it may have been installed partly for that purpose. Subsequently the entire leading and lagging capability of the compensator may be used in damping power swings (Chapter 3), and the TCR controls may incorporate special feedback signals to enhance this effect.

Power Losses. The power losses in a compensator are an important consideration. Because of the high cost of energy, the capitalized value of the losses can be comparable to the capital cost of the compensator. In a TCR compensator with a fixed shunt capacitor the losses increase with conduction in the TCR as shown in Figure 22 as a percentage of the capacitive reactive power rating. They include resistive losses in the reactor,

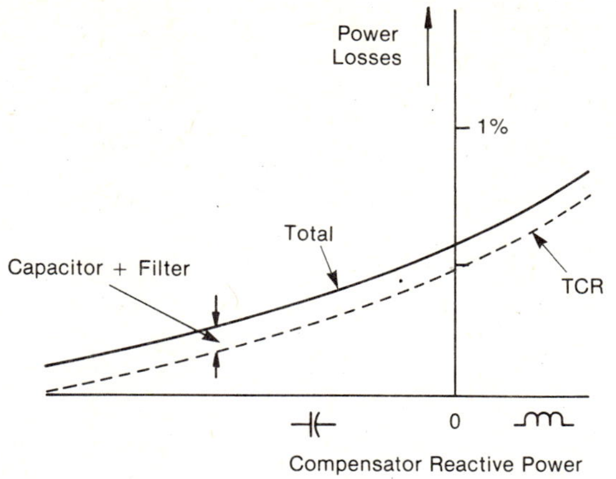

FIGURE 22. Power loss characteristic for TCR compensator with fixed shunt capacitor/filter.

and conduction, switching, and other losses in the thyristor controller. (Transformer and auxiliary losses are not included in Figure 22.) The losses increase with decreasing leading reactive power and with increasing lagging reactive power. The opposite trend is obtained in the hybrid compensator where the capacitors are switched in groups and the TCR is smaller. The loss characteristic for this type of compensator is of the form shown in Figure 23. This example is for thyristor-switched capacitors. With mechanically switched capacitors the characteristic has the same general form but the losses in the leading (capacitive) regime are much lower.

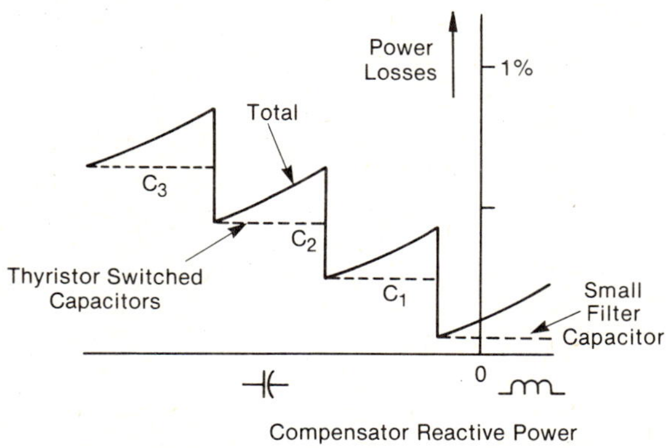

FIGURE 23. Power loss characteristic for TSC/TCR hybrid compensator.

4.3. THE THYRISTOR-SWITCHED CAPACITOR*

4.3.1. Principles of Operation

The principle of the thyristor-switched capacitor (TSC) is shown in Figures 24 and 25. The susceptance is adjusted by controlling the number of parallel capacitors in conduction. Each capacitor always conducts for an integral number of half-cycles. With k capacitors in parallel, each controlled by a switch as in Figure 25, the total susceptance can be equal to that of any combination of the k individual susceptances taken 0,1,2,... or k at a time. The total susceptance thus varies in a stepwise manner. In principle the steps can be made as small and as numerous as desired, by having a sufficient number of individually switched capacitors. For a given number k the maximum number of steps will be obtained when no two combinations are equal, which requires at least that all the individual susceptances be different. This degree of flexibility is not usually sought in power-system compensators because of the consequent complexity of the controls, and because it is generally more economic to make most of the susceptances equal. One compromise is the so-called binary system in which there are $k-1$ equal susceptances B and one susceptance $B/2$. The half-susceptance increases the number of combinations from k to $2k$.

The relation between the compensator current and the number of capacitors conducting is shown in Figure 26 (for constant terminal voltage). Ignoring switching transients, the current is sinusoidal, that is, it contains no harmonics.

4.3.2. Switching Transients and the Concept of Transient-Free Switching

When the current in an individual capacitor reaches a natural zero-crossing, the thyristors can be left ungated and no further current will flow. The reactive power supplied to the power system ceases abruptly. The capacitor, however, is left with a trapped charge (Figure 27a). Because of this charge, the voltage across the thyristors subsequently alternates between zero and twice the peak phase voltage. The only instant when the thyristors can be gated again without transients is when the voltage across them is zero (Figure 27b). This coincides with peak phase voltage.

---

* Material in this section is reproduced by kind permission of the Institution of Electrical Engineers. (See Miller, T.J.E. and Chadwick, P., "An Analysis of Switching Transients in Thyristor-Switched-Capacitor Compensated Systems," IEE Conference Publication No. 205, "Thyristor and Variable Static Equipment for AC and DC Transmission," London, November/December 1981.)

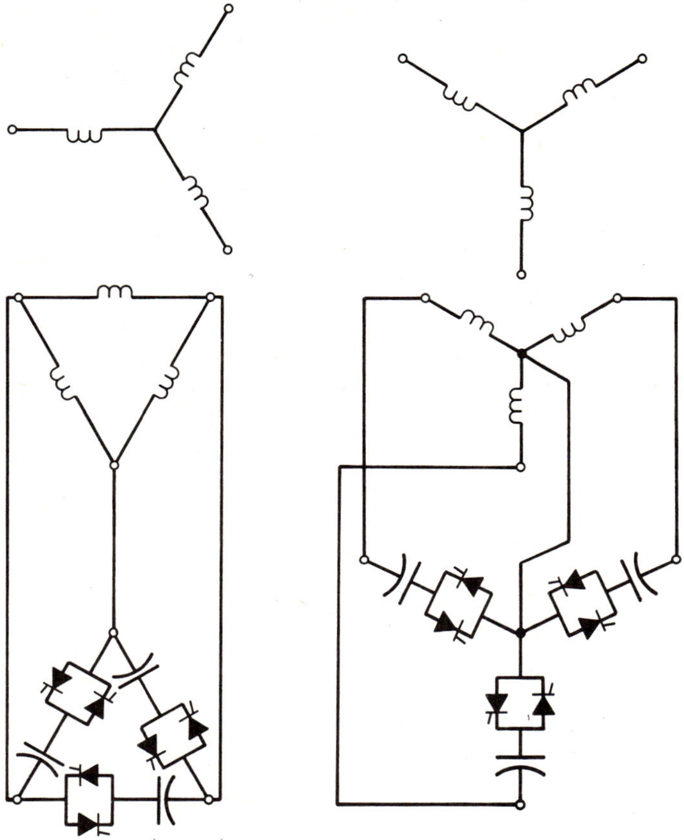

(a) (b)

FIGURE 24. Alternative arrangements of three-phase thyristor-switched capacitor. (*a*) Delta-connected secondary, delta-connected TSC; (*b*) Wye-connected secondary, wye-connected TSC (4-wire system).

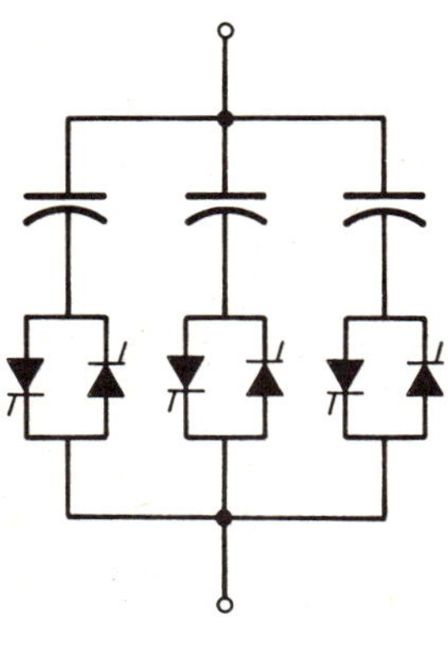

FIGURE 25. Principle of operation of TSC. Each phase of Figure 24 comprises a parallel combination of switched capacitors of this type.

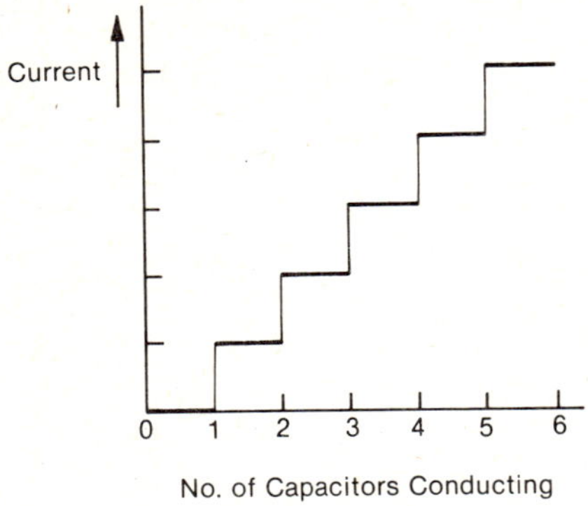

FIGURE 26. Relationship between current and number of capacitors conducting in the TSC.

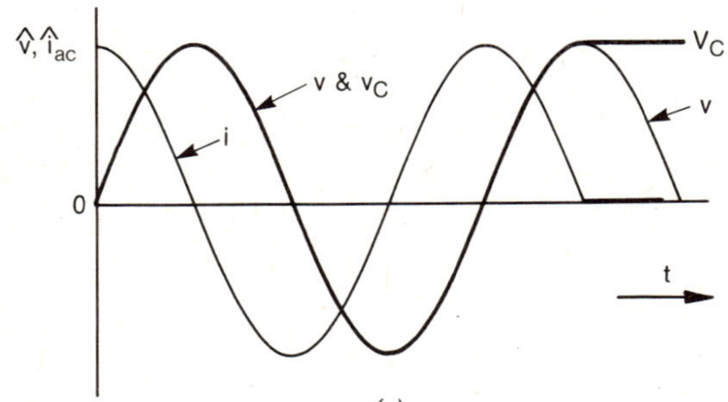

(a)

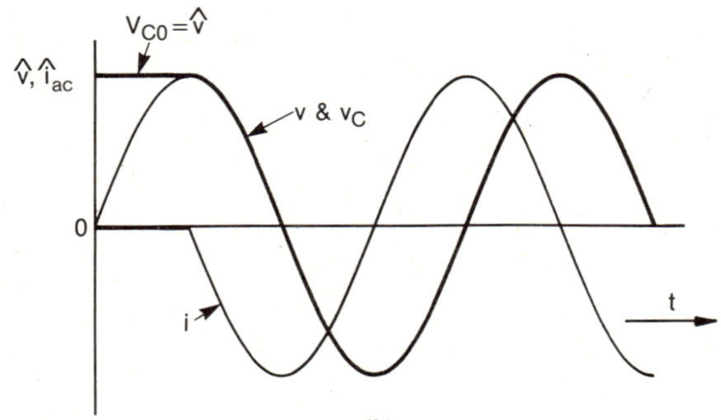

(b)

FIGURE 27. Ideal, transient-free switching waveforms. (*a*)Switching on. (*b*)Switching off.

Ideal Transient-Free Switching. The simple case of a switched capacitor, with no other circuit elements than the voltage supply, is used first to describe the important concept of transient-free switching. Figure 28 shows the circuit.

With sinusoidal ac supply voltage $v = \hat{v} \sin (\omega_0 t+\alpha)$, the thyristors can be gated into conduction only at a peak value of voltage, that is, when

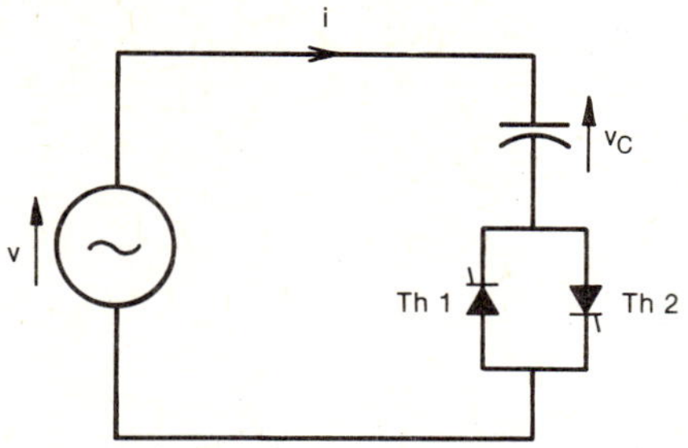

FIGURE 28. Circuit for analysis of transient-free switching.

$$\frac{dv}{dt} = \omega_0 \hat{v} \cos(\omega_0 t + \alpha) = 0. \tag{9}$$

Gating at any other instant would require the current $i = C\,dv/dt$ to have a discontinuous step change at $t = 0+$. Such a step is impossible in practice because of inductance, which is considered in the next section. To permit analysis of Figure 28, the gating must occur at a voltage peak, and with this restriction the current is given by

$$i = C\frac{dv}{dt} = \hat{v}\omega_0 C \cos(\omega_0 t + \alpha), \tag{10}$$

where $\alpha = \pm\pi/2$. Now $\omega_0 C = B_c$ is the fundamental-frequency susceptance of the capacitor, and $X_c = 1/B_c$ its reactance, so that with $\alpha = \pm\pi/2$

$$i = \pm\hat{v}B_c \sin\omega_0 t = +\hat{i}_{ac} \sin\omega_0 t, \tag{11}$$

where $\hat{i}_{ac}$ is the peak value of the ac, $\hat{i}_{ac} = \hat{v}B_c = \hat{v}/X_c$.

In the absence of other circuit elements, we must also specify that the capacitor be precharged to the voltage $V_{c0} = \pm\hat{v}$, that is, it must hold the prior charge $\pm\hat{v}/C$. This is because any prior dc voltage on the capacitor cannot be accounted for in the simple circuit of Figure 28. In practice this voltage would appear distributed across series inductance and resistance with a portion across the thyristor switch.

With these restrictions, that is, $dv/dt = 0$ and $V_{c0} = \pm\hat{v}$ at $t = 0$, we have the ideal case of transient-free switching, as illustrated in Figure 27. This concept is the basis for switching control in the TSC. In principle, once each capacitor is charged to either the positive or the negative system peak voltage, it is possible to switch any or all of the capacitors on or off for any integral number of half-cycles without transients.

Switching Transients in the General Case. Under practical conditions, it is necessary to consider inductance and resistance. First consider the addition of series inductance in Figure 28. In any practical TSC circuit, there must always be at least enough series inductance to keep di/dt within the capability of the thyristors. In some circuits there may be more than this minimum inductance. In the following, resistance will be neglected because it is generally small and its omission makes no significant difference to the calculation of the first few peaks of voltage and current.

The presence of inductance and capacitance together makes the transients oscillatory. The natural frequency of the transients will be shown to be a key factor in the magnitudes of the voltages and currents after switching, yet it is not entirely under the designer's control because the total series inductance includes the supply-system inductance which, if known at all, may be known only approximately. It also includes the inductance of the step-down transformer (if used), which is subject to other constraints and cannot be chosen freely.

It may not always be possible to connect the capacitor at a crest value of the supply voltage. It is necessary to ask what other events in the supply-voltage cycle can be detected and used to initiate the gating of the thyristors, and what will be the resulting transients.

The circuit is that of Figure 29. The voltage equation in terms of the Laplace transform is

$$V(s) = \left[Ls + \frac{1}{Cs}\right] I(s) + \frac{V_{c0}}{s}. \tag{12}$$

The supply voltage is given by $v = \hat{v} \sin(\omega_0 t + \alpha)$. Time is measured from the first instant when a thyristor is gated, corresponding to the angle α on the voltage wave. By straightforward transform manipulation and inverse transformation we get the instantaneous current:

$$i(t) = \hat{I}_{ac} \cos(\omega_0 t + \alpha) - nB_c \left[V_{c0} - \frac{n^2}{n^2-1}\hat{v}\sin\alpha\right] \sin\omega_n t$$

$$- \hat{I}_{ac} \cos\alpha \cos\omega_n t\,, \tag{13}$$

where ω_n is the natural frequency of the circuit,

$$\omega_n = 1/\sqrt{LC} = n\omega_0, \tag{14}$$

and

$$n = \sqrt{X_c/X_L}\,. \tag{15}$$

n is the per-unit natural frequency.

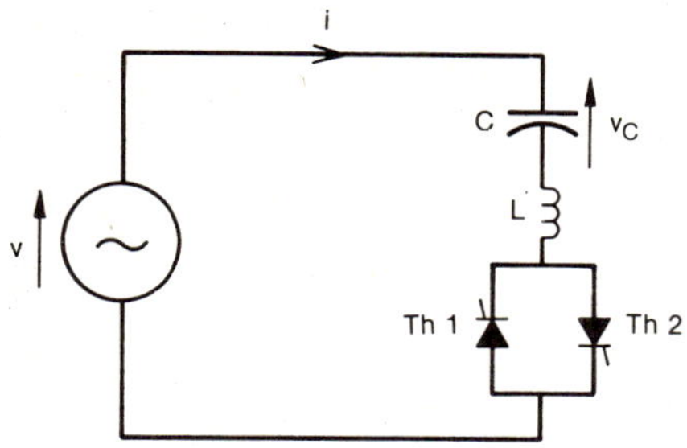

FIGURE 29. Circuit for analysis of practical capacitor switching.

The current has a fundamental-frequency component i_{ac} which leads the supply voltage by $\pi/2$ radians. Its amplitude $\hat{i}_{ac}$ is given by

$$\hat{i}_{ac} = \hat{v}B_c \frac{n^2}{n^2-1} \tag{16}$$

and is naturally proportional to the fundamental-frequency susceptance of the capacitance and inductance in series, that is, $B_c n^2/(n^2-1)$. The term $n^2/(n^2-1)$ is a magnification factor which accounts for the partial series-tuning of the L-C circuit. If there is appreciable inductance, n can be as low as 2.5, or even lower, and the magnification factor can reach 1.2 or higher. It is plotted in Figure 30.

The last two terms on the right-hand side of Equation 13 represent the expected oscillatory components of current having the frequency ω_n. In practice, resistance causes these terms to decay. The next section considers the behavior of the oscillatory components under important practical conditions.

(a) Necessary Conditions for Transient-Free Switching. For transient-free switching, the oscillatory components of current in Equation 13 must

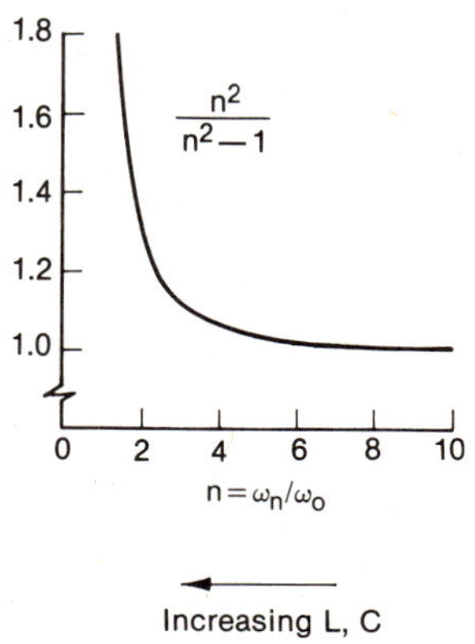

FIGURE 30. Voltage and current magnification factor $n^2/(n^2-1)$.

be zero. This can happen only when the following two conditions are simultaneously satisfied:

$$\text{(A)}\ \cos\alpha = 0\ \text{(i.e., } \sin\alpha = \pm 1\text{)} \tag{17}$$

$$\text{(B)}\ V_{c0} = \pm\hat{v}\,\frac{n^2}{n^2-1} = \pm X_c\,\hat{i}_{ac}\,. \tag{18}$$

The first of these means that the thyristors must be gated at a positive or negative crest of the supply voltage sinewave. The second means that the capacitors must also be precharged to the voltage $\hat{v}n^2/(n^2-1)$ with the same polarity. The presence of inductance means that for transient-free switching the capacitor must be "overcharged" beyond $\hat{v}$ by the magnification factor $n^2/(n^2-1)$. With low values of n, this factor can be appreciable (see Figure 30).

Of the two conditions necessary for transient-free switching, the precharging condition B is strictly outside the control of the gating-control circuits because V_{c0}, n, and $\hat{v}$ can all vary during the period of nonconduction before the thyristors are gated. The capacitor will be slowly discharging, reducing V_{c0}; while the supply system voltage and effective inductance may change in an unknown way, changing n. In general, therefore, it will be impossible to guarantee perfect transient-free reconnection.

In practice the control strategy should cause the thyristors to be gated in such a way as to keep the oscillatory transients within acceptable limits. Of the two conditions A and B, A can in principle always be satisfied. Condition B can be approximately satisfied under normal conditions. For a range of system voltages near 1 pu, condition B will be nearly satisfied if the capacitor does not discharge (during a nonconducting period) to too low a voltage; or if it is kept precharged or "topped up" to a voltage near $\pm\hat{v}n^2/(n^2-1)$.

(b) Switching Transients Under Nonideal Conditions. There are some circumstances in which conditions A and B are far from being satisfied. One is when the capacitor is completely discharged, as for example when the compensator has been switched off for a while. Then $V_{c0} = 0$. There is then no point on the voltage wave when both conditions A and B are simultaneously satisfied.

In the most general case V_{c0} can have any value, depending on the conditions under which conduction last ceased and the time since it did so. The question then arises, how does the amplitude of the oscillatory component depend on V_{c0}? How can the gating instants be chosen to minimize the oscillatory component?

Two practical choices of gating angle are (a) at the instant when $v = V_{c0}$, giving $\sin\alpha = V_{c0}/\hat{v}$; and ($b$) when $dv/dt = 0$, giving $\cos\alpha = 0$.

The first of these may never occur if the capacitor is overcharged beyond $\hat{v}$. The amplitude $\hat{i}_{osc}$ of the oscillatory component of current can be determined from Equation 13 for the two alternative gating angles. In Figures 31 and 32 the resulting value of $\hat{i}_{osc}$ relative to $\hat{i}_{ac}$ is shown as a function of V_{c0} and n, for each of the two gating angles.

From these two figures it is apparent that if V_{c0} is exactly equal to $\hat{v}$, the oscillatory component of current is non-zero and has the same amplitude for both gating angles, whatever the value of the natural frequency n. For any value of V_{c0} less than $\hat{v}$, gating with $v = V_{c0}$ always gives the smaller oscillatory component whatever the value of n.

The conditions for transient-free switching appear in Figure 32 in terms of the precharge voltage required for two particular natural frequencies corresponding to $n = 2.3$ and $n = 3.6$.

Switching a Discharged Capacitor. In this case $V_{c0} = 0$. The two gating angles discussed were (a) when $v = V_{c0} = 0$ and (b) when $dv/dt = 0$ ($\cos \alpha = 0$). In the former case only condition B (Equation 18) is satisfied. From Equation 13 it can be seen that in the second case (gating when $dv/dt = 0$) the oscillatory component of current is greater than in the first case (gating when $v = V_{c0} = 0$). An example is shown in Figures 33 and 34. The reactances are chosen such that $\hat{i}_{ac} = 1$ pu and the natural frequency is given by $n = X_c/(X_s + X_t) = 3.6$ pu. In case (a), the amplitude of the oscillatory component of current is exactly equal to $\hat{i}_{ac}$. In case (b), the oscillatory component has the amplitude $n\hat{i}_{ac}$ and much higher current peaks are experienced. The capacitor experiences higher voltage peaks and the supply voltage distortion is greater.

4.3.3. Voltage/Current Characteristics

In a TSC for transmission system application the cost of the thyristor switches and other complications make it desirable to minimize the

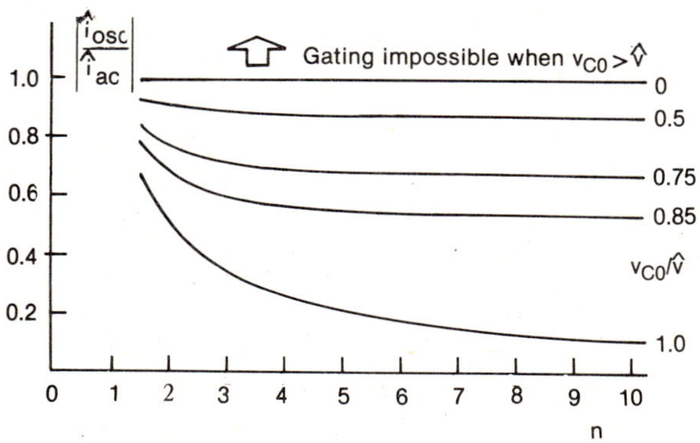

FIGURE 31. Amplitude of oscillatory current component. Thyristors gated when $v = V_{c0}$.

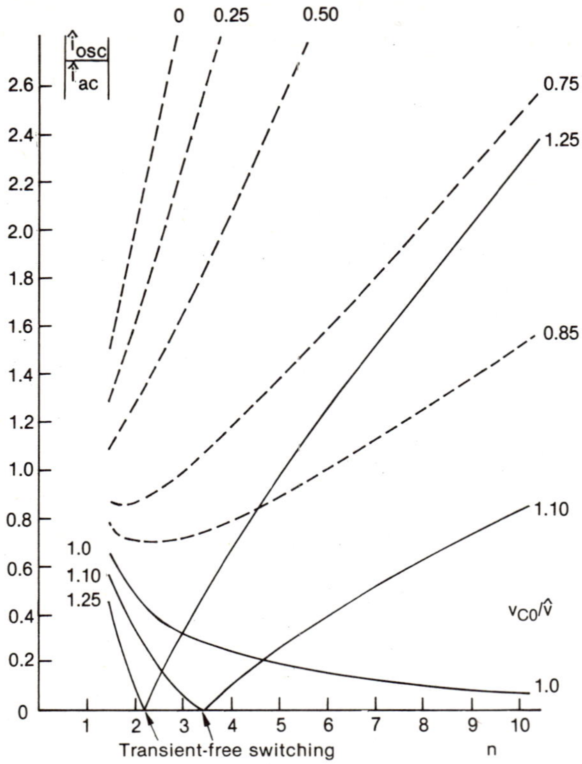

FIGURE 32. Amplitude of oscillatory current component. Thyristors gated when $dv/dt = 0$.

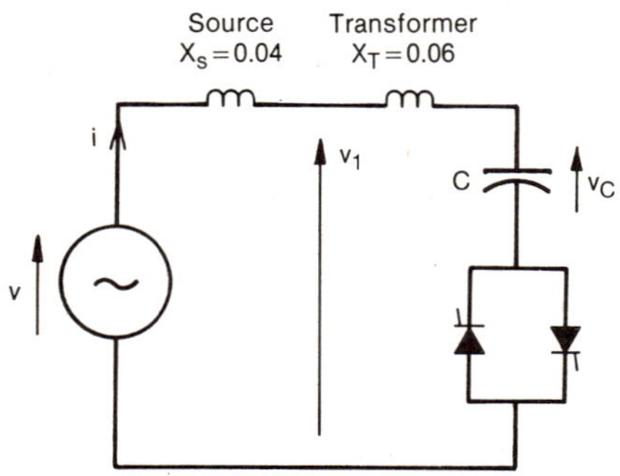

FIGURE 33. Switching a discharged capacitor; circuit diagram.

number of parallel capacitor units. A figure of 3 or 4 is typical of existing or presently planned installations. With so few capacitors a smooth voltage/current characteristic is unobtainable, and a stepped characteristic is obtained (Figure 35). The capacitor characteristics 1, 2, and 3 intersect the system voltage/current characteristic at discrete points, and operation can be at any of these points depending on the number of capacitors con-

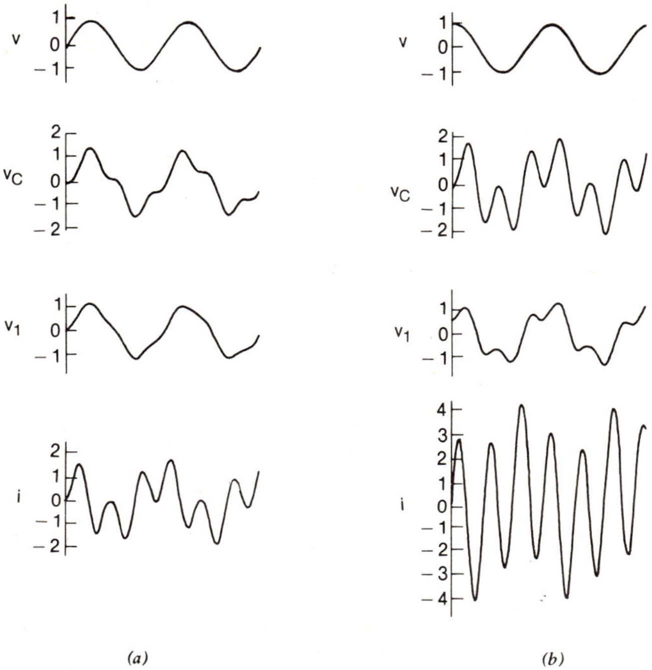

FIGURE 34. Switching transients with discharged capacitor. (*a*) Gating when $v = V_{c0} = 0$. (*b*) Gating when $dv/dt = 0$.

ducting. With two capacitors conducting, operation would be at point A.

In order to obtain a smoother voltage/current characteristic it is usual to have a parallel-connected TCR which "interpolates" between the capacitor characteristics. If the TCR characteristic has a small positive slope, the resultant characteristic is shown by the heavy-line segments in Figure 35. This construction shows that the TCR current rating must be a little larger than that of one capacitor bank at rated voltage, otherwise deadbands arise as shown by the shading in Figure 35. The increase in TCR rating enables the heavy segments to be extended to the left through the deadband.

With fixed TCR controls (i.e., fixed knee voltage and slope) the voltage/current characteristic of this hybrid TSC/TCR is still stepped, giving rise to the possibility of bistable operation. It is therefore necessary to adjust the TCR knee voltage and slope by a small amount every time a capacitor is switched in or out. This can in principle be done either by open-loop or closed-loop (current feedback) modification of the control system giving a continuous V/I characteristic as in Figure 14. A further sophistication in the control system is to incorporate a hysteresis effect so that the capacitors are switched in at a lower voltage than that at which they are switched out. This helps to prevent a "hunting" instability

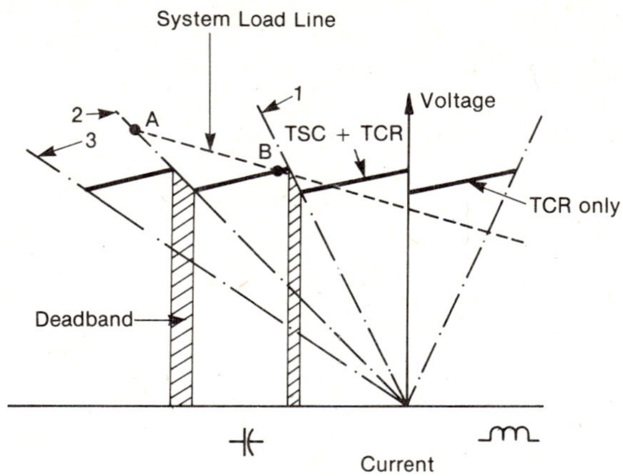

FIGURE 35. Effect of paralleling a TSC and TCR before control systems are properly coordinated.

which can arise if the system characteristic intersects the compensator characteristic near the junction of two segments.

4.4. SATURATED-REACTOR COMPENSATORS

A variety of saturating-iron devices have been used for voltage stabilization. At least four basic principles have been applied, but only one has been developed for transmission-system applications, the remainder having been employed only in smaller sizes and at lower voltages. This is the so-called polyphase, harmonic-compensated, self-saturating reactor. It is closely related to, and was developed from, the phase-multiplying type of frequency multiplier, of which many different versions were developed both in the United States and in Europe since about 1912.

Other classes of saturated-reactor voltage stabilizer include the ferroresonant constant-voltage transformer, the transductor, and the tapped-reactor/saturated-reactor compensator. The ferroresonant constant-voltage transformer is manufactured in the United States only in kVA sizes. The transductor has dc control windings and works as an adjustable susceptance. Its speed of response compares unfavorably with other types of compensator. The tapped-reactor/saturated-reactor compensator is essentially single-phase, and has a series as well as a shunt element. It is therefore suitable only for load compensation (e.g., small arc furnaces), and cannot be connected to a power-system busbar for voltage stabilization.

4.4.1. Principles of Operation

The principle of the saturated reactor is shown in Figure 36. (It is interesting to compare this with Figure 5 for the TCR). The controlling

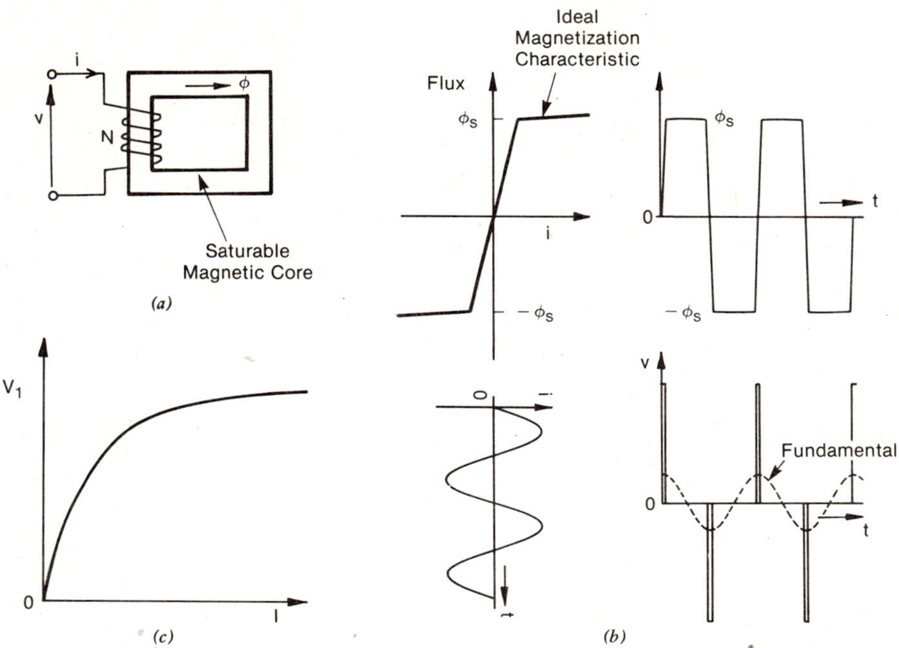

FIGURE 36. Principle of elementary saturated reactor. (*a*) Elements of saturated reactor. (*b*) Flux, current, and voltage waveforms. (*c*) Variation of fundamental voltage with current (assumed sinusoidal).

element is the saturable magnetic core, whose idealized magnetization characteristic is shown in terms of flux and current (Figure 36*b*). Assume that a sinusoidal current flows in the winding. Then the flux and voltage waveforms ($v = N\mathrm{d}\phi/\mathrm{d}t$) are as shown in Figure 36*b*. The flux waveform is approximately square, the flux alternating between the saturation levels $\pm\phi_S$. The voltage is almost a series of impulses, but because the flux waveform is nearly independent of the current the fundamental component of the voltage is *constant*. In practice the magnetization characteristic is not perfectly flat, but is nearly linear above ϕ_S with a slope proportional to the permeability of free space. The result is a fundamental voltage/current characteristic of the form shown in Figure 36*c*, with a small positive slope. The constant fundamental voltage property follows directly from the saturation transitions of the core, each of which induces a fixed voltage impulse (volt-seconds) in the winding, no matter how rapidly the transitions take place. The fundamental voltage lags 90° behind the current and reactive power is absorbed.

The plain saturated reactor just described is unsuitable for use in transmission systems because the voltage, or the current, or both, are too distorted. In transmission-system compensators the harmonics are reduced to an extremely low level by internal compensation. The techniques are in principle the same as used in phase-multiplying transformers or magnetic frequency multipliers, and a simple explanation is given in terms of Figure 37, which shows a three-phase saturated reactor having an additional winding in the form of a closed delta. In the absence of the secondary winding, the primary currents are highly nonsinusoidal. However, if the secondary winding is closed, by three-phase symmetry it will be found that most of the necessary triplen harmonic components of the magnetomotive forces in each limb can be provided by the *secondary* current. The primary current is free of these harmonics. The secondary currents are predominantly third-harmonic, and the circuit is thus an elementary frequency-tripler, which shows the generic relationship between this type of voltage stabilizer and the magnetic frequency multiplier. (In frequency multipliers the high-frequency power is transferred to a load either by direct series insertion in the secondary circuit, or by transformer coupling.)

The primary currents, under balanced conditions, contain no harmonics lower than 5th and 7th. It was found in the 1930s that by inserting a reactor of suitable value in the secondary circuit, even these harmonics could be reduced. This phenomenon of partial harmonic compensation has been exploited particularly by Friedlander.

The harmonic performance and the voltage/current characteristics of the plain frequency tripler are not good enough for application in power systems, and much better characteristics are obtained with frequency multipliers of up to nine times, such as for example the treble-tripler reactor. This shares with the plain tripler the open-mesh secondary winding, which carries predominantly 9th harmonic, but in order to generate this from a balanced three-phase system a phase-multiplying arrangement is necessary, as shown in Figure 38. There are nine limbs in the magnetic core, and in common with other magnetic frequency multipliers only one is unsaturated at a time. Each limb saturates alternately in both directions, giving a total of 18 unsaturations per cycle. In the terminology of the TCR, this corresponds to 18-pulse operation and helps to explain the high speed of response of the reactor by itself. The lowest-order characteristic harmonics of the treble-tripler reactor are the 17th and 19th, but

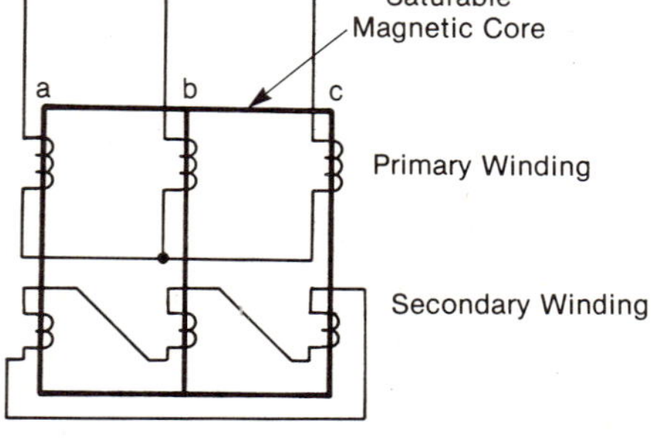

FIGURE 37. Elementary frequency tripler having approximately constant voltage characteristic.

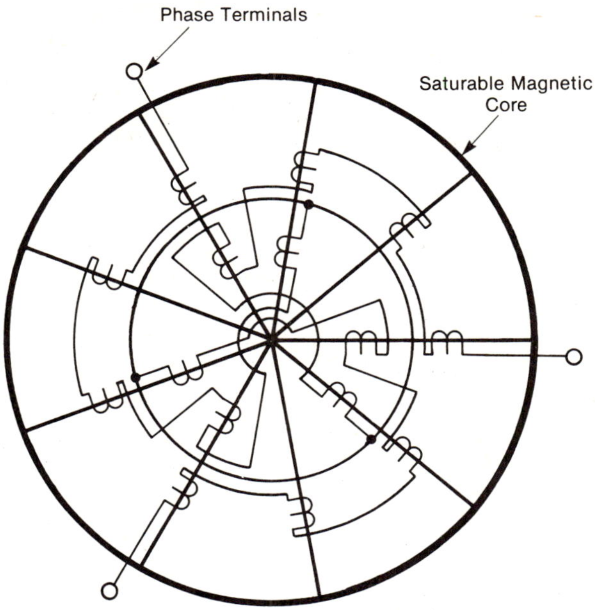

FIGURE 38. Polyphase treble-tripler reactor. Arrangement of magnetic core and windings (part of the winding is omitted for simplicity).

by appropriate design of an inductive load in the 9th harmonic mesh circuit, these can be reduced to around 2% by partial harmonic compensation. Plain shunt capacitors can normally be used to absorb these residual harmonics.

4.4.2. Voltage/Current Characteristics

The voltage/current characteristic of the saturated reactor by itself is shown in Figure 39. There is a slope of between 5 and 15% (based on the reactor rating), which depends on the reactor design and in particular on the after-saturation inductance of the windings. A lower slope reac-

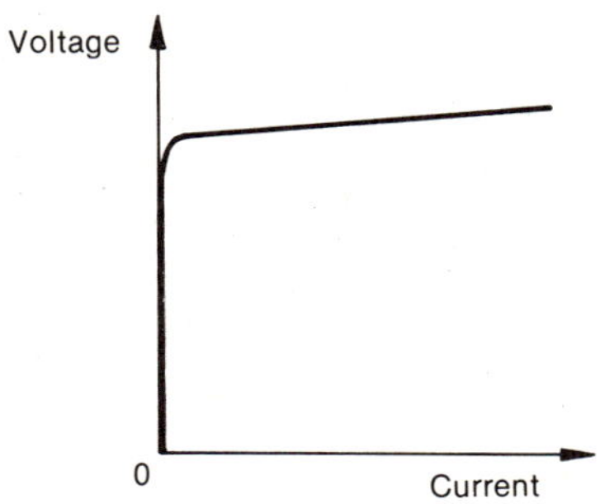

FIGURE 39. Voltage/current characteristic for polyphase saturated reactor.

tance requires a more expensive and larger reactor design. The characteristic is very linear above about 10% of rated current.

A lower slope reactance is sometimes obtained by connecting a capacitor in series with the saturated reactor. This slope-correcting capacitor can be sized to make the slope zero or even negative, but typical slopes are the same as would be specified for a TCR. The effect of the slope-correcting capacitor is shown in Figure 40.

Just as in the TCR compensator, the voltage-stabilized operation can be biased into the leading power-factor region by means of shunt capacitors. These may be designed as filters if the system resonances require it, although this is usually not necessary for the characteristic harmonics of the reactor.

A step-down transformer is normally used for EHV connection, because it is uneconomic or impractical to design the reactor for direct connection above about 132 kV. The transformer may have a load tap-changer that can effectively alter the knee-point voltage of the compensator on the high voltage side. It is possible to insert capacitors in series with the transformer as well as the reactor, to obtain flat stabilization of both the EHV and the compensator busbars. In other respects the step-down transformer would be similar to the one used with the TCR compensator. A general arrangement is shown in Figure 4.

The slope-correcting capacitors can make the saturated reactor compensator susceptible to subharmonic instability especially on weak systems, and it is normal practice to have a harmonic damping filter in parallel with the capacitors. This filter as well as the capacitors must have a voltage rating compatible with any transients which might occur during energization or other system disturbances, and their overload capability limits the otherwise substantial overload capability of the saturated-reactor compensator, unless the capacitor is bypassed by a spark gap or some other device. The slope-correcting capacitor also slows the response of the compensator.

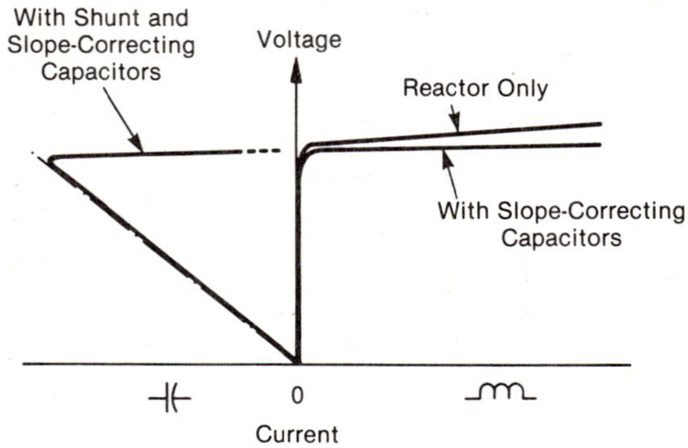

FIGURE 40. Voltage/current characteristics for saturated reactor compensator with series (slope-correcting) and shunt capacitors.

Energization is by direct closure of the compensator circuit breaker, although the shunt capacitors need not be energized simultaneously. Losses are comparable to those of similarly rated transformers, and their variation with current is not unlike that of the TCR. Noise levels can be quite high near the reactor because of high-frequency magnetostrictive forces, and a brick enclosure is sometimes used. Reliability is comparable with that of similarly rated transformers.

4.5. SUMMARY

Table 3 summarizes the comparative merits of the main types of compensator. Note that some specialized types are omitted. For example, the thyristor-controlled transformer has many properties in common with the thyristor-controlled reactor. The dc controlled transductor is probably not competitive because of its slow response. Breaker-switched capacitors are also omitted. They are less flexible than the TSC, although they have been more widely used and they certainly have lower losses.

The comparisons in Table 3 are not hard and fast, because each type of compensator can be designed with a wide range of properties. There is therefore no general way of deciding that one type is better than another. Comparisons can be realistically made only in specific circumstances and with respect to specific operating criteria. For example, it is meaningless to say that the TSC has lower losses than the TCR, unless the reactive power is specified at which the losses are evaluated. This reactive power should be representative of the average operating point of the compensator throughout its lifetime.

Again, it is often pointed out that the synchronous condenser has a slow speed of response. But the speed of response can be increased by increasing the exciter ceiling voltage: this has actually been done on some synchronous condensers to the point where they have sufficient speed of response to perform comparably with static compensators. Other criteria, such as cost and reliability, then determine which type of compensator should be used.

For the ultimate flexibility of control, with present-day technology, the TCR compensator is the best type. The TSC, which may be combined with the TCR, generally results in a loss characteristic which is lower in the lagging regime, and may enhance the control flexibility. The TCR compensator can also be designed to have some useful overvoltage-limiting capability which is not available in the plain TSC.

For an absolute minimum of maintenance the saturated reactor compensator has advantages, but it has almost no control flexibility and it may require appreciable expenditure on damping circuits to avoid any possibility of subharmonic instability. It does have overload capability which is useful in limiting overvoltages, although the full exploitation of

TABLE 3
Comparison of Basic Types of Compensator

Quality	Synchronous Condenser	TCR (with shunt capacitors where necessary)	TSC (with TCR where necessary)	Polyphase Saturated Reactor
Construction	Rotating machine	Thyristor equipment with static reactors and capacitors	Thyristor equipment with static capacitors	Transformer type with shunt capacitors
Reactive Power Capability	Leading/Lagging	Lagging/Leading (indirect)	Leading/Lagging (indirect)	Lagging/Leading (indirect)
Control	Continuous	Continuous	Discontinuous (continuous with vernier TCR)	Continuous
Response Time	Slow	Fast, but system-dependent	Fast, but system-dependent	Fast, but system-dependent and slowed by slope-correction capacitors
Harmonics	Very good	Filters may be necessary depending on system conditions	Good, but filters may be necessary with TCR	Good, up to 17th and 19th

Losses	Moderate	Good, but increase with lagging current	Good, but increase with leading current	Good, but increase with lagging current
Flexibility (=programmability of control)	Good within limitations of response speed	Excellent	Good; excellent if TCR is added	Poor
Phase-balancing Ability	Limited	Good	Limited	Limited
Overvoltage Limitation	Good	Moderate	None (or very limited with vernier TCR)	Good, within limitations of slope-correction capacitors
Rotating Inertia	Yes	No	No	No
Accuracy of Compensation	Good	Excellent	Good; excellent if TCR is added	Good
Direct EHV Connection	No	Reactor – No Capacitors – Yes	No	Saturated reactor – no Capacitors – yes
Starting	Slow	Fast, with minimal transients	Fast, with some transients	Fast, with some transients

this capability may involve bypassing the series slope-correcting capacitor with a spark gap, or other device.

The synchronous condenser also has a useful overload capability and good harmonic performance. In addition, because of its inertia, it has a measure of waveform stability which is useful in some applications.

4.6. FUTURE DEVELOPMENTS AND REQUIREMENTS

Although the demand for electric power in the United States is increasing only slowly, the trends in worldwide generation and transmission suggest that reactive power compensation will become increasingly important. This follows from a wide commitment to ac transmission, together with economic factors which necessitate the maximum utilization of generation and transmission facilities. In addition, there is no evidence that large, irregular loads like arc furnaces will become less common.

As far as bulk transmission is concerned, the continued development of remote hydro resources worldwide should continue to provide a demand for EHV compensators. Among the competitive issues in this field are power losses and harmonics, and innovations that improve performance in these areas are greatly to be desired. Overvoltage performance is another area where improvements are desirable. Most static compensators, including the polyphase saturated reactor, are limited in their ability to hold down the system voltage under emergency conditions. Developments in HVDC transmission also generate reactive power control requirements which compensators can help to fulfill.

Chapter 5

DESIGN OF THYRISTOR CONTROLLERS

R. L. ROFINI

5.1. THYRISTORS

The application of solid-state technology to reactive power control (compensation) was made possible by the increase in the voltage and current ratings of thyristors as shown in Table 1.

The disciplines used in applying thyristors to reactive power control are related to those used in thyristor valves for high-voltage dc transmission. (In the HVDC field, thyristor valves have completely replaced mercury-arc valves in new installations.)

5.2. THE THYRISTOR AS A SWITCH; RATINGS

It is not the intent of this text to provide an in-depth description of the thyristor. Such in-depth descriptions are readily available from application notes and descriptive manuals issued by the thyristor manufacturers. This section is written to enable the reader who lacks previous knowledge of the thyristor to understand its application in the thyristor controller.

As would be expected, the most important component in the thyristor controller is the thyristor itself. The thyristor, when introduced in the late 1950s, was known as a "silicon-controlled rectifier" (SCR). The device's correct name, as later determined by industrial committees, is "reverse-blocking triode thyristor." But the name in common usage is "thyristor."

The thyristor is a three-terminal semiconductor device. Symbolically, it is as shown in Figure 1. It comprises four layers of alternately p- and n-doped semiconducting silicon, having three p-n junctions.(1) Functionally, the thyristor is a unidirectional switch. In general it will conduct

TABLE 1
Progress in Thyristor Ratings

Year[a]	Wafer Diameter (mm)	Peak Repetitive Voltage Rating (V)	Average Current Rating (A)
1958	14	200	80
1960	24	700	150
1963	28	1300	300
1963	28	1800	225
1965	33	1800	550[b]
1969	40	1800	750
1969	40	2600	600
1973	53	2600	1000
1975	53	2900	1000[c]
1977	77	3800	1600
1980	77	4500	1350
1985[d]	100	6000	—

[a] Year of availability for commercial application.
[b] Introduction of double-sided cooling.
[c] Introduction of neutron-irradiated silicon.
[d] Estimated.

current (a closed-switch function) only when both the anode and the gate are made positive with respect to the cathode. Once the thyristor is in conduction, the gate has no further control — the thyristor will cease conduction only when the current flowing through the external circuit is reduced to zero.

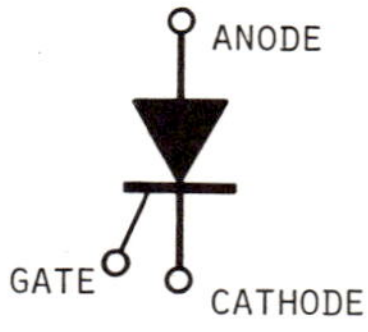

FIGURE 1. Thyristor circuit symbol.

Figure 2 shows the voltage-current characteristics of a thyristor when the gate–cathode voltage is zero. The figure shows two regions of the voltage/current characteristics that are not within the normal operating range. One region is that of the reverse avalanche area where the voltage/current product (watts dissipation in the thyristor) becomes excessive. The other area is the off-state breakover voltage area where the thyristor went into the "closed switch" function without the supply of gate current. Operation in either of these two regions can cause degradation or failure.

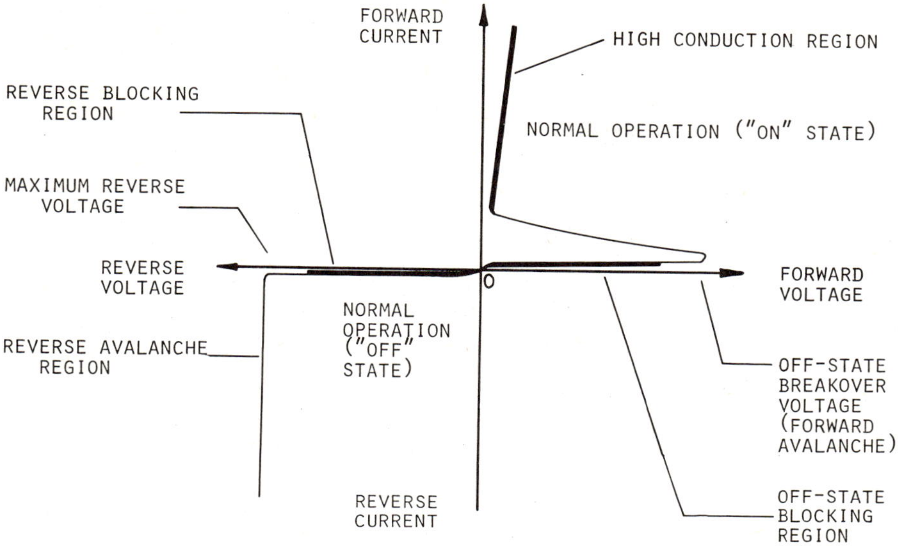

FIGURE 2. Thyristor characteristics without applied gate voltage.

The reverse voltage blocking capability will be at least equal to the minimum forward breakover voltage. The thyristor has a voltage rating that corresponds to the peak off-state blocking voltage — a voltage magnitude at which the thyristor will not go into conduction unless gated. The blocking voltage capability is related to gate circuit conditions, junction temperature, and rate of change of applied voltage. These relationships are discussed in detail in the literature.[1,2] If voltage is applied to the gate while the thyristor is in the off-state (i.e., forward-biased) blocking region, the thyristor state will change from the off-state blocking region to the high-conduction region. The applied gate voltage causes the thyristor to change function from an open switch to a closed switch. In typical applications, the voltages applied to the thyristor are well below its off-state and reverse voltage ratings.

Thyristor manufacturers issue detailed specification sheets for their commercially available thyristors. In special circumstances a manufacturer may provide a "custom" thyristor that is particularly suited for a thyristor controller application.

As the ratings of thyristors have increased over the years, it has become common to find a thyristor described not only by its voltage rating, but also by a dimensional parameter that is an indication of the size of the silicon wafer. There is no uniform definition of this dimensional parameter, but typical parameters used by manufacturers are (1) the diameter of the silicon wafer, (2) the diameter of the thyristor *package*, and (3) the conducting area of the silicon wafer. The dimensions of the wafer determine its current rating, and a typical relationship is shown in Table 1.

Figure 3 is an outline of a typical thyristor used in thyristor controllers. The two surfaces defined as "mounting surfaces" are placed against a surface capable of being cooled. The heat-dissipating element within the ceramic housing is the silicon wafer. Each side of the silicon wafer dissipates heat through metal slugs to the mounting surfaces.

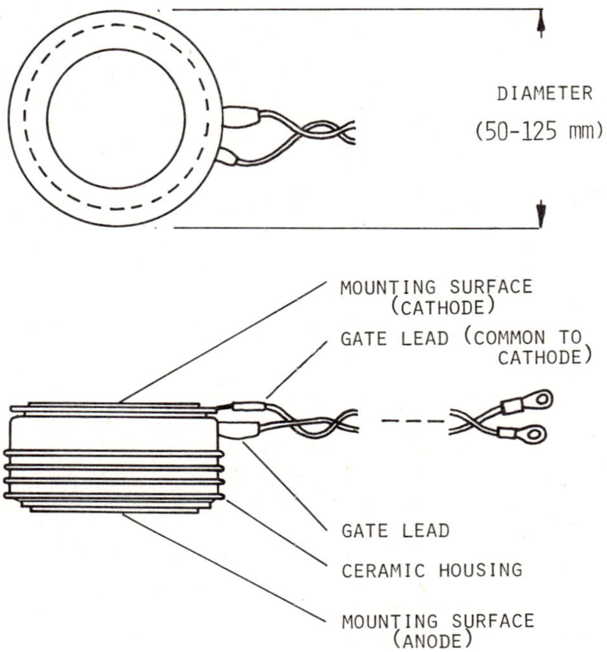

FIGURE 3. Typical thyristor outline.

5.3. THERMAL CONSIDERATIONS

The thyristor specifications describe the electrical characteristics at certain junction temperatures. The word *junction* refers to a theoretical boundary layer within the silicon wafer, and the expression *junction temperature* is commonly used to describe the "hot spot" temperature of the layer within the silicon wafer.

Figure 4 represents a simple equivalent thermal circuit that allows calculation of the junction temperature. The watts magnitude is the total rate of heat generation within the thyristor. The product of watts and the thermal resistances expressed in °C/watt results in a temperature difference. Given the thermal resistances from the junction to the case of the thyristor, the thermal resistance from the case to the heat sink, the

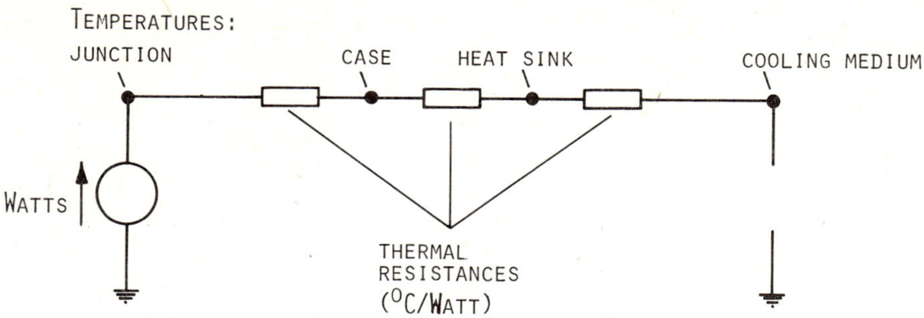

FIGURE 4. Thyristor thermal circuit—steady-state conditions.

thermal resistance from the heat sink (or heat dissipator) to the cooling medium, and the cooling-medium temperature, it is possible to calculate the junction temperature for any operating condition.

The equivalent thermal circuit of Figure 4 is for a steady-state condition. The combination of the thyristor and its heat sink does have a heat storage capacity; in other words, the combination has a thermal time constant that allows overload currents for brief periods of time. Typical thermal time constants for air-cooled thyristors are of the order of 3 min. For a fluid-cooled thyristor, the thermal time constant is generally less. The thermal time constant, of course, is influenced by the heat sink mass. This thermal time constant allows the thyristor controller designer to provide for overloads lasting for milliseconds or even seconds (especially with an air-cooled thyristor heat sink). The designer will consider the operating condition (current flow through the thyristor controller) to determine the wattage dissipation in the thyristor controller. The cooling system will determine the thermal resistance from the heat sink to the cooling medium. By a proper selection of the cooling system parameters, the junction temperature can be kept within predefined limits during normal and abnormal operation.

The operating junction temperature must be less than the maximum specified by the thyristor manufacturer. As the junction temperature increases, the thyristor's capability to withstand voltage during its nonconducting state decreases. This is just one example of a degradation of performance due to an increased junction temperature. Also, an increased junction temperature causes an increase in the thermal stress within the thyristor assembly, owing to differences in thermal expansion of the materials in it. The designer must choose a low enough junction temperature to ensure reliable operation. Typical maximum operating junction temperatures for thyristor controller applications may be up to 120°C or more at maximum design current with maximum cooling-medium temperature.

5.4. DESCRIPTION OF THYRISTOR CONTROLLER

5.4.1. General

The voltage and current ratings required in a complete thyristor controller are usually greater than those of even the largest individual thyristor. In addition, for ac operation it is necessary to connect oppositely-poled thyristors in parallel to permit conduction in both directions. The required voltage and current ratings are achieved respectively by series and parallel connection. A typical thyristor controller circuit used in a TCR system is shown in Figure 5. Only one phase of a three-phase delta is shown. The thyristors are connected in several series "voltage levels," each with two or more thyristors in parallel. One phase of the thyristor controller is sometimes called a thyristor switch.

The functions of the components are as follows:

1. L1 and L1A are the air-cored reactors, external to the thyristor controller. Together they make up the total reactance of one phase; they are split into two to protect the thyristor switch against certain types of electrical fault.

2. Q1, Q1A, Q2, ... and QNA are the thyristors. They are connected anti-parallel for the bi-directional current flow. Figure 5 shows two thyristors per voltage level (one in each direction). The current through the external reactors L1 and L1A is controlled by causing the thyristors to conduct current during a portion of the cycle.

3. L2 and L2A are small reactors. They are physically located near the terminals of the thyristor switch, and their function is to delay the discharging of the stray capacitances through the thyristor switch when the thyristors are gated. This limits the initial di/dt (rate of rise of current) in the thyristors.

4. The capacitances C_{stray} are the stray capacitances (such as bus-to-ground, bushing and/or standoff insulator capacitances, and terminal-to-terminal capacitances). Depending on the current rating, more than one thyristor may be used to carry current in one direction. As many as eight thyristors per voltage level (four carrying current in one direction connected in parallel with four carrying current in the opposite direction) have been used in high-current TCR systems.

5. The series circuits of resistors and capacitors (R1, R2, ... RN, and C1, C2, ... CN), one across each thyristor level, limit the transient voltage that appears across the thyristor switch when the thyristor ceases current conduction. The R-C circuit is called a "snubber" and is a common feature of thyristor and diode circuits.[3,4] At power-frequency the series of R-C circuits also equalizes the voltage distribution among the thyristors.

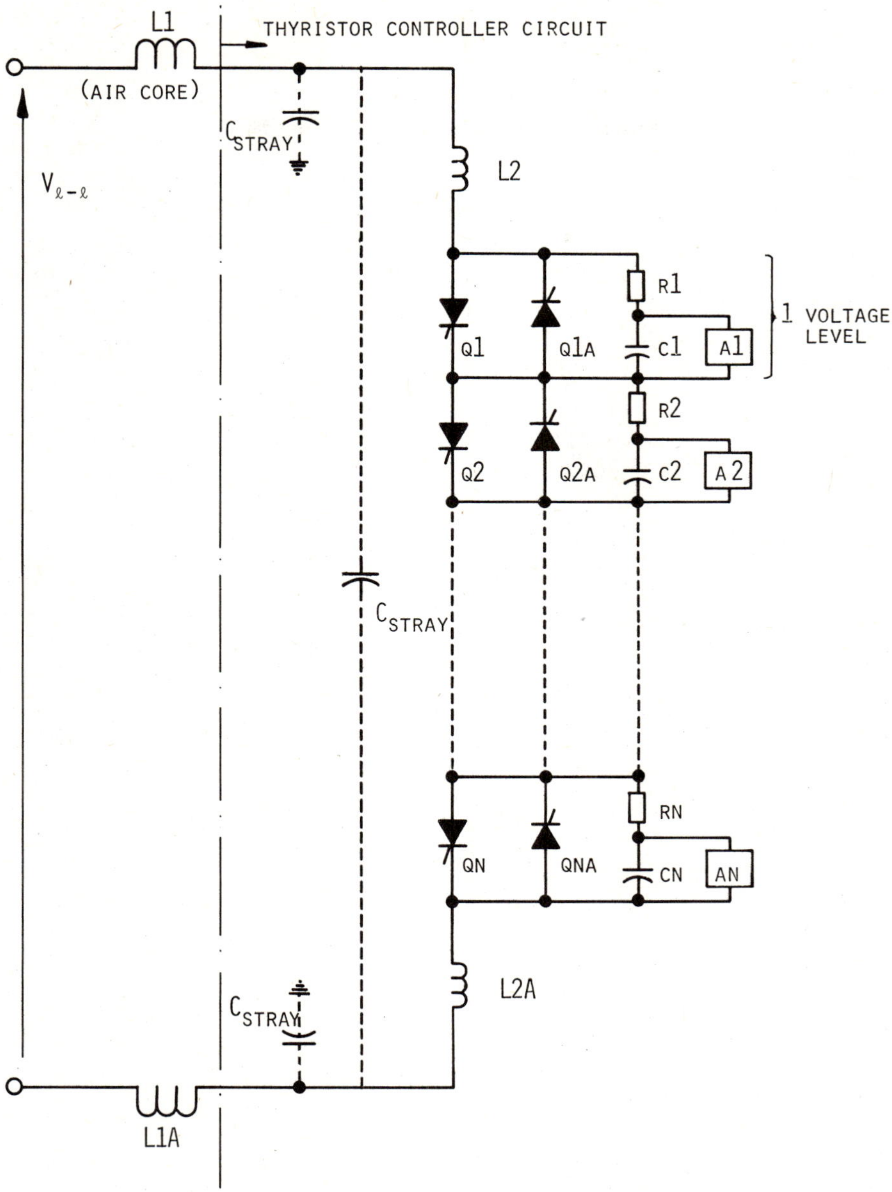

FIGURE 5. Typical thyristor controller circuit in a TCR system—gating circuits not shown.

6. When a thyristor fails, it becomes a short circuit. It is necessary to have an indication of such a failure. Its effect is to increase the voltage on all the remaining thyristors. The design parameters of the thyristor controller will be such that a small number (usually fewer than 10%) of short-circuited thyristors can be tolerated. One way to monitor failed

thyristors is to connect a status lamp circuit across a voltage level (or across the snubber capacitor). Such circuits draw currents only of the order of milliamperes. Since a short-circuited thyristor cannot withstand voltage, its failure extinguishes the corresponding status lamp. The status lamp circuit is, of course, at the same high voltage to ground as is the whole thyristor switch.

In some instances a remote status indication is desired. Remote indication can be obtained by fiber optics. The initial light source to the fiber optic is energized by a portion of the voltage existing across the thyristor level. The fiber optic isolates the remote indication circuits from the high voltage of the thyristor switch. One way to obtain the energy for the status monitoring circuit is to use the snubber capacitor as a source of voltage. This would be the function of circuits A1, A2, ... AN.

7. There are additional control circuits for gating the thyristors into conduction. These circuits, not shown in Figure 5, are discussed later.

5.4.2. R-C Snubber Circuit

An important characteristic of thyristors and diodes is recovery current. Recovery current occurs during the time when the thyristor is functionally changing from a closed switch to an open switch. Typical thyristor current and voltage waveforms during this transition time are shown in Figure 6. The recovery current is a transient phenomenon and can be ascribed to the removal of charge-carriers from the thyristor when it is changing from the conducting to the nonconducting state.

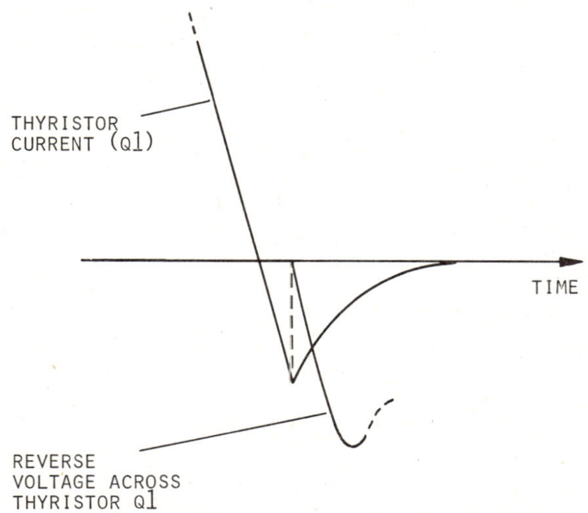

FIGURE 6. Thyristor recovery current and reverse voltage.

The product of the inductance of the circuit and the di/dt (rate of change) of the recovery current results in a transient voltage across the thyristor switch. In order to control the magnitude and waveform of the transient voltage, it is the practice to place a series resistor-capacitor circuit across the individual voltage levels.

The resultant equivalent circuit is shown in Figure 7.

The usual choice of the resistor and capacitor is such that the circuit is almost critically damped, giving the typical voltage waveform shown in Figure 6.

There are numerous articles (see, for example, Reference 4) written on choosing the values of the capacitor and resistor. The thyristor controller designer will base a choice of capacitor and resistor on the overshoot magnitude, the rate of change of the voltage across the thyristor, and the desirability of having a minimum value of snubber capacitance. The reason for desiring a minimum capacitance is that the wattage losses of the snubber resistor are due to the currents that charge and discharge the capacitor. There are differences in the peak recovery currents among the thyristors installed in a thyristor controller. These differences tend to cause bias voltage differences among the snubber capacitors. This is readily apparent by considering the current loops shown in Figure 8 in a circuit of just two series thyristors. The loops i_{rec1}

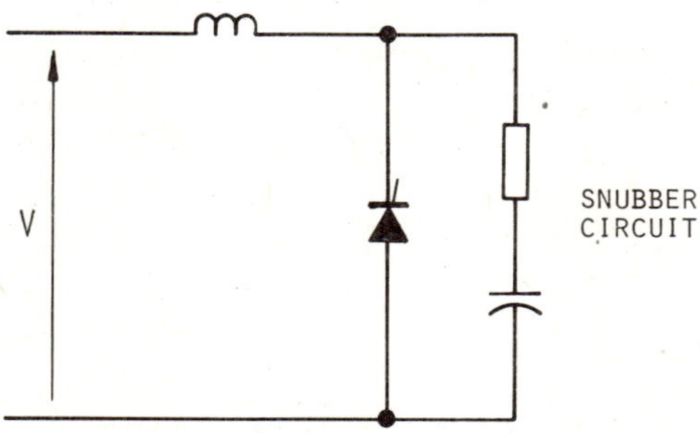

FIGURE 7. Equivalent circuit of snubber.

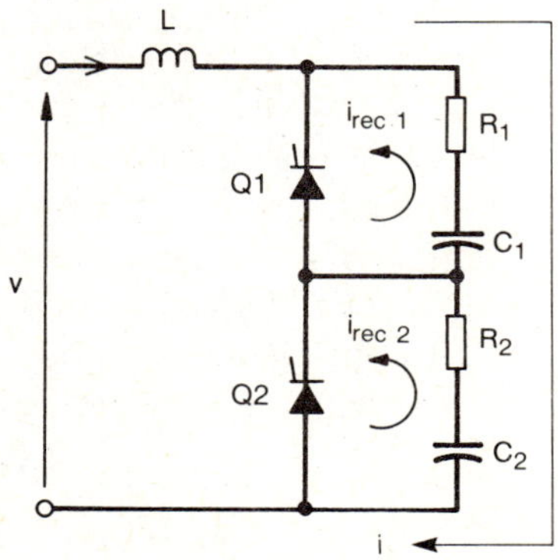

FIGURE 8. Circuit of two series thyristors showing recovery currents.

and i_{rec2} represent the recovery currents of the individual thyristors Q1 and Q2. The differences between the recovery currents i_{rec1} and i_{rec2} will cause a corresponding difference in the voltages on C1 and C2 and will lead to a deviation from an ideal voltage distribution between the series thyristor voltage levels. It is the usual practice to minimize the recovery differences between the thyristors by careful manufacture and by the matching of thyristors with closely similar characteristics.

In addition, the tolerance on the capacitances of the snubber capacitors must be allowed for in determining the maximum deviation from equal voltage-sharing between the voltage levels at power-frequency.

5.4.3. Gating Energy

The gating of the thyristor controller into conduction is initiated by control signals from circuits which are at ground potential. The usual method of transmitting the gating signal from the control circuit to the thyristor controller (which is usually at 13.8 kV or higher potential) is by a pulse of light transmitted by fiber-optic light guides. Light guides provide a high isolating impedance between the thyristor controller and ground potential.

The thyristors used in most thyristor controller applications require an *electrical* gating signal. Therefore most thyristor controllers contain a low-voltage electronic gating circuit that provides the gating signal to the thyristors in response to a light pulse. The power required by the low voltage electronic circuit can be obtained from three sources:

1. The voltage across the thyristor controller when the thyristor controller is not conducting.
2. The current through the thyristor controller when the thyristor controller is conducting.
3. Auxiliary transformers from ground potential.

One type of circuit is shown in Figure 9. Thyristors Q1 through QNA represent a portion of the total series thyristor circuit for one phase. Capacitor C1 is capable of being charged from a single-phase full-wave bridge circuit through resistor R1 when voltage exists across the nonconducting thyristors. Also, C1 will be charged from the current transformer T1 when the thyristors are conducting. The function of the zener diode across C1 is to regulate the voltage on capacitor C1. The charged capacitor C1 is used as the low-voltage energy source for the gating circuits. These types of circuit lend themselves to redundancy in that the gating

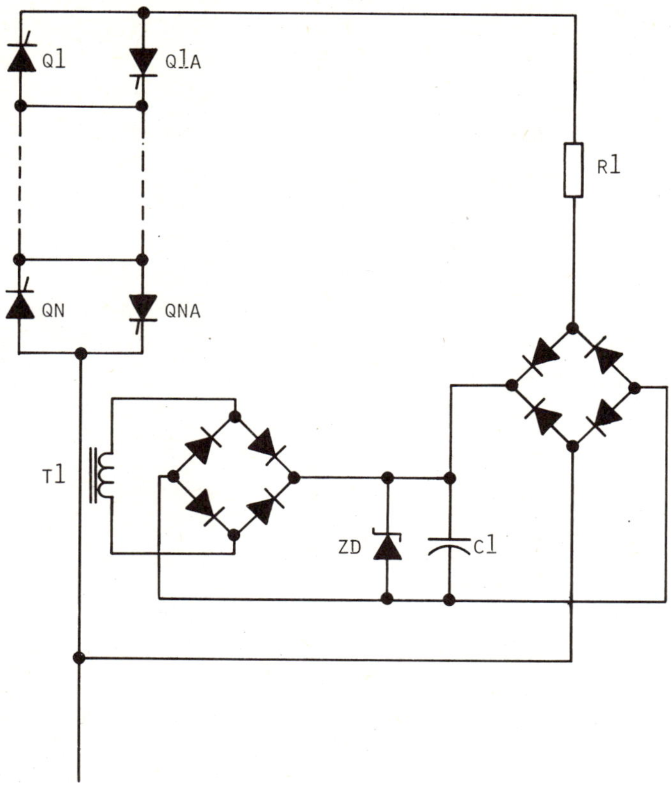

FIGURE 9. Gating energy circuit.

circuits of a series circuit of thyristors can be interrelated so that a failure of a portion of the gating circuit does not result in the elimination of the gating control function. The circuits providing the gating energy dissipate only a small portion of the total thyristor controller losses.

Present thyristor development will eliminate the need for such gating energy circuits as individual thyristors will be gated directly by light pulses via the fiber-optic light guides.

5.4.4. Overvoltage Protection

Voltages greater than the voltage rating of the thyristor may cause the thyristor to fail (becoming short-circuited). Although the controller is designed so that the normal applied voltage is well within the thyristor rating, it is usual to provide an overvoltage protection circuit as an integral part of the thyristor controller. The overvoltage protection circuit is activated when abnormal voltages (due to switching transients and so forth) occur across the thyristor controller.

The typical operating sequence of the overvoltage protective circuit is as follows:

1. An overvoltage occurs across the thyristor controller.
2. The overvoltage reaches a magnitude (below the voltage capability of the thyristor) that activates the overvoltage protective circuit.
3. The protective circuit initiates a gating signal to the thyristors. Thus, the thyristors are caused to go into conduction. This rapidly reduces the voltage across the thyristors to the on-state value of 1 to 2 volts and prevents damage.

There are two approaches in implementing the protective circuit. One approach is to monitor the voltage across the total series circuit of thyristors of the phase. When the voltage reaches the protective level, a gating signal is initiated to cause the series circuit of thyristors to go into conduction. In the other approach, each thyristor level has its own individual overvoltage protective circuit. The first approach has to allow for the possibility of unequal voltage distribution among the individual thyristor levels particularly when a transient voltage appears across the thyristor switch. Otherwise the transient voltage across an individual voltage level may exceed the thyristor capability before the overvoltage protective circuit across the total thyristor switch is activated.

5.4.5. Variation of Thyristor Controller Losses during Operation

The voltage and current waveforms of a thyristor controller for a TCR system depend on the duration of current conduction during the half cycle. The waveforms shown in Figure 10 are for minimum, maximum, and an intermediate length of current conduction.

The maximum loss dissipation within the thyristor controller occurs during maximum conduction. The snubber circuit losses are then also at their peak because voltage swings on the snubber capacitors are at their greatest magnitude. For the minimum-current condition (Figure 10), both the thyristor and the snubber circuit losses are reduced.

During dynamic operation of the thyristor controller, the losses will vary from half-cycle to half-cycle. In quoting the design parameters of the thyristor controller, a steady-state operating condition at maximum current conduction is often (but not always) specified.

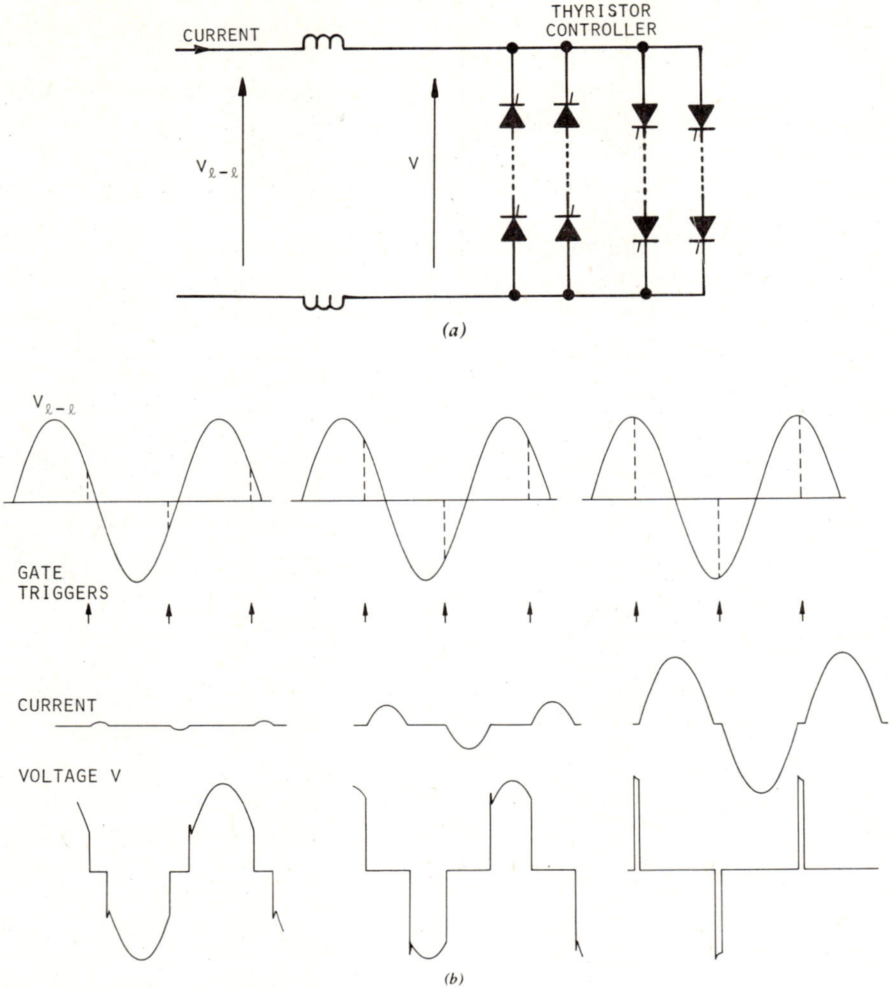

FIGURE 10. (*a*) Circuit for Figure 10*b*. (*b*) Thyristor controller voltage and current waveforms.

5.5. COOLING SYSTEM

The function of the thyristor controller cooling system is to maintain the thyristors and other components within their temperature limits. The cooling medium can be either air or liquid. The cooling arrangement can be varied according to the particular installation requirements. If possible, the cooling system should have redundancy: that is, a failure of a portion of the cooling system should automatically bring into operation standby equipment so that the thyristor controller may continue operation.

The following are descriptions of various cooling systems that have been used on thyristor controllers.

5.5.1. Once-through Filtered Air System

A once-through filtered-air system is a cooling system wherein outside air is drawn through a filter and then through the thyristor controller, and the heated air then exhausted to the outside. Figure 11 shows an air schematic of the cooling system.

The following is a summary of the schematic and the functions of the components.

1. Outside air is drawn through a wall-mounted louver into the building.
2. The air passes through a roll-type filter and then through a bag-type filter (both of these filters are widely used in industrial plants).
3. The air is then drawn into the intake of the operating blower. Only one of the two blowers is normally operating, the other blower being on standby.
4. Air is exhausted by the blower through a counter-balanced backdraft damper into the thyristor controller.
5. The heated air leaves the thyristor controller and is exhausted from the building through a counter-balanced backdraft damper to the outside.

5.5.2. Variations of the Once-through Filtered Air System

When ambient temperatures are well below freezing during the winter season, a set of temperature-controlled dampers can be used to allow some of the exhausted heated air from the thyristor controller to reenter the intake duct of the blower. The controlling temperature can be the intake plenum temperature or the thyristor controller room temperature. This variation allows the thyristor controller room to be held at a comfortable temperature for personnel as well as limiting the minimum temperature for the thyristor controller. It is possible to use a portion of the exhausted heated air from the thyristor controller to heat adjacent rooms through a small wall-mounted damper and fan.

When using the partially recirculated air method, a choice has to be made on the nonenergized position of the dampers. The damper motors should be equipped with a spring return that automatically brings the damper to the nonenergized position when the motor control power is removed. One practice is to have the nonenergized damper position be

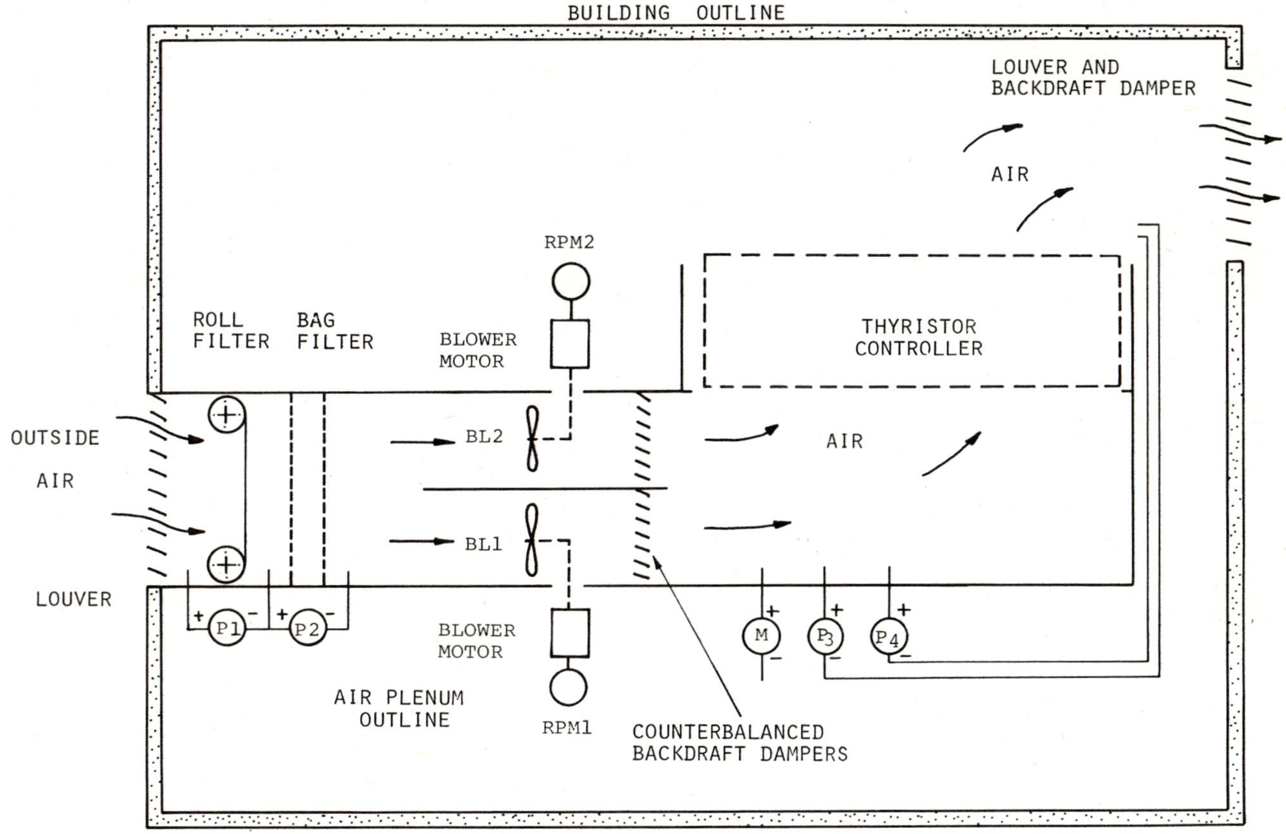

FIGURE 11. Once-through filtered air system.

that for which no recirculation occurs. Therefore, if the control power is inadvertently disconnected during operation, the system is designed such that excessive temperatures do not occur in the thyristor controller room under any abnormal operating conditions.

5.5.3. Recirculated Air System

Thyristor controllers can be built with a completely recirculated air system. The heat is removed from the circulating air by water coils. A make-up air system is often used to maintain a positive air pressure within the thyristor controller room with respect to the outside, to prevent excessive dust leaking into the room.

5.5.4. Liquid Cooling System

Thyristor controller cooling can be done by liquid cooling. The thyristors and other heat-dissipating components are mounted on liquid-cooled heat sinks. A standby coolant pump should be available to allow continuous operation in case the running pump malfunctions. Automatic changeover from the running pump to the standby pump can be done by using check valves. The cooling media most usually considered are water and Freon. The use of liquid cooling requires the controller design to be safe in the unlikely event of a coolant leak. However, this technology has been successfully developed and it is likely that in future the trend will be towards the use of liquid cooling.

5.5.5. General Comments on Cooling Systems

The choice of the type of cooling system is influenced by the ambient conditions and the user's preference. An air-to-water heat exchanger is often used when the ambient air conditions would require frequent changing or cleaning of the filters. A liquid-cooled thyristor controller offers a lower cooling-system power loss than the air-cooled system. The effect of possible liquid leakage within the high-voltage thyristor controller must be considered, however. The advantage of an air-cooled system is its simplicity in operation and maintenance.

5.6. AN EXAMPLE OF A THYRISTOR CONTROLLER

Figure 12 is a photograph of an operating air-cooled thyristor controller. The light guides are shown near the center of the structure. The cooling air is drawn in from a plenum below the structure and the heated air is exhausted from the top of the structure. The heated air leaves the room through the dampers shown at the end of the room.

Figure 13 is a photograph of a similar thyristor controller with one of

FIGURE 12. Air-cooled thyristor controller, 13.8 kV, 40 MVAr.

FIGURE 13. Air-cooled thyristor controller with power module partly withdrawn.

its power modules partially withdrawn. The air-cooled thyristors are shown near the center of the module.

REFERENCES

1. *General Electric SCR Manual,* 6th Ed., General Electric, Auburn, New York, 1979.
2. S. K. Gandhi, *Semiconductor Power Devices,* Wiley, New York, 1977.
3. V. R. Stefanovic, "Thyristor Rectifiers and Synchronous Inverters," IEEE Tutorial *Power Electronics Applications in Power Systems,* Publ. 78EH0135-4-PWR, 1978.
4. W. McMurray, "Optimum Snubbers for Power Semiconductors," *IEEE Trans. Industry Appl.* **8**, 593–600, (September/October 1972).
5. S. G. Dewan and A. Straughen, *Power Semiconductor Circuits,* Wiley, New York, 1975.
6. Gentry, et al., *Semiconductor Controlled Rectifiers — Principles and Application of p-n-p-n Devices,* Prentice-Hall, New Jersey, 1964.

Chapter 6

AN EXAMPLE OF A MODERN STATIC COMPENSATOR

R. W. LYE

6.1. INTRODUCTION

This chapter provides a detailed description of a modern, thyristor-controlled static compensator, the Rimouski compensator installed on the transmission network of Hydro Québec at 230 kV. The compensator is typical of many such installations on high-voltage transmission systems, but many of its design features are reproduced in load compensators also, particularly in supplies to electric arc furnaces. The thyristor controllers, reactors, and capacitors are essentially the same in both cases. The main differences are in the control strategy and the system voltage.

The Hydro-Québec system has many long-distance, high-voltage transmission lines. Prior to 1978 synchronous condensers were installed to provide reactive compensation. The Baie James project, a major extension of the system exploiting hydropower resources in the north of the province, will involve the transmission of more than 11,000 MW of power at 735 kV over a distance of about 1000 km, and it will have dynamic shunt compensation in the form of static compensators at five locations. Planning studies, which considered various alternative forms of compensation, led to the decision to install two static compensators for performance evaluation, at locations not on the Baie James system. One of these was installed near Rimouski, Québec, on the 230-kV system of the Gaspé region. It was commissioned in 1978 and serves as a representative example of a transmission system compensator.

6.2. BASIC ARRANGEMENT

The compensator is a thyristor-controlled reactor (TCR) with a fixed capacitor. Figure 1 is a simplified one-line diagram of the main com-

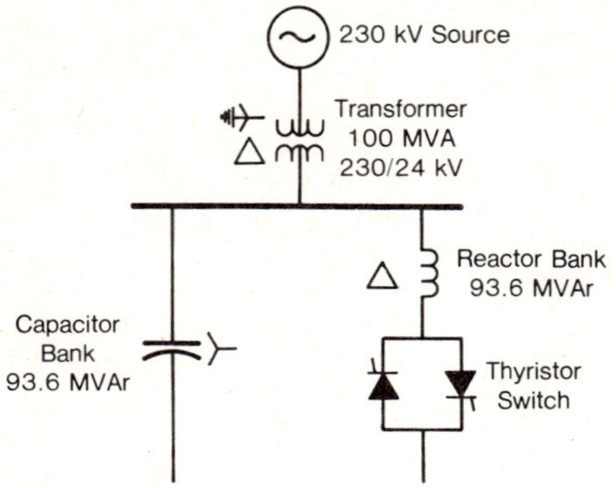

FIGURE 1. Basic circuit of TCR/FC compensator.

ponents. The compensator is connected to the system through a 100-MVA 230/24-kV step-down transformer. On the secondary side, a three-phase wye-connected capacitor bank rated 93.6 MVAr is paralleled with the delta-connected TCR. The rating of the reactor bank is also 93.6 MVAr.

The voltage/current characteristic of the compensator at the 230-kV bus is shown in Figure 2. The slope of the control range is nominally 3%. This means that a voltage change of −3% produces the rated capacitive reactive power of 84 MVAr. For a linear voltage/current characteristic, a

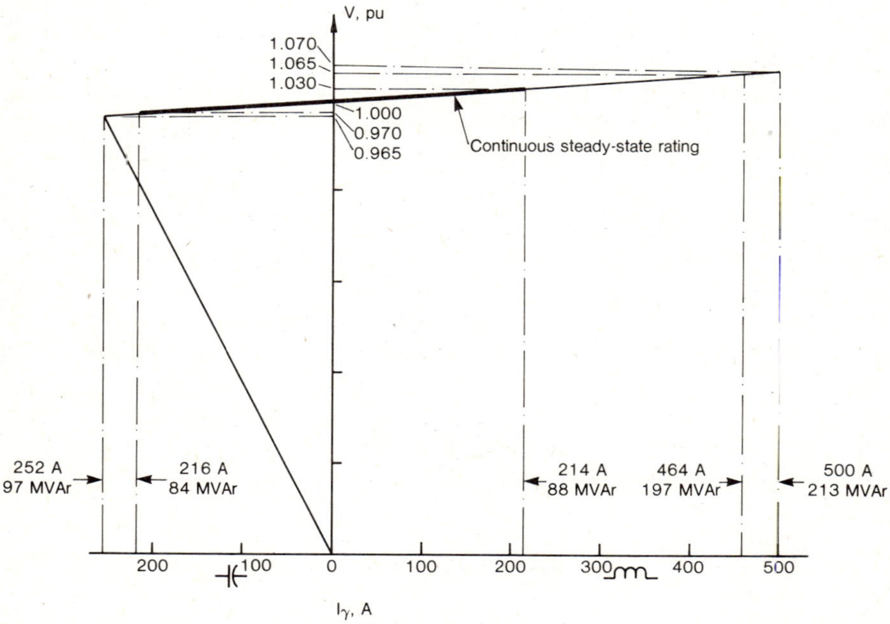

FIGURE 2. Voltage/current characteristic.

voltage change of +3% produces an inductive reactive power of 88 MVAr (Figure 2). This slope was chosen as a result of system studies.

The nominal rating of the compensator is summarized in Table 1, and shown graphically in Figure 3 in terms of a hypothetical loading cycle which would utilize the compensator's full capability. This loading cycle was derived from system performance studies including load rejection and fault recovery studies.

TABLE 1
Nominal Rating of Rimouski Compensator

Reactive Power + = Generation − = Absorption	230-kV Bus Voltage (pu)	Duration
+84	0.97	Continuous
0	1.00	Continuous
−120	1.04	5 min
−200	1.065	1 sec

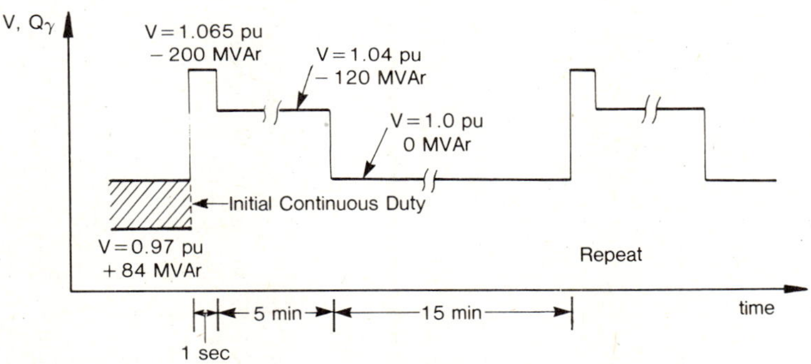

FIGURE 3. Compensator rated loading cycle.

6.3. DESCRIPTION OF MAIN COMPONENTS

The compensator layout in the substation is shown in the plan drawing, Figure 4, and in the photograph, Figure 5.

Main Power Transformer. This is rated 60/80/100 MVA with ONAN/ONAF/ONAF cooling; 65°C temperature rise; 230-kV wye, 24-kV delta; three-phase; 60 Hz. The impedance is 11%. A load tap-changer is connected in the HV neutral and provides ±10% adjustment in 33 steps. The tap-changer automatically regulates the 24-kV bus voltage with a response time of about 3 sec per step.

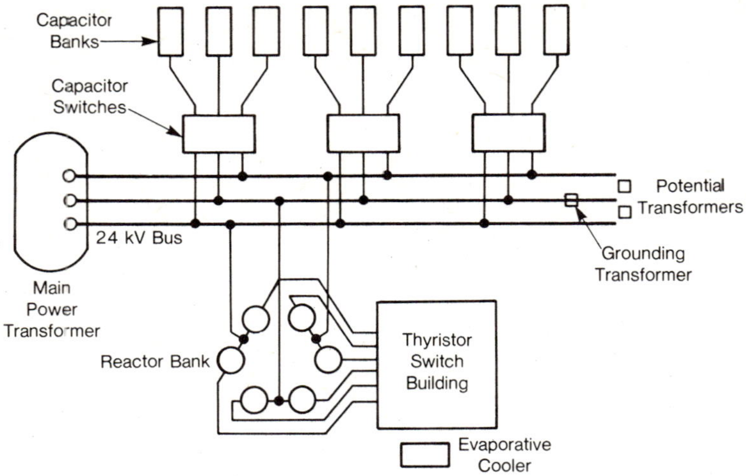

FIGURE 4. Station layout.

FIGURE 5. Rimouski station showing static compensator building on right, with reactor bank in foreground.

Capacitor Banks. The capacitor bank of 93.6 MVAr is divided into three 3-phase groups, each of 31.2 MVAr, separately switched and connected at 24 kV in ungrounded double-wye (Figure 6). A CT is con-

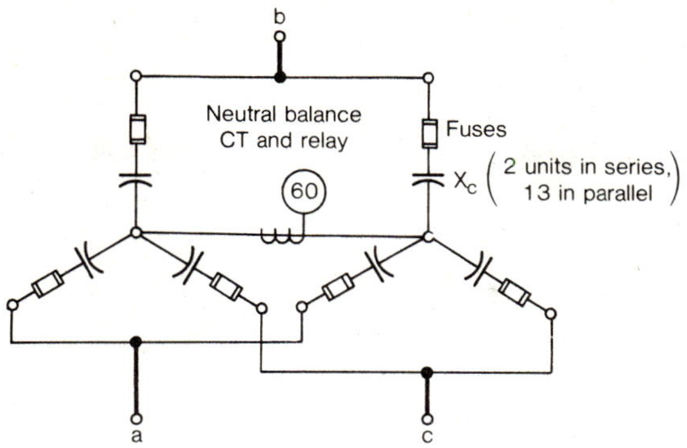

FIGURE 6. 31.2-MVAr capacitor bank.

nected between the neutral points of the two wyes to detect unbalance. Each phase of each wye consists of 26 capacitor units arranged in two series groups, each having 13 units in parallel. Each unit is rated 200 kVAr, 6930 V.

Capacitor Switches. Each wye of each capacitor bank is switched with a three-phase, 1000-A, 34.5-kV vacuum switch. A reactor of 30 μH is connected in series with each pole to limit di/dt. The switches are operated manually, or by certain protective relays, but are never operated by the compensator voltage regulator (as would be the case in a hybrid compensator; see Chapter 4).

Reactor Bank. The main power reactor bank consists of six air-core reactors shown in Figure 5 and connected as in Figure 7. They are laid out in a hexagonal plan (Figure 4). This arrangement is designed to equalize the magnetic coupling between the reactors. Each reactor has a reactance of 2.49 Ohms at 60 Hz and is rated 1300 A continuously; 3800 A for 1 sec; and 2790 A for 5 min. These are the currents required for the rated loading cycle (Figure 3). The values are rms values for the fundamental component. The reactors also carry the harmonic components 3rd, 5th, 7th, and so on (see Chapter 4).

Thyristor Controller. The thyristor controller is of the type discussed in Chapters 4 and 5 (see Figure 7) and the current rating is the same as that of the reactor bank. Each phase of the controller consists of four thyristors in parallel for each polarity and 36 in series. The thyristors are mounted on insulating panels, together with heat sinks, resistor-capacitor damper circuits (snubbers) and gate drive circuits. A lamp monitors the voltage across each series voltage level. The thyristor panels are all mounted in a steel cubicle which forms a support structure, an air plenum chamber, and a "deadfront" grounded safety barrier.

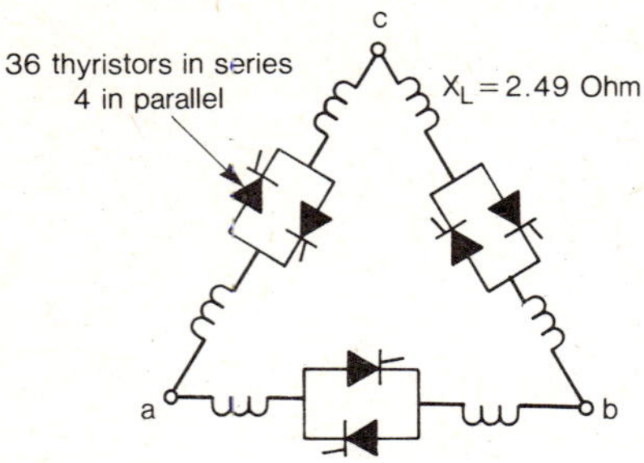

FIGURE 7. Reactor bank connections.

Thyristor Controller Building. This contains the three-phase thyristor controller together with its control system and protection; the station service supplies; and the thyristor cooling system (except for the evaporative cooler, which is located outdoors nearby). The building is metal-sheathed, thermally insulated, and can be electrically heated when necessary. Inside dimensions are 40 ft × 37 ft × 25 ft high.

Thyristor Switch Cooling. The thyristors are forced-air cooled. The closed-cycle air recirculates after passing through air-glycol heat exchangers. The glycol is in turn recirculated through an evaporative cooler. Figure 8 shows the cooling system schematically. The total air flow required is 100,000 CFM which is supplied by six fans, each rated for 20,000 CFM. One of the six fans is redundant, for reliability. Each

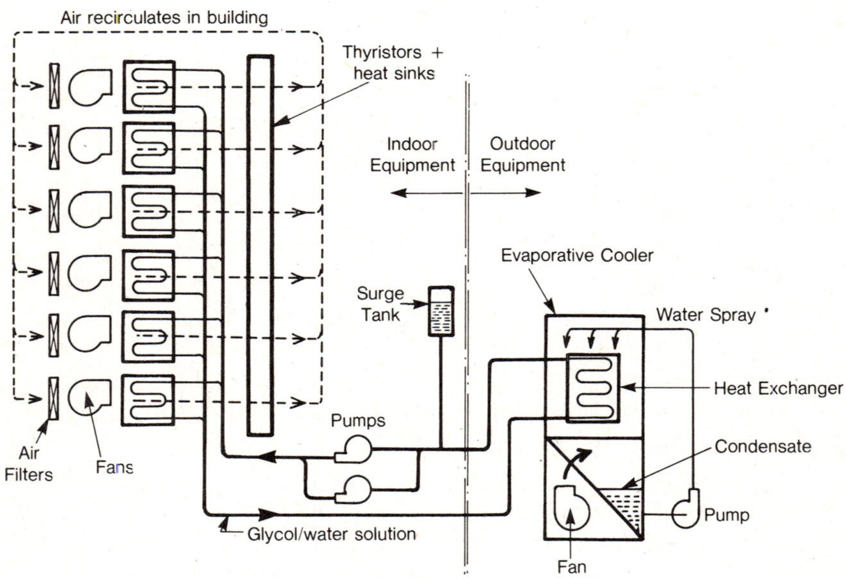

FIGURE 8. Schematic of cooling system.

fan has its own associated air-glycol heat exchanger and one of these also is redundant. In case the glycol system or evaporative cooling system must be shut down for any reason, an emergency system is provided whereby outside air is blown through the building by wall fans. This emergency system is satisfactory for outside air temperatures of up to about 30°C.

Compensator Protection (Figure 9). The 230-kV and 24-kV buses are protected by voltage differential relays with overcurrent back-up. The capacitor banks employ neutral unbalance current detection to avoid capaci-

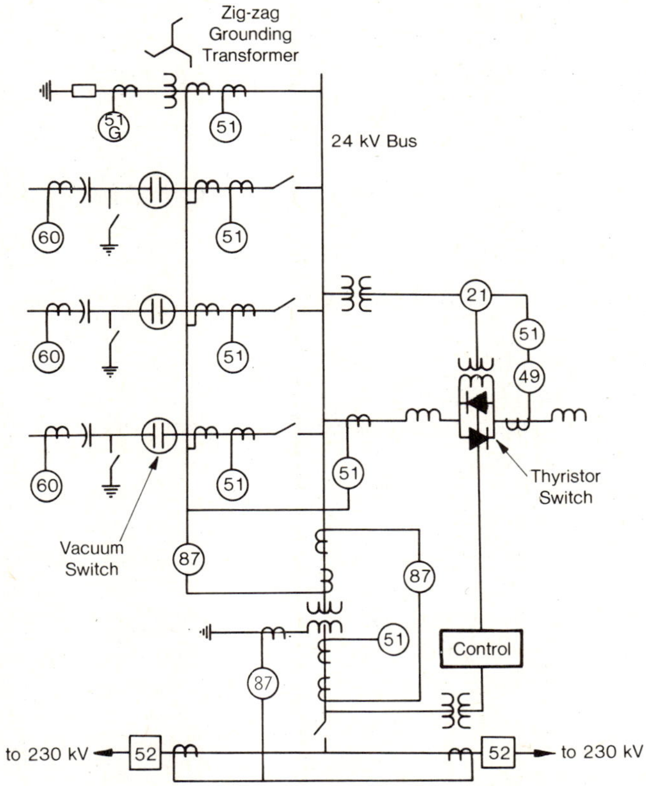

FIGURE 9. One-line diagram showing protective relay system. Protective relay device function numbers from ANSI Standard C37.2-1970 (as used in this figure): 21, distance relay, functions when circuit impedance increases or decreases beyond predetermined limits; 49, thermal relay, functions when the temperature of a power rectifier exceeds a predetermined value; 51, ac time overcurrent relay; 51G, ac time overcurrent relay monitoring ground current; 52, ac circuit breaker; 60, current balance relay, functions on a given difference in current between two circuits; 87, differential protective relay, functions on a difference of two currents or some other electrical quantity.

tor unit overvoltages due to failed capacitor elements. Fault protection is provided by inverse time/overcurrent relays.

Ground faults on the entire 24-kV system will be detected in the neutral of the zig-zag grounding transformer, and if not cleared by one of the primary differential systems will result in a shutdown via the ground overcurrent relays.

Fault currents associated with the power reactors and thyristor controller are detected by inverse overcurrent relays, and if the thyristor controller is unable to turn itself off and eliminate the fault, the compensator will be tripped off-line by the 230-kV circuit breakers.

Overload protection for the thyristor controller includes continuous monitoring of glycol temperature and flow, air temperature and flow, and a special thermal overload relay which monitors thyristor junction temperature. This relay is programmed with the thyristor heating/cooling characteristics and can calculate junction temperature from a measurement of thyristor current.

6.4. CONTROL SYSTEM OF THYRISTOR CONTROLLER

Referring to the block diagram in Figure 10, the control system consists basically of a voltage regulator, a current-limit, and thyristor gating circuit.

The *voltage regulator* is a conventional closed-loop system, with feedback from potential transformers on the 230-kV bus. The feedback is three-phase ac rectified, and so represents an average of the three-phase 230-kV system voltage.

The *current-limit* circuit is also closed-loop, with feedback from current transformers in the reactor circuits. Three-phase currents are summed and rectified to represent average reactor current. The value of the current limit is set by a *load cycle control* circuit. The load cycle control ensures that the time-current rating of the thyristor controller is not exceeded. To perform this function, electronic circuits monitor the ampere-second integrals applied to the thyristors. The load cycle control should ensure that the allowed thyristor junction temperature is never exceeded. However, a thermal overload relay is also monitoring the thyristor junction temperatures. If a malfunction of the load cycle control allows the junction temperature to exceed the allowed value, then the overload relay will cause a shutdown of the compensator.

The thyristor gating circuit begins with an error signal from the voltage regulator, which determines at what point in time the thyristors are to be gated, and ends with current pulses being delivered to the gates of the thyristors.

The phase control block shown in Figure 10 compares the voltage error signal to timing ramps generated by the voltage across the thyristors. At

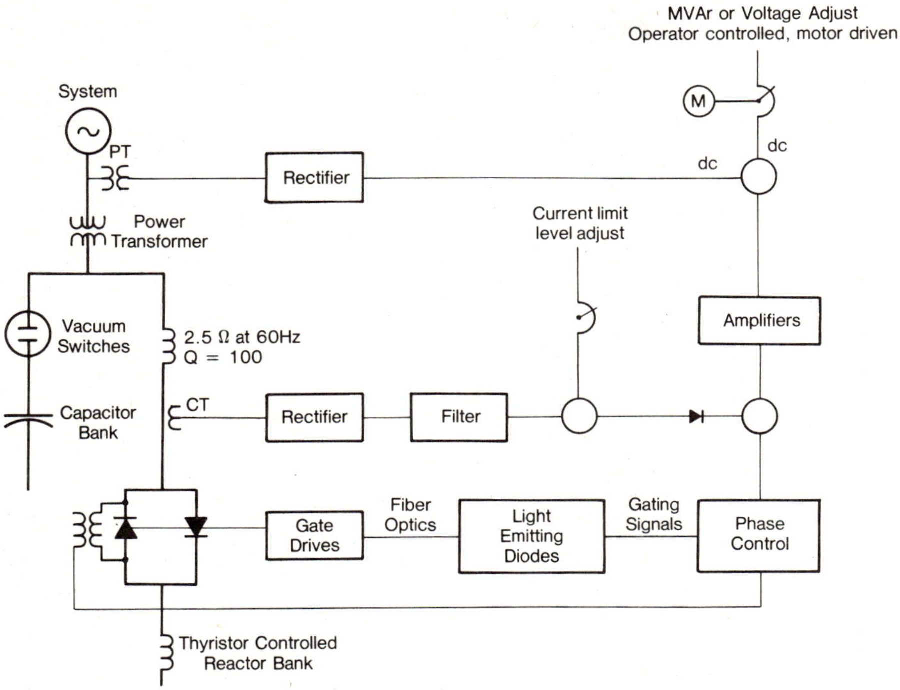

FIGURE 10. Control block diagram.

the crossover of the ramp and error signal a gating signal is generated. This energizes light emitting diodes (LED's) which transmit light through fiber optics (light guides) to the gate drives. The light guides form the insulation between the control circuits and the thyristors. The gate drives convert the light signal to a gate current pulse. Energy for the gate drives is obtained from either voltage across the thyristors or current in the thyristors, whichever is available at a given time (see Chapter 5).

The voltage/current slope of 3% is obtained by setting the proper relationship between the voltage error signal and the gating instants of the thyristors. The thyristor gating instants determine the current in the reactors and thus the total reactive current of the compensator. The compensator current is not a linear function of the gating angle, so linearizing circuits are included (see Chapter 4). The result is a linear relationship between system voltage change (voltage error) and compensator current.

Start-Stop Control.

(a) Starting. The compensator can be started either from the thyristor controller building (Local) or from the substation control room (Remote). An automatic starting sequence is provided for starting from ei-

ther location (Auto Start). In addition all the starting functions can be performed individually by control switches in the thyristor controller building (Manual Start). The normal starting mode is Remote–Auto. The Local–Auto or Local–Manual modes are used only for maintenance procedures or in an emergency.

The automatic starting sequence has two main sections. The first starts all the auxiliary motors in an orderly timed sequence. The motors operate fans and pumps associated with the thyristor cooling system. The second section connects the compensator to the 230-kV bus in a manner which causes the least disturbance to the system.

The starting sequence can be summarized as the following steps:

1. Start auxiliary motors.
2. Close 230-kV breakers.
3. Remove the gating suppression signal from the thyristors. The reactor will be energized and if the operator's setting called for the existing system voltage, the voltage regulator will call for zero current in the reactor.
4. Close the capacitor bank switches. For switching purposes the three sections of the capacitor bank are energized in sequence. As each section is energized, the voltage regulator automatically adjusts the reactor current to maintain zero net current from the compensator.

(b) Stopping. A complete stop, or shut-down, can be accomplished by operator control from either Local or Remote locations, or by various protective relays. Whether the stop is initiated manually or by a protective relay, the following occurs in quick sequence:

1. Trip 230-kV circuit breakers.
2. Trip all capacitor bank vacuum switches.
3. Suppress gating of thyristors.
4. Stop cooling system motors.

6.5. PERFORMANCE TESTING

An extensive series of tests was made, during and after commissioning, to check the performance of the compensator. These tests included measurements of:

Regulator transfer functions.

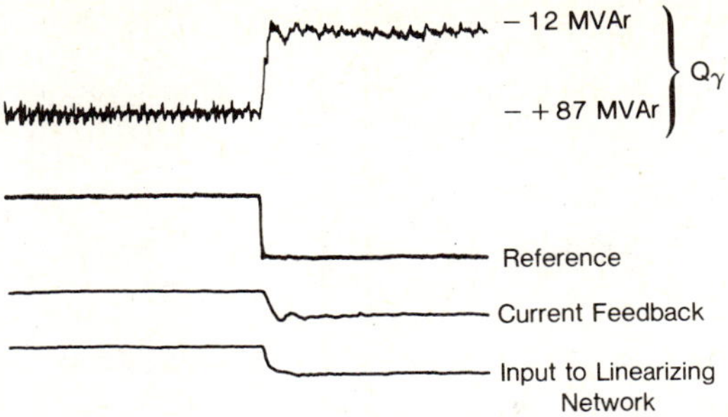

FIGURE 11. Compensator performance. Sudden change of −99 MVAr in response to a step change in reference signal.

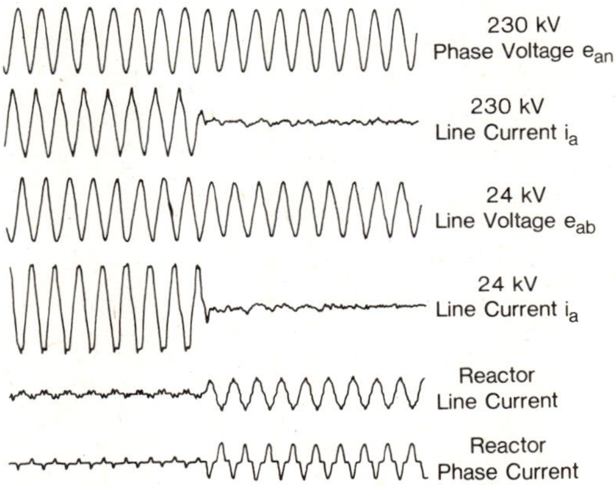

FIGURE 12. Compensator performance. Voltage and current waveforms corresponding to Figure 11.

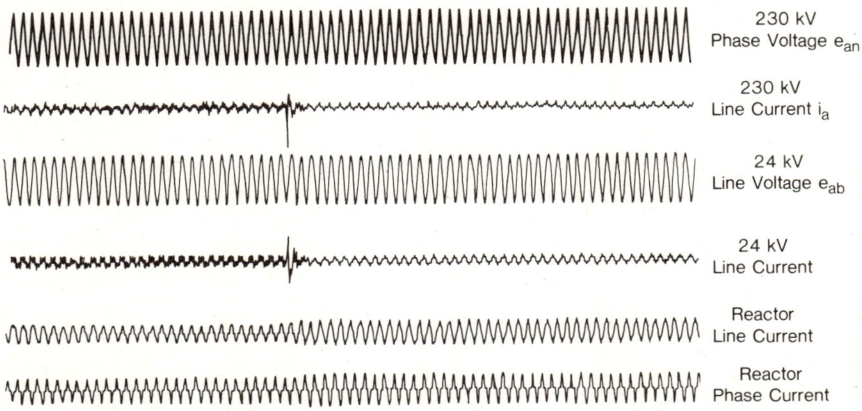

FIGURE 13. Compensator performance. Energizing No. 3 capacitor bank, producing a sudden change of 31.2 MVAr.

Accuracy of phase-to-phase spacing of thyristor gating (commonly known as "tracking").

Slope of V/I and V/Q characteristics.

Harmonic content of reactor current, 230-kV and 24-kV bus voltages, 230-kV and 24-kV line currents to compensator.

Oscillograms were taken of a number of response tests including:

Injection of sinusoidal reference signals at frequencies from 0.01 to 10 Hz.

Injection of step changes of reference. These were of various magnitudes and from various starting points, to fully test the response of the compensator. Some representative examples are shown in Figures 11 through 13. It can be observed that the response time of the reactor current and the reactive power of the compensator is about 1 cycle. Also the response is well damped with very little overshoot.

Finally the loading cycle of Figure 3 was performed. This was done by injecting a continuous high-reference signal into the control causing the compensator to go to the −200 MVAr current-limit. The load cycle control allows this for 1 sec, then allows −120 MVAr for 5 min, then imposes a limit at the continuous rated current.

Chapter 7

SERIES CAPACITORS

L. E. BOCK

7.1. INTRODUCTION

Rights of way for transmission lines have become increasingly difficult and costly to obtain. This, coupled with concerns for our energy resources and spiraling energy costs, forces utilities to operate transmission systems at maximum possible efficiency.

Both series and shunt capacitor banks are useful tools in improving system efficiency and power transfer capability. Both have had growth rates significantly greater than the rate of growth of active power generation. Shunt capacitors generate reactive power and help to reduce the amount of reactive power that flows through the network. For maximum effectiveness in reducing line losses and voltage control, they are typically installed near the load. Series capacitors are applied to compensate for the inductive reactance of the transmission line. They may be installed remote from the load, as for example at an intermediate point on a long transmission line. Their benefits include:

1. Improved system steady-state stability.
2. Improved system transient stability.
3. Better load division on parallel circuits.
4. Reduced voltage drops in load areas during severe disturbances.
5. Reduced transmission losses.
6. Better adjustment of line loadings.

The theory of series capacitors as a means of "line length compensation" has been developed in Chapter 2. In this chapter we focus attention on the design, application, and performance of series capacitors.

7.2. HISTORY

One of the first installations of a series capacitor was in 1928 at a New York Power & Light substation in Ballston Spa, New York. This was a 1.25 MVAr bank on a 33-kV circuit using capacitor units rated 10 kVAr each, and it was applied to control the division of load between parallel circuits.

Since that time, successful installations have been made at line voltages up to 550 kV, with bank ratings of up to 800 MVAr. Capacitor unit ratings likewise have increased, so that ratings essentially the same as those for shunt applications can be provided. Figure 1 shows the growth in installed capacitor unit size in series capacitor installations since 1920.

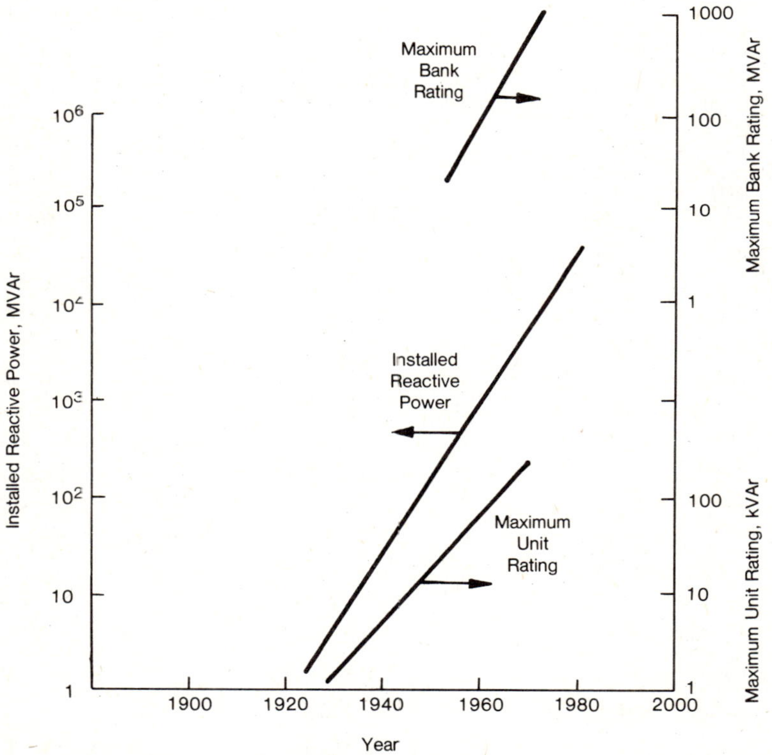

FIGURE 1. Series capacitor trends since 1920, showing sustained growth in installed capacity, maximum bank size, and unit size.

7.3. GENERAL EQUIPMENT DESIGN

7.3.1. Capacitor Units

Capacitor design economics still dictate the use of individual *units* assembled in appropriate series and parallel connected groups to obtain the desired bank voltage and reactive power ratings, both in shunt and series capacitor equipments. Figure 2 is a cutaway view of a typical power capacitor unit. Series capacitor duty usually requires that a unit designed for a series application be more conservatively rated than a shunt unit, but basically there is no significant construction difference.

Since the first production of power capacitors in 1914, the construction has undergone many stages of improvement. The introduction of thin Kraft paper to replace linen, and askarel to replace oil in the early 1930s, made possible individual unit ratings up to 15 kVAr. By the early 1960s the 100-kVAr rating was introduced, evolving from a series of comparatively small, very costly refinements in the basic paper/askarel dielectric. In 1965 General Electric designed the Magvar unit, a 150-kVAr unit using a paper/polypropylene film/askarel dielectric system. Further refinements led to single unit ratings up to 600 kVAr, although the most economical unit rating today is in the 200–300 kVAr size. Replacement of askarel in 1976 with non-PCB fluids did not have an immediate significant effect on unit sizes or ratings. The newer all-polypropylene film dielectric units offer distinct advantages in reduced losses and probability of case rupture, as well as some improvements in unit size and rating. Table 1 illustrates the effects of these changes on the capacitor unit ratings, volume, density, and price. Without the advent of these larger capacitor unit ratings, the physical equipment size and site area requirements would make the larger shunt and series banks impractical or uneconomical.

7.3.2. Fusing

In the United States most series as well as shunt capacitor banks have been designed using external fuses, one for each unit. This contrasts with the widespread use of internally fused units in many other places. Internally fused units have the advantage that failure of a single element or "roll" within the unit does not cause the entire unit to fail. However, since there is no visual indication of a blown fuse, frequent maintenance checks involving detailed and accurate capacitance measurement are

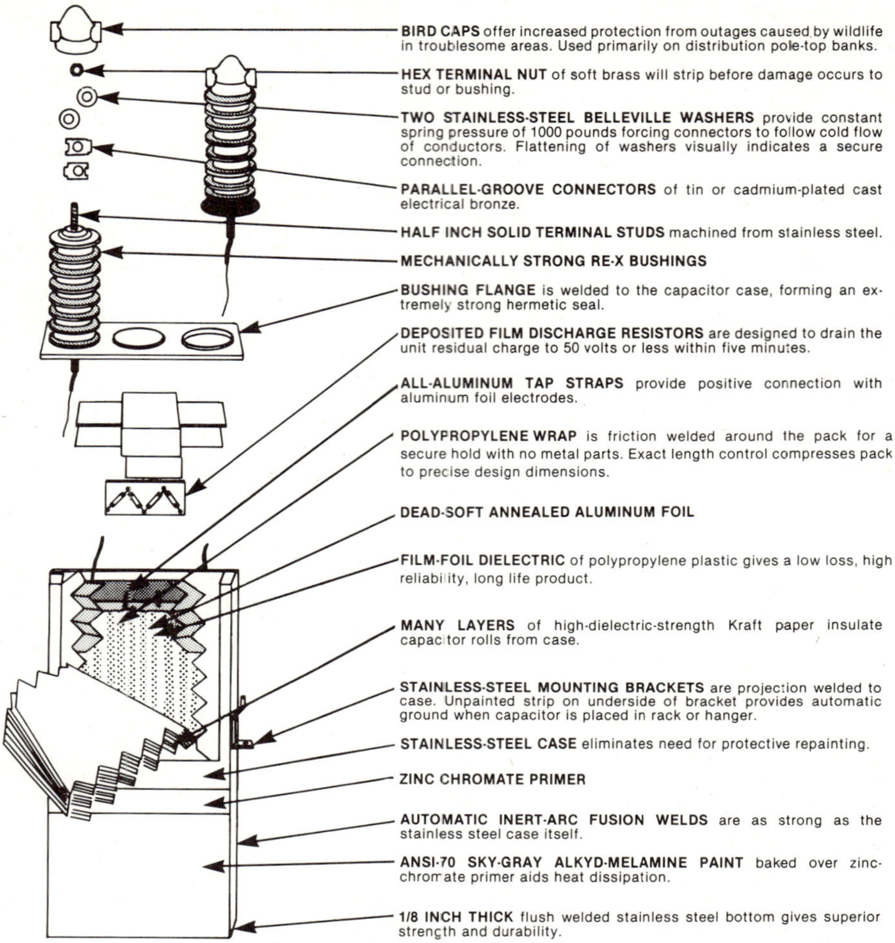

FIGURE 2. Typical power capacitor unit, for either shunt or series application.

TABLE 1
Effect of Design Improvements on Capacitor Unit Characteristics

Date	Maximum Unit Rating (kVAr)	Volume (in^3/kVAr)	Density (lb/kVAr)	Price ($/kVAr)
1929	10	130	8.6	18.70
1930	10	65	4.4	18.20
1931	15	43	3.2	16.00
1959	50	32	2.0	3.30
1962	100	19	1.2	1.50
1965	150	11	0.7	1.70
1977	300	10	0.6	1.70
1980	300	8	0.4	1.80

required. These are usually performed annually. Use of internal fuses also limits the voltage rating of the capacitor, typically to 5 kV or less, so that there is less flexibility in the overall bank design. This voltage restriction is due to limits in the number of parallel·connected elements within the unit to keep the overvoltage on remaining elements reasonably low when one element fails. There are, however, several 500-kV series capacitor equipments with internal fuses that have been in service successfully for over 10 years.

External current-limiting fuses, with a physical disconnect to avoid trackover and provide a visual indication of operation, are commonly specified. Suitable ratings of this type of fuse for series capacitor application have been available since the early 1970s. These fuses must be capable of clearing at voltages up to the maximum gap sparkover levels, often three times rated voltage or more, and must be sized to withstand all transient overload and discharge duties without damage.

7.3.3. Compensation Factors

Line compensation becomes economical for distances typically more than 200 mi, although series capacitors can be found in shorter sections of line. Economic loading for longer distances is in the neighborhood of the surge impedance loading of the line (see Chapter 2 and Reference 1).

If resistance is neglected, maximum ac power transfer is given by:

$$P_{\max} \sim \frac{E_1 E_2}{X_L - X_C}, \tag{1}$$

where E_1 and E_2 are the line terminal voltage magnitudes. X_L is the inductive reactance between the terminal voltages; X_C is the reactance of the compensating capacitors. Typical EHV applications are in the 25–70% compensation range. There are other factors that may place limits on the compensation level, such as subharmonic stability, subsynchronous resonance, and switching transients.

7.3.4. Physical Arrangement

Series capacitor banks are built in one or more discrete segments or steps per phase, each with several components mounted on a platform insulated from ground. Small distribution class banks may use ground-mounted hardware, but the larger transmission class equipments are all platform-mounted.

Figure 3 is a photograph of an EHV series capacitor equipment, rated approximately 318 MVAr. The bank is located on the Pacific NW-SW Intertie at Pacific Gas and Electric's Tesla substation near Tracy, California, and was energized in January 1968. The dimensions of the bank are about 95 × 124 × 41 feet high. With the improvements previously men-

FIGURE 3. 318-MVAr, 500-kV transmission-line series capacitor.

tioned in capacitor and fuse design, it is now possible to approximately double the rated bank reactive power in the same space. Each phase has four separate segments, with two platforms per phase.

Prior to an earthquake in February 1971 that caused considerable damage to the Sylmar substation of the Los Angeles Department of Water and Power, specifications for series capacitor equipments usually required no more than a 0.2g static withstand. Since then, however, requirements have become more stringent for equipments located in seismically active areas, particularly in the western United States.

Specifications for such equipments now include a ground motion response spectrum and dynamic analysis, with a base acceleration of typically 0.5g. EHV equipments cannot practically be designed as rigid structures with natural resonant frequencies above about 25 Hz, so the designer must take into account frequency and damping in the support structure construction. Requirements on vertical acceleration have also been added, usually specified as 70–80% of the horizontal value. These more demanding specifications substantially influence the cost of the structure.

7.4. PROTECTIVE GEAR

Since the cost increases roughly in proportion to the square of the current rating, series capacitor equipment cannot be economically designed to withstand all abnormal voltages produced by all possible excessive line and fault currents. The equipment is therefore designed to withstand certain less severe abnormal voltages and currents that occur during some system disturbances, particularly when series capacitors are functionally necessary at the time, (for example, when they are necessary to maintain transient stability). Although there are industry standards for series capacitors (ANSI C55.2 and IEC-143), each application is sufficiently different that equipment design is specifically tailored for each case after thorough studies on the transient network analyzer or by computer.

Equipment is provided to automatically and instantaneously bypass segments when design voltage levels are exceeded, as might occur during a line fault. Traditionally this has been accomplished by sparkover or triggering of a gap in parallel with the capacitor. New zinc-oxide varistor technology offers improvements over this method of overvoltage protection, and is discussed later. After fault clearing, control circuits initiate the reinsertion of the capacitor. For equipments in unfaulted line sections this means that the protective gear must interrupt current and withstand without sparkover a *combination* of voltages generated by reinsertion of the bypassed capacitor bank and the electromechanical system disturbance.(2)

For a bank which has been installed primarily to improve transient stability, rapid reinsertion after fault clearing and rapid fault clearing time are both essential for substantially increased line loading capability, as illustrated in Figure 4.(3) The results shown are based on two parallel 500-kV lines with 50% series compensation. For the case where the fault is cleared in five cycles, the effects on line loading capability for reinsertion delays of five and eight cycles are also shown. [Line loading capability here means the maximum prefault power transfer for which the system has transient stability (Chapter 3) following the fault indicated in Figure 4].

Figure 5 is a schematic diagram of a typical segment of a series capacitor bank. The capacitor groups are arranged in two parallel strings. A sensing circuit is included to compare the currents in the two strings and detect any failed capacitor units. Often two levels of pickup are used in a differential circuit, one to give an alarm when only a slight unbalance exists, and one to close the bypass switch when the overvoltage on the remaining units is excessive. If possible, an alarm is signaled for one blown fuse, and a bypass is initiated for overvoltages usually above 10%. Where there is a large number of parallel units, control circuits may not be sensitive enough to detect a single blown fuse, but then these conditions also give smaller overvoltages. Transmission class equipments typically have 40–80 total parallel connected units per group. An alternative unbalance protective scheme is to use potential transformers to compare voltages on the capacitors arranged in two series groups, rather than comparing currents in two parallel strings.

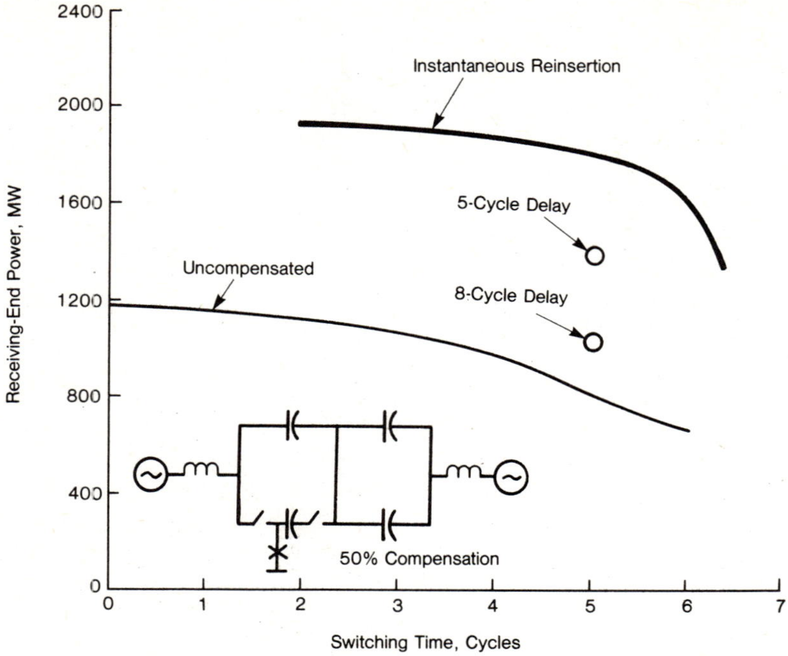

FIGURE 4. Increase of line loading capability with rapid reinsertion.

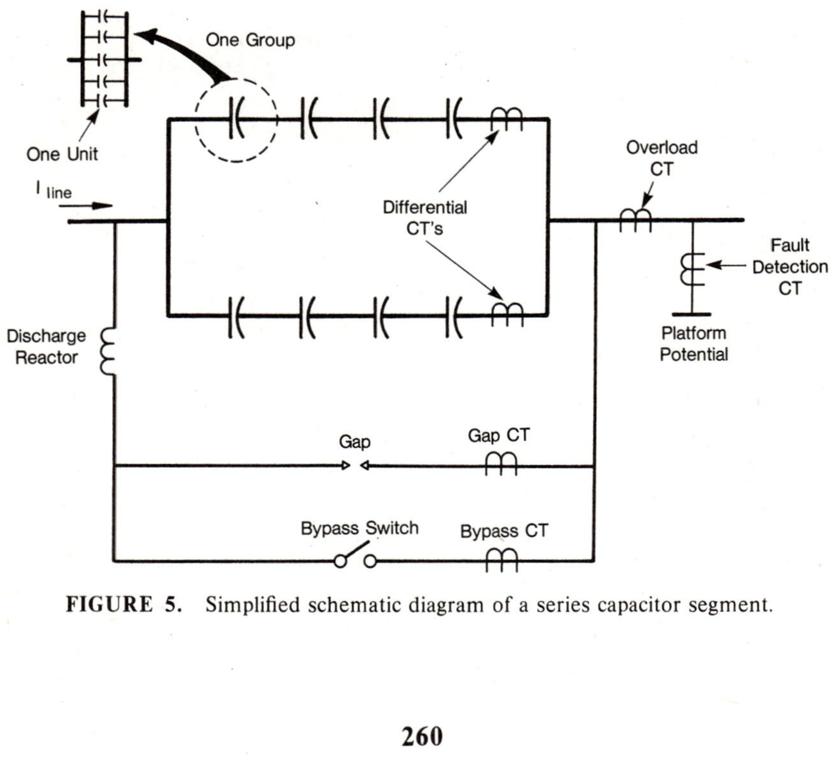

FIGURE 5. Simplified schematic diagram of a series capacitor segment.

The main gap is set to spark over at a preset level to protect the capacitors during a line fault. Usual protective values are 2.5–4 times normal operating voltages. A typical control sequence is for the presence of gap current to be detected by the gap CT, and a signal sent to automatically close the bypass switch. When the switch closes, current is diverted from the gap, and when the line current as measured by the switch CT returns to normal a signal is given to automatically open the switch, so reinserting the capacitor into the line. The above sequence is used for a nonself-clearing gap. A variation for an air-blast gap is to monitor line current with the gap CT, and turn on the compressed air to extinguish the gap when the line fault is cleared. Reinsertion time following fault clearing is dependent on switch operating speeds, control circuit timing, and gap deionization and recovery withstand time.

For air-blast gaps, faster reinsertion is obtained if the air is turned on as soon as gap current is sensed. The disadvantage of this procedure is that the gap current will extinguish at every current zero, and reignition at voltages above the sparkover setting usually follows due to increased air pressure in the gap. This is a more severe duty on the capacitor than a single sparkover followed by a single reinsertion.

The discharge reactor is necessary to limit the magnitude and frequency of the current through the capacitor when the gap sparks over. This prevents damage to the capacitor units and fuses as well as easing the duty on the gap. In Figure 5 the reactor is shown in the bypass circuit. It is sometimes located in series with the capacitor such that it still limits the discharge, but does not have to withstand the full power frequency fault current.

Overload protection is usually provided to protect the capacitor from a sustained overload below the gap sparkover setting. Frequently this is a CT with several current relays and appropriate timers to approximate the time-current withstand capability of the capacitor. Industry standards specify overload capabilities for series capacitors (higher than for shunt units), but again each application generally has unique requirements. Operation of this circuit signals the bypass switch to close and protect the capacitor. The switch will automatically open to reinsert the capacitor when the overload circuit senses that the current has returned to a safe level. A thermal analog to simulate capacitor unit temperature is sometimes used in addition to the shorter-time/higher-current protection of the timers and relays. This then protects for relatively low overcurrents (less than about 150%), taking ambient temperature into account. Overload circuits are normally provided on only one phase, and operation causes bypass of the complete bank. This assumes that phase currents will be essentially balanced during system conditions leading to sustained overcurrents.

One end of the capacitor segment is usually connected to the platform. This keeps the platform from floating at some unknown potential, and also allows the use of CT's rated at lesser voltage, as well as reduced BIL on one side of the platform-mounted gap and switch. A fault detection CT located in the connection to the platform is used to sense an insulation failure or sparkover from any point on the equipment to the platform. Operation of this circuit will signal the bypass switch to close. Insulation between components on the platform must be carefully selected to assure withstand at all transient voltage levels, particularly the maximum gap sparkover level. IEC Standards relate all insulator BIL ratings to gap sparkover, whereas U.S. standards only specify values for capacitor units.

Many of the operations previously discussed require that a signal be sent from the platform to the ground control circuits, either to signal an alarm or to relay a command to close the corresponding bypass switches in the other phases to maintain system balance. Communication between platform and ground is usually made through a signal column, whose channels can be solenoid-operated insulated rods, coiled compressed air tubing, or fiber optics. Equipments are in service with all three types of communications.

Usually a source of power on the platform is also required to operate the bypass switch and the platform control circuits. This can be compressed air with a compressor at ground level, or a platform-mounted battery with a charging system using line current as a primary source. A prototype system using solar power to charge the battery is also in service. It is normal practice to detect low platform power and close the bypass switch. Often an intermediate power level is also detected to send an alarm to ground.

Except for response to a gap sparkover, bypass switch commands are normally made on a three-phase basis to maintain system balance. This would include operation due to capacitor differential, overload, platform fault, loss of platform power, and remote manual operation. Automatic switch closings, except those due to gap sparkover and overloads, are usually designed to lock the switch in the closed position until the problem is corrected and the switches are manually reset.

Figure 6 is a copy of an oscillograph taken during a field test on a series capacitor equipment at Pacific Gas & Electric Company's Table Mountain substation in 1971. The capacitor current when the line circuit breakers close into a fault is sufficient to initiate a gap sparkover, causing immediate gap conduction. The switch travel indicator shows that the bypass switch starts to close a few milliseconds after gap conduction, and is fully closed less than 16 msec after sparkover. Note the high frequency discharge current, and its transfer from the gap to the switch. The switch

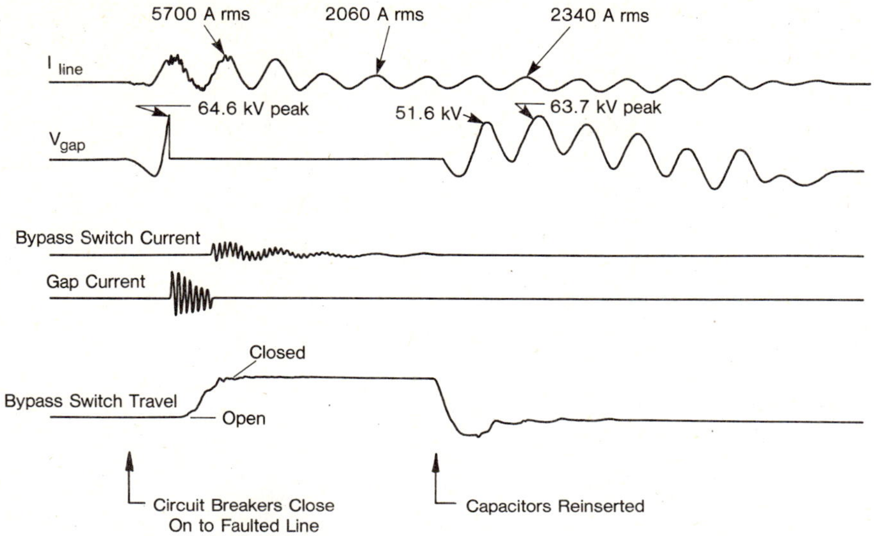

FIGURE 6. Series capacitor field test. Capacitor rating = 9 Ohms, 1800 A.

current trace also shows the 60-Hz current superimposed on the high-frequency discharge current. After the line current is reduced from about 5700 to 2060 A rms, the control circuit opens the switch to reinsert the capacitor after 2.5 cycles. Of particular interest is the transient voltage upon reinsertion, at one point being completely offset. The non-60 Hz component is the subharmonic oscillation of the series capacitors and the system, and this is almost large enough to cause the gap to spark over again, even though the steady state 60 Hz voltage is well below the gap setting. Switched capacitor shunting resistors can be used to substantially reduce the reinsertion transient voltages. Note too the increase in line current from 2060 to 2340 A rms when the capacitor is reinserted, owing to the change in net impedance of the test system.

In a system having two parallel lines such as shown in Figure 4, severe swing currents and voltages, overload currents and reinsertion transient voltages can occur when capacitor segments are bypassed due to a fault in a parallel line section. Upon clearing the faulted line section, these equipments must carry the full system swing current following reinsertion against more than double the normal line current. Sustained overload current following the swing may be about twice the prefault loading. These overload current requirements often determine the required continuous current rating of the bank. A typical line current profile for a series capacitor bank in a parallel line section of a two-parallel line system is shown in Figure 7.

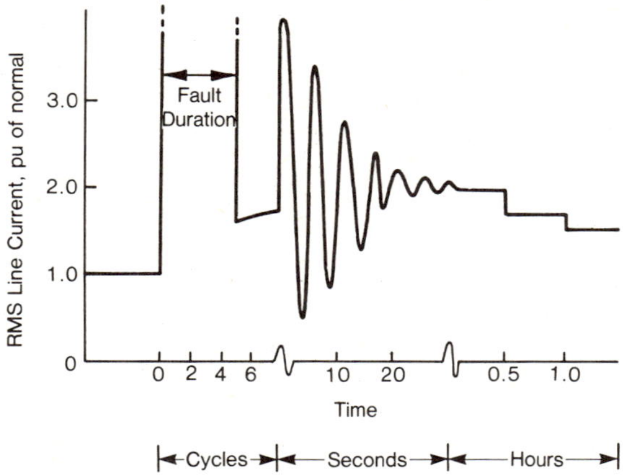

FIGURE 7. Typical swing and overload current profile on parallel line section, with faulted section held open.

7.5. REINSERTION SCHEMES

As previously discussed, the time delay following fault clearing until the capacitor is reinserted into the line can be critical to proper system operation. The shorter the delay, the higher the transient stability limit as shown in Figure 4. This reinsertion time delay is dependent on control relaying speed, bypass switch opening time, and the protective gap voltage recovery characteristics.

A slow-speed reinsertion system typically uses a nonself-clearing gap, and starts to close the bypass switch when sparkover occurs. After fault clearing, sufficient delay is provided to deionize the gap and allow it to recover its dielectric strength, then the switch is reopened. Reinsertion times on slow systems are usually one second or more.

A high speed reinsertion system is one that reinserts the capacitor within a few cycles following fault clearing. This can be achieved with high-speed switches and controls and gaps that rapidly recover dielectric strength, such as a vacuum gaps or self-clearing air gaps. A self-clearing gap could use either high- or low-pressure air with air flow initiated when the fault is cleared. Reinsertion times with self-clearing gaps are a function of the fault current magnitude and duration since these parameters affect the recovery characteristics.

Instantaneous reinsertion has been achieved by initiating air flow to the gap when sparkover occurs, to quickly build up insulation strength. The gap will then recover and attempt to insert the capacitor at every current zero during the fault, until capacitor voltage falls below the gap withstand. This has the disadvantage of subjecting the capacitor units to repeated high voltage pulses every time the gap reignites, and in fact may allow the capacitor voltage to rise dangerously above the sparkover setting while the air flow is on.

7.6. VARISTOR PROTECTIVE GEAR

Satisfactory equipment operation has been achieved with all of the different types of protective gear components previously discussed, including nonself-clearing gaps, high- and low-pressure air blast gaps, vacuum gaps, and several types of switches. Each new generation of equipment has offered improvements in characteristics over previous designs.

A new development in series capacitor protective equipment utilizes zinc-oxide varistors, a form of nonlinear resistor, to limit the voltage across the capacitor.[4] The exceptionally nonlinear resistance characteristic of the varistor is used to limit the voltage on the capacitor and offers all the advantages of practically *instantaneous* reinsertion while eliminating the need for high speed switching and rapid recovery gap characteristics. Figure 8 shows a typical varistor voltage/current profile.

The basic circuit for this new type of protective gear is shown in Figure 9. The varistor is connected directly in parallel with the capacitor. The stability of zinc oxide allows the varistor to withstand continuous energization below the knee voltage without significant deterioration, and considering also the exceptional nonlinearity, the need for a complicating series gap is avoided.

There is a triggered air gap shown in the diagram, but in this equipment the gap is not used to limit capacitor voltage—the varistor performs that function. Rather, the firing of this gap is initiated by control logic which monitors the energy absorbed by the varistor. Therefore for certain extreme system faults where the varistor current magnitude or duration is excessive, the gap is triggered to short circuit the varistor and protect it from further duty. A bypass switch is still necessary in this protec-

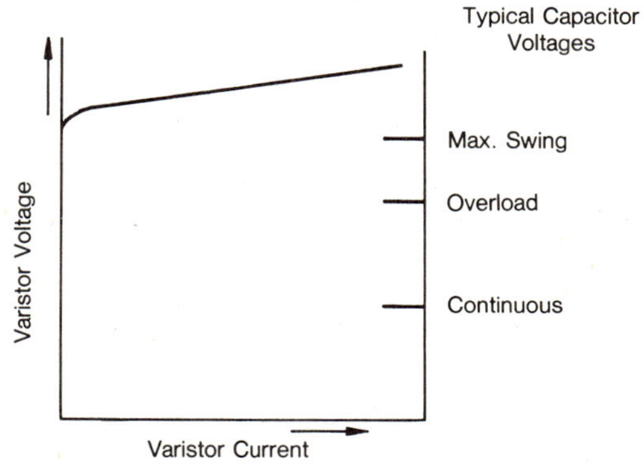

FIGURE 8. Voltage/current characteristic of varistor. © 1980 IEEE.

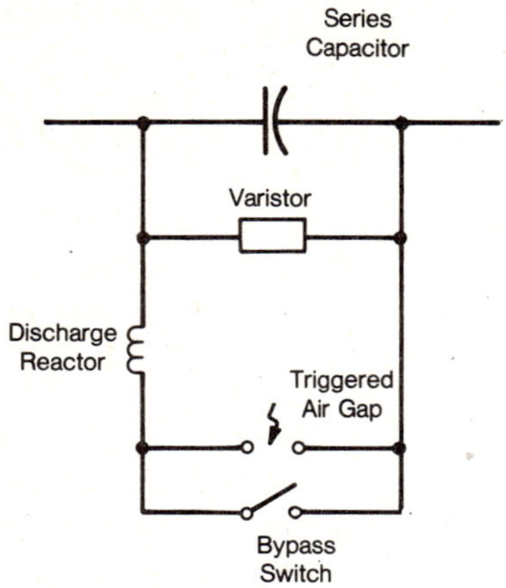

FIGURE 9. Series capacitor protection using varistor.

tive system to allow manual control of insertion, as well as to bypass the capacitor for abnormal system or equipment conditions. The discharge reactor is likewise still needed to limit the magnitude and frequency of the capacitor current when either the gaps spark over or the bypass switch closes.

Typically the varistor will have insignificant conduction at voltage levels up to the maximum system swing condition, but will still limit voltage to a value within the capacitor capability. The varistor characteristic (Figure 8) is determined by the number of zinc-oxide disks and their series-parallel connections. For a particular application the major considerations for the varistor are its required voltage and energy ratings. The voltage rating and the related voltage/current characteristic are established by the maximum nonfault voltages that are expected across the capacitor.

The essential features of the zinc-oxide varistor protective system can be described by reviewing simulations of its operation on the 500-kV, 60-Hz idealized system illustrated by the one-line diagram in Figure 10*a*. The varistor indicated for each phase has a protective level of 2 pu (crest) of the rated capacitor voltage for a fault current of 25 kA (crest). The capacitor rated current was assumed to be 1600 A rms. The system at each end of the compensated line section was represented by an equivalent inductive reactance of 16 Ohms.

Under normal system conditions, line current flows through the series capacitor and negligible current flows in the varistor. The bypass switch is open and the gap is not conducting. This condition is illustrated in the first cycle of the waveforms shown in Figure 10*b*.

A fault on the system increases the capacitor current and voltage. If the capacitor voltage rises enough, the varistor conducts and limits further voltage increase. If the fault is external to the line section in which the capacitor is installed, the line fault current is modest because it is limited by the impedance of the total line. The action of the varistor for

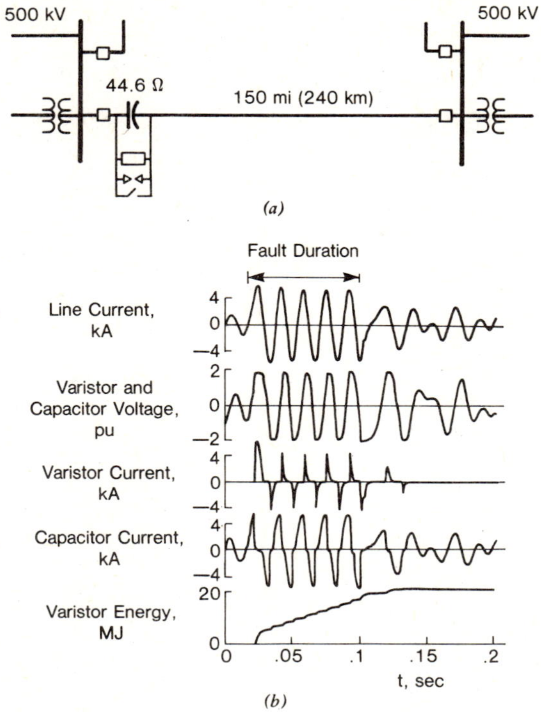

FIGURE 10. (*a*) 500-kV circuit with a three-phase fault external to the compensated line section. (*b*) Voltage and current waveforms for a five-cycle fault on the circuit of (*a*). © 1980 IEEE.

such a case is illustrated in Figure 10*b* for a three-phase fault. These waveforms were calculated on the digital computer using a simulation program. As shown, the varistor limits the voltage on each half-cycle and the current alternates between the capacitor and the varistor. When the varistor clamps the voltage, dv/dt is reduced which reduces the capacitor current to a low level. The capacitor-varistor shared conduction continues until the fault is cleared by the system breakers (Figure 10*a*) as directed by their own relaying. The line current then drops to the post-fault level, reducing the capacitor voltage and causing the varistor to virtually cease conduction. Thus the line current flow is fully restored to the series capacitor. This reinsertion or restoration is instantaneous and automatic, and can markedly increase power system transient stability and power transfer as was previously discussed.

A prototype of a complete protective gear system using zinc-oxide varistors was installed on an existing series capacitor at the North John Day substation on the Bonneville Power Administration system. A plan view of the protective gear platform and a photo of the installed prototype are shown in Figures 11 and 12. A series of staged fault tests were successfully conducted in April and May 1979 on this prototype installation.

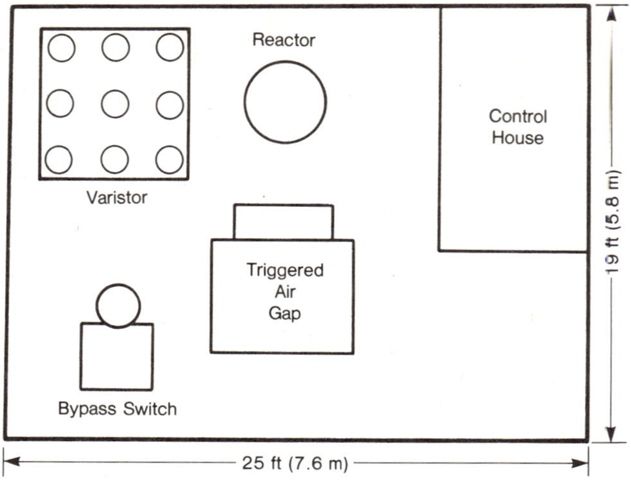

FIGURE 11. Plan view of prototype platform. © 1980 IEEE.

FIGURE 12. Prototype installation. © 1980 IEEE.

7.7. RESONANCE EFFECTS WITH SERIES CAPACITORS

A capacitor in series with the inductance of the transmission line forms a series-resonant circuit with a natural frequency given by

$$f_e = \frac{1}{2\pi\sqrt{LC}} = f\sqrt{\frac{X_{C\gamma}}{X_l}}, \tag{2}$$

where $X_{C\gamma}$ is the reactance of the capacitor in each phase and X_l is the total reactance of the line at the power frequency f. Since the degree of compensation, $X_{C\gamma}/X_l$, usually is in the range 25–70%, f_e is usually less than the power frequency, and we say that the system has a subharmonic resonance or "mode." X_l must include the equivalent series reactance of the generators and loads connected at the ends of the line. In practice these have complex frequency-response characteristics, as does the line itself, and for accurate estimation of the resonance phenomena a more detailed circuit model of the power system must often be used.

The first effect of a subharmonic resonance is that during any disturbance, transient currents are excited at the subharmonic resonant frequency f_e. These currents are superimposed on the power-frequency component, and are usually damped out within a few cycles by the resistances of the line and of the loads and generators connected to it. An example is shown in Figure 6. In this case the damping is said to be positive, and the subharmonic mode is stable. It should be noted that the subharmonic mode is only one of the natural modes of a power system. Other natural modes have resonant frequencies above the power frequency, and these can be troublesome if they occur at or near integer-order harmonic frequencies when sources of harmonic currents (e.g., large rectifiers) are connected (see Chapter 10). In general any disturbance, including any switching operation, will excite all the natural modes of the system, in different degrees. Usually all the resulting transient currents are positively damped, again in varying degrees.

Under certain conditions the subharmonic mode associated with series capacitors can experience a destabilizing influence from polyphase ac rotating machines. In extreme cases it can even become unstable in the absence of corrective measures. The destabilizing influence shows itself as a negative resistance in the equivalent circuit of synchronous and induction machines. If saliency is neglected, the per-phase equivalent circuit of a synchronous machine is of the form shown in Figure 13. The generated (power-frequency) emf has been omitted from this circuit

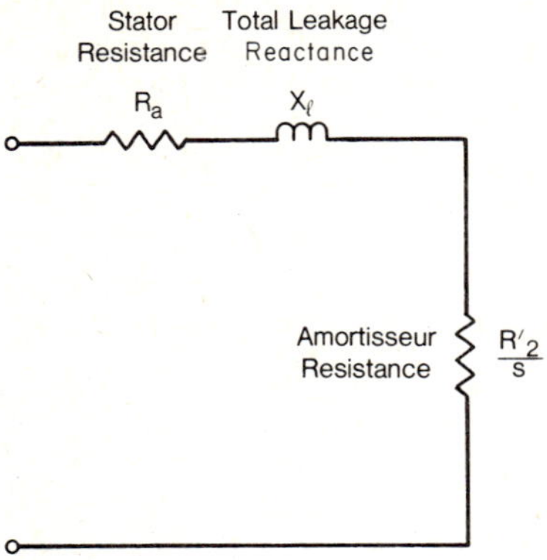

FIGURE 13. Simplified equivalent circuit of synchronous machine, for explaining subharmonic effects.

while we consider the subharmonic frequency alone. Also the field winding has been omitted for simplicity. Suppose that subharmonic currents have been transiently excited by a disturbance in the external system. In general they will be unbalanced between the three phases, but if we resolve them into symmetrical components the positive-sequence components flowing in the machine stator will set up a magnetic flux which rotates in the same direction as the rotor, but at the angular velocity $2\pi f_e$ electrical radians per second. The rotor is rotating at $2\pi f$ elec rad/sec, faster than the subharmonic field. It is said to be slipping relative to this field, the slip s being given by

$$s = \frac{f_e - f}{f_e}. \tag{3}$$

Since $f_e < f$, the slip is negative, and the rotor behaves much like that of an induction motor running above synchronous speed. The resistance of the amortisseur (or solid rotor in the case of high-speed synchronous generators) referred to the stator is R'_2/s, as in the induction motor equivalent circuit, and this is a negative quantity which contributes negative damping or "undamping"[(7)] when added in series with the stator and external system resistances. The machine is therefore capable of converting mechanical energy into electrical energy associated with the subharmonic mode.

If s is very small, as can happen when the degree of series compensation is large and f_e approaches f, then R'_2/s can become large enough to overcome the positive resistances in the system. The subharmonic mode is then unstable, and will grow to dangerous levels of current and voltage as a result of the smallest disturbance. This situation rarely arises in practice, but when it does, corrective measures are necessary. These can include any or all of those which are applied to prevent subsynchronous resonance (discussed below).

The electrical subharmonic natural mode is rarely troublesome except where *subsynchronous* resonance (SSR) can occur. As it rotates in the backward direction relative to the rotor and the main field, the subharmonic field produces an alternating torque on the rotor at the frequency $f - f_e$. If this difference-frequency coincides with one of the natural torsional resonances of the machine's shaft system, torsional oscillations can be excited. This condition is known as subsynchronous resonance. SSR is a combined electrical/mechanical natural mode or resonance. Like the purely electrical subharmonic mode, it can be stable or unstable, depending on the degree of damping. Although the negative resistance effect in synchronous machines can have a destabilizing influence, instability of the subsynchronous mode is more likely to be a result of phase shifts in the circuit external to the generator whose shaft is oscillating. The oscillation produces a frequency modulation of the power frequency with subharmonic and harmonic sidebands, and the subharmonic sidebands may be made unstable by these phase shifts.

The machines which are most susceptible to SSR are large multiple-stage steam turbines, which typically have four or five torsional modes in the frequency range 0–60 Hz. The lowest torsional frequency is the "swing" frequency in which the entire system of turbine cylinders and the generator oscillates about synchronous speed as one inertia. Torsional resonances at higher frequencies involve the twisting of the shaft in different mode shapes, and resonant frequencies can extend up to hundreds of Hz. The damping of these modes is generally extremely small.

The consequences of an SSR condition can be dangerous in the short term, if the oscillations are unstable and build up sufficiently to break the shaft. But even if the oscillations are relatively well damped, disturbances (like switching, fault clearing, etc.) can use up the fatigue life of the shaft. This slow deterioration is called "low-cycle fatigue," and in recent years considerable effort has been made to understand it quantitatively.

The corrective measures for SSR are:

1. Tripping sections of line, or bypassing series capacitors, using protective relays sensitive to very small incipient levels of subharmonic current.
2. The installation of special subharmonic filter circuits. These can take the form of blocking filters (parallel-resonance type) in series with the power line; or damping circuits in parallel with the series capacitors.
3. The use of excitation control (modulation of field current) in turbine-generators phased so as to provide increased positive damping at the subharmonic frequency.

4. The use of a static compensator whose reference voltage is modulated with such a phase as to provide increased positive damping at the subharmonic frequency.

In some severe cases a combination of methods (1) through (3) has been used successfully with as many as four parallel-tuned blocking filters to provide damping for each of four subsynchronous resonances in a system equipped with a large number of series capacitors, deployed at many points throughout the network.

7.8. SUMMARY

Series capacitors have been and continue to be an economic means of enhancing the technical feasibility and efficiency of power transmission on EHV systems. The applications are not simple, and a great deal of study is required to assure proper operation. Continuing advances in both capacitor unit and protective gear designs have played an important role in maintaining the growth rate of series capacitors, and this line compensation technique should be a key factor in future EHV transmission systems.

REFERENCES

1. L. E. Bock and G. R. Mitchell, "Higher Line Loadings with Series Capacitors," *Transmission,* March 1973.
2. I. S. Benko, S. H. Gold, and W. N. Rothenbuhler; and L. E. Bock, I. B. Johnson, and J. R. Stevenson, "Internal Overvoltages and Protective Devices in EHV Compensated Systems – Series Capacitors and Shunt Reactors," *CIGRE 33-05,* 1976.
3. I. B. Johnson, *Capacitor Banks for Transmission System Compensation,* Missouri Valley Electr. Assoc., April 1973.
4. ©1980 IEEE Reprinted with Permission from "A Zinc Oxide Varistor Protective System for Series Capacitors," by J. R. Hamann, S. A. Miske, Jr., I. B. Johnson, and A. L. Courts, IEEE Power Eng. Soc. Summer Meeting July 13–18, 1980, Minneapolis, Minnesota.
5. D. N. Walker, C. E. J. Bowler, and D. H. Baker, "Torsional Dynamics of Closely Coupled Turbine Generators," *Trans. IEEE, Power Appar. Syst.* **97**(4), 1458–1466 (1978).
6. L. A. Kilgore, D. G. Ramey, and M. C. Hall, "Simplified Transmission and Generation System Analysis Procedures for Subsynchronous Resonance Problems," in IEEE Power Eng. Soc. Winter Meeting, 1976.
7. C. E. J. Bowler, "Understanding Subsynchronous Resonance," IEEE Publ. 76CH1066-0-PWR, Symposium on Analysis and Control of Subsynchronous Resonance, 1976.

Chapter 8

SYNCHRONOUS CONDENSERS

P. G. BROWN

8.1. INTRODUCTION

Synchronous condensers have played a major role in voltage and reactive power control for more than 50 years. They have been connected at both subtransmission and transmission voltage levels to improve stability and to maintain voltages within desired limits under varying load conditions and contingency situations.[1,2] For economic reasons applications in the subtransmission area have been largely supplanted by switched shunt capacitor banks. Synchronous condensers have an inherent advantage over capacitors for emergency voltage support in maintaining or increasing their output at reduced voltage. This has prompted their continued installation at transmission voltage levels where large ratings may be required. During the past decade major increases in condenser sizes have occurred, as shown in Figure 1. This growth has been spurred by the continuing increase in transmission voltage levels with correspondingly higher bulk power flow capability per circuit. At the same time, utilities have increasingly been faced with postponement or cancellation of planned circuits. This has created potentially more critical emergency situations calling for large amounts of emergency reactive power support.

Another primary application of synchronous condensers is with high voltage dc transmission (HVDC),[3] where they supply a portion of the converter reactive power requirements and provide necessary system reinforcement where the ac receiving system short-circuit capacity is low.

In this chapter the basic characteristics and performance of condensers are reviewed.

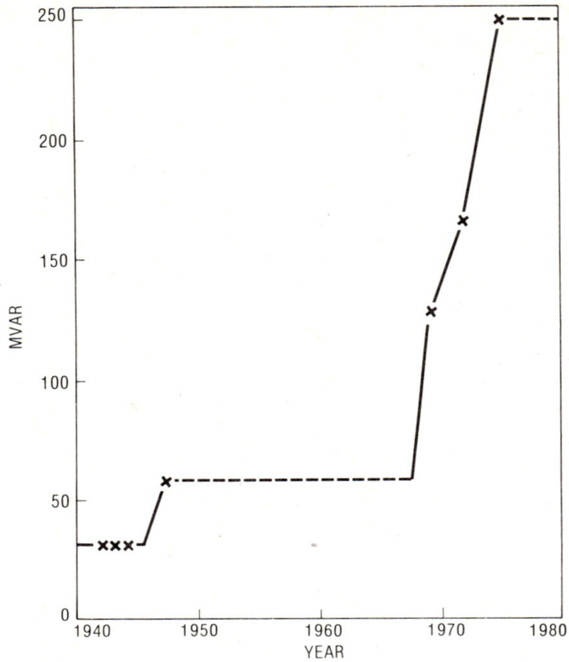

FIGURE 1. Growth in hydrogen-cooled condenser ratings.

8.2. CONDENSER DESIGN FEATURES

Functionally, a synchronous condenser is simply a synchronous machine that is brought up to speed and synchronized to the power system. After the unit is synchronized the field is controlled to either generate or absorb reactive power as needed by the power system. The synchronous condenser falls into the class of active shunt compensators discussed in Chapter 2.

The majority of synchronous condenser installations are outdoor design and operate unattended with automatic controls for startup, shutdown, and on-line monitoring. Historically both air-cooled and hydrogen-cooled condensers have been used extensively; however, nearly all of the large sizes in the United States are hydrogen-cooled. There is in addition a large (345 MVA) water-cooled unit in service.[2]

Figure 2 shows a 167-MVAr hydrogen-cooled condenser together with major auxiliary components. In addition to the automatic control and protection equipment, the control building houses the excitation control equipment and motor control equipment.

The condenser is contained entirely within a gas-tight enclosure, with no running seals, which makes it inherently suitable for outdoor installation. All leads required outside the shell are through individual bushings.

FIGURE 2. 167-MVAr, 900-rpm hydrogen-cooled synchronous condenser. Control building and cooling tower in background.

The slip rings for excitation power and the starting motor, if one is used, are contained in a separate compartment within the main shell, with an inflatable seal for use when the unit is at rest. This permits brush-changing without the necessity of purging the entire condenser.

The condenser pit area directly below the condenser, houses a number of auxiliaries, including the lube oil system, high-pressure bearing lift pump for reducing friction during starting, hydrogen accessory equipment, carbon-dioxide purging equipment and the neutral grounding equipment.

Figure 3 shows the salient-pole type of rotor construction generally used. In addition to providing damping of rotor oscillations, the amortisseur bars carry the rotor circulating currents during startup where reduced voltage starting is used. For that type of starting "complete amortisseurs" are required, with interpole connections of the amortisseur bar groups as shown in Figure 3.

A number of fundamental design advances, including higher operating speeds and higher hydrogen pressures, have accompanied the extension of ratings to the larger sizes. Large condensers of the type shown in Figure 2 are typically rated 900 rpm and operate with 30-psig hydrogen pressure.

Figure 4 shows the essential elements of the condenser installation. These include connections to the system and auxiliary systems, discussed later. A key element is the excitation control equipment which to a large

FIGURE 3. Synchronous condenser rotor.

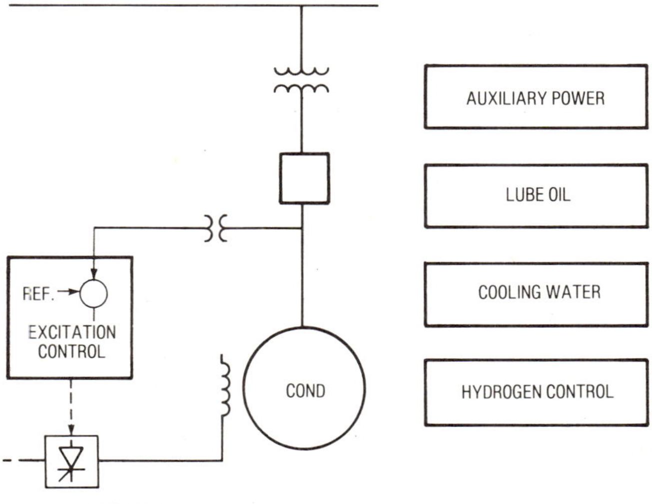

FIGURE 4. Condenser major auxiliary systems.

measure determines the condenser performance on the power system. This is done simply by voltage regulator, as in the excitation control of most generating stations. A major system disturbance results in abnormal voltage which the condenser and its excitation control respond to correct. Voltage sensing may be either from the condenser terminals, as shown, or from a transmission bus. Where direct sensing of transmission bus voltage is used, some droop in the control circuit will generally be required to provide stable operation. In Figure 4 this droop is provided by the step-up transformer reactance, and can be increased or decreased by the control.

The total full-load losses of the condenser, including its auxiliary systems, are of the order of 1% of the condenser rating, about two-thirds of which are a function of the loading.

8.3. BASIC ELECTRICAL CHARACTERISTICS

8.3.1. Machine Constants

Typical values of the major electrical constants are given in Table 1, along with those of steam turbine-generators and hydro generators for comparison. The reactances and time constants are similar to those of 4-pole

TABLE 1
Typical Machine Constants

Constants	Turbine-Generator 2-Pole	Turbine-Generator 4-Pole	Hydro	Synchronous Condenser
X_d	2.0	2.0	1.0	2.0–2.5 pu
X'_d	0.22	0.33	0.32	0.35–0.45 pu
X''_d	0.16	0.22	0.21	0.25–0.30 pu
T'_{d0}	4	9	5–10	9–10 sec
H	2.5	4	3–6	1.2 kW-sec/kVA
X_q	2	2	0.6	1.2–1.6 pu
X_P (Potier reactance)				0.30–0.40

Note: transient and subtransient reactance values are for rated voltage.

turbine-generator units. The inertia constant *H*, which is a relatively unimportant characteristic for a condenser, is much smaller because there is no connected turbine.

8.3.2. Phasor Diagram

The phasor diagram for a condenser is very simple because, except for a very small loss component which can be neglected, all of the current is in quadrature with the terminal voltage either leading or lagging. Figure 5 shows the phasor relationships corresponding to the two directions of reactive power flow. From a no-load condition, an increase of excitation (i.e., "overexcited" operation) generates reactive power into the system (Figure 5*a*). The corresponding phasor relationship based on the convention normally applied for generators is also shown. Reactive power flows into the system with overexcited operation, just as that which is produced by a shunt capacitor. A decrease in excitation from the no-load condition (i.e., "underexcited" operation) absorbs reactive power from the system with the phasor relation shown in Figure 5*b*, just as that which is absorbed by a shunt reactor.

Figure 6 shows a more complete phasor diagram for overexcited operation. The voltage $\mathbf{E}_1$ represents the per-unit field current, neglecting saturation, and $\mathbf{E}_f$ the total field current derived as shown. Since the condenser current is only reactive, the conventional generator reactive capability curve is not applicable and other curves such as the V-curve are used to portray condenser operation.

8.3.3. V-Curve

A synchronous condenser provides stepless adjustment of the reactive power in both the overexcited and underexcited regions. In the overexcited region there is both a continuous "nameplate" capability and short-

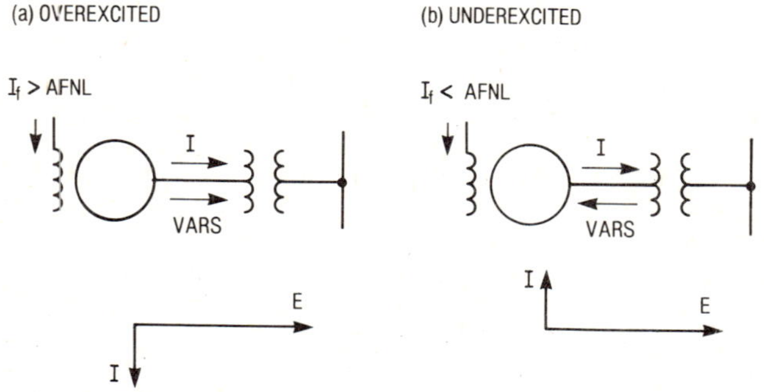

FIGURE 5. Basic definitions of excitation and reactive power flow. (AFNL is no load excitation.)

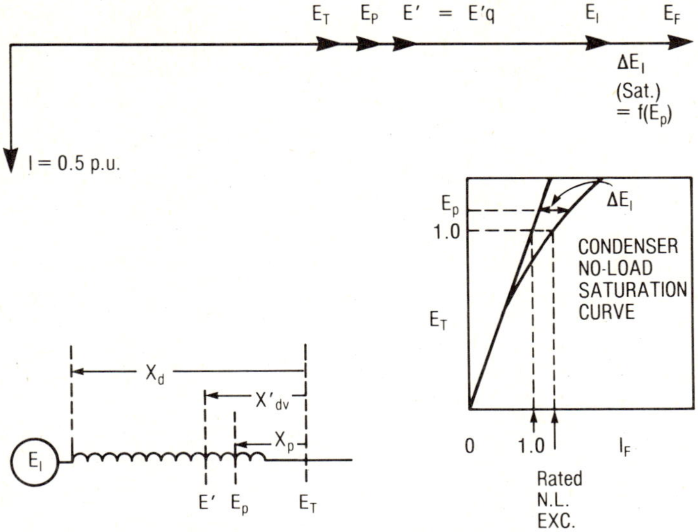

FIGURE 6. Phasor diagram for overexcited operation.

time overload capabilities. Steady-state operating characteristics are shown by a V-curve, such as that of Figure 7. The right hand portion of the V-curve represents overexcited operation, as for a shunt capacitor bank. The left hand portion of the V-curve represents underexcited operation, in which the machine is absorbing reactive power from the system, as for a shunt reactor bank.

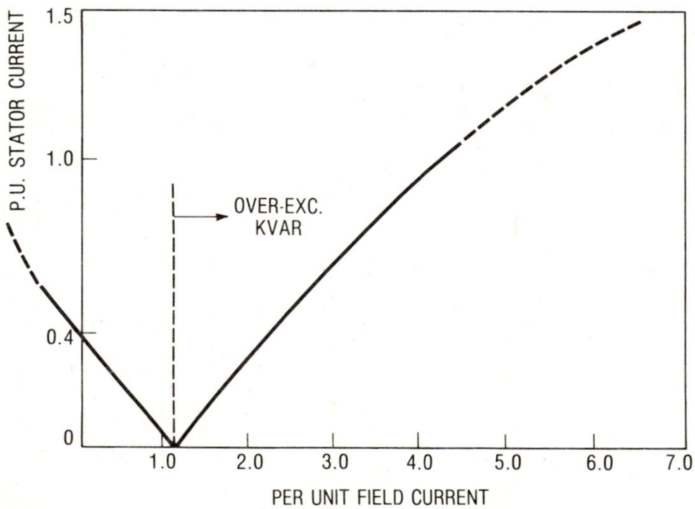

FIGURE 7. Condenser V-curve. Base (1.0 pu) field current as defined in Figure 6.

Normal continuous operation may be at any point on the curves below 1.0 pu or rated stator current. The dashed portion to the right represents short-time overload operation obtained by increased excitation.

The condenser underexcited capability is obtained by operating at field currents below no-load excitation. As indicated in Figure 7 this operation may theoretically extend even into the negative field current range. Operation at negative field currents (within limits) is possible without pole slipping because of the reluctance torque associated with the salient-pole rotor construction. The limiting condition where a pole would be slipped in steady-state operation occurs in Figure 7 at approximately 0.6 pu underexcited reactive current, corresponding to E_T/X_q. A dc commutator type exciter may have its polarity reversed to provide negative field currents and many condensers in service inherently have that capability. A conventional static excitation system (transformer-fed thyristor) cannot provide reversed current even though it will have full voltage-inverting capability for transient field forcing. This static system permits operation down to a minimum exciter voltage of about 10% of the no-load excitation requirement. Typically, this underexcited capability as shown in Figure 7 is about 35–40% of rating. A static exciter with negative current capability can be provided by using additional thyristor bridges and associated controls, so that the underexcited capability of a condenser can be increased to its stability limit.

8.3.4. Simplified Equivalents

While the published V-curve usefully describes normal operation, it is not very helpful for unusual operating conditions where for short periods of time voltages are not normal. Transient stability computer programs should be used for such studies. For estimating, however, a simple equivalent circuit consisting of a voltage behind a machine reactance as in Figure 8 can be useful in understanding the machine performance in contingency situations, and in providing checks of computer results.

For example, with the condenser initially running at half rated output as in Figure 7, consider a sudden system emergency arising from the loss

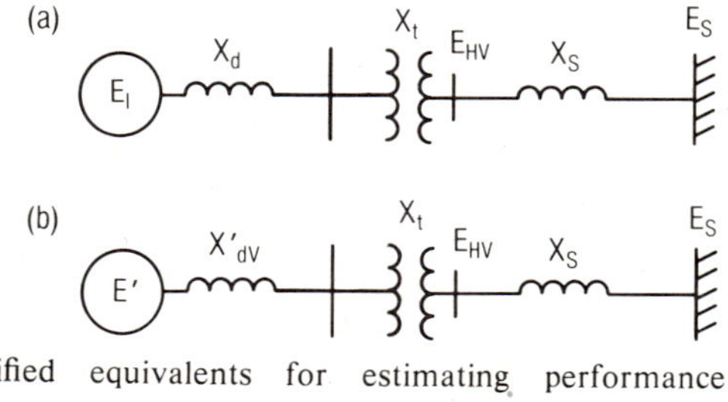

FIGURE 8. Simplified equivalents for estimating performance. (*a*) Steady State. (*b*) Transient.

of critical lines or generation, which reduces the transmission bus voltage at the condenser location to 70%. Assuming the lower-band reactance values of Table 1 and a 10% transformer reactance, the immediate response of the condenser (neglecting the first few cycles) would be a step increase in reactive current output to 113% of condenser rated current. Subsequent response is a function of the excitation system action, as discussed in the following section.

For the same voltage disturbance shunt capacitors would experience a sharp reduction in output current. Assuming a corresponding case with the same available nominal capacity of switched capacitors the reactive current would be only 80% of rated. This shows that in comparing alternate means of emergency reactive power supply, it is important to consider equipment ratings derived from system studies that demonstrate equivalent improvement, rather than simply the nominal ratings.

8.4. CONDENSER OPERATION

In recent years the major applications of synchronous condensers have been for the following purposes:

1. Power system voltage control.
   (a) normal voltage control.
   (b) emergency voltage control.
2. HVDC applications.

8.4.1. Power System Voltage Control

"Normal" operation of a power system is characterized by a continual change in transmission line loading, between light load and peak load periods; under the changing demands of pool interchange agreements; and as a result of network changes arising from maintenance outages of lines, and so on. Such changes in loadings translate into similar continual changes in the reactive power requirements throughout the system. In the distribution and subtransmission areas the changing reactive requirement are generally satisfied by switched capacitor banks supplemented with load tap-changing transformers and feeder voltage regulators for voltage control. On the transmission network wide variations in reactive requirements exist between the case of light line loadings (where lines behave like reactive sources), and heavy line loadings (where lines appear as reactive loads). The net result may be an inability to maintain correct voltage levels, or to adequately control reactive power interchange with neighboring utilities. Switched shunt capacitor banks and shunt reactor banks are commonly used for transmission voltage control, and their ability to accomplish this is a function of the size of switching steps.

Synchronous condensers have a technical advantage in this role in that they:

1. provide a continuously adjustable (stepless) reactive power which enables close control of transmission voltages,
2. have the capability to provide both capacitive and inductive reactive power in accomplishing the requirements of (1).

8.4.2. Emergency Reactive Power Supply

The ability to provide emergency voltage control during major system disturbances is probably the primary incentive for most recent utility condenser applications. Such needs arise for contingencies such as the occurrence of a fault and sudden loss of major transmission or major generation. In extreme cases this can result in system breakup, or islanding.

Major system upsets are generally characterized by abnormal voltages, at least initially. Voltage extremes in either direction may occur, particularly where an area becomes islanded. Synchronous condensers under voltage regulator control automatically change their output in a direction to alleviate the voltage excursion. Figure 9 shows the short-time emergency reactive power output with reduced voltage available at one installa-

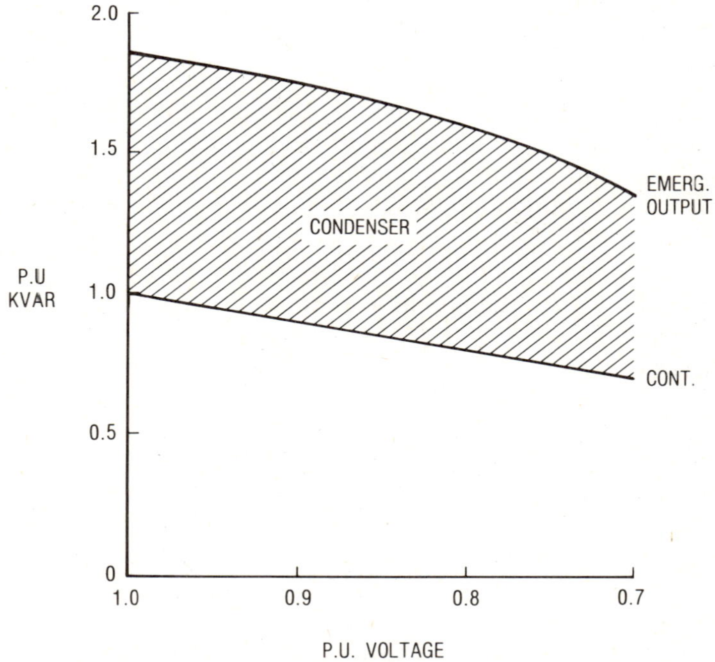

FIGURE 9. Emergency reactive power support.

tion. As shown, condensers have short-time capabilities well in excess of their nameplate ratings. Typically they may have a short-time rating of 150% of rated current for 1 min. In actual operation, the condenser output is, of course, a function of the system voltage and excitation ceiling available rather than assigned ratings. Figure 10 illustrates the effect of excitation system design ceiling on the condenser response to a depressed system voltage. In this example a step change to 0.8 pu of the HV bus voltage is assumed. For extreme system conditions very high condenser output occurs automatically and it is therefore important that the control design provide for adequate stator winding as well as rotor winding overload protection for a wide range of conditions. Reduction of the field current to its rated value by a maximum excitation limit is not sufficient to guarantee protection of the stator winding which could still be considerably overloaded.

The time period during which the condenser overload capability is available during a system emergency is adequate to allow reclosing of transmission circuits and action by some prime-mover controls. For example, gas turbines and some hydro units in a spinning reserve mode can pick up their full capability in a matter of less than a minute. Automatic line reclosing is usually initiated within 10–20 sec.

The importance of providing reactive support in a load area, on the receiving end of the transmission system, is illustrated by Figure 11, which is taken from Reference 4. This illustrates the importance of reactive support at the receiving end to improve the capability of the transmission network to import power into a generation-deficient area.

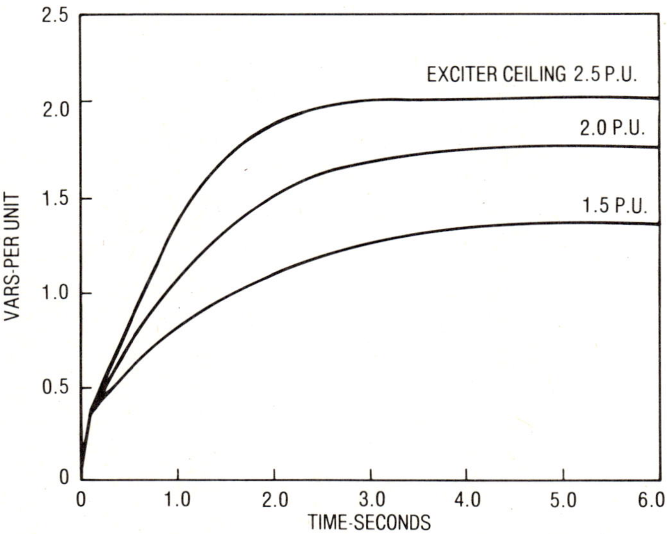

FIGURE 10. Synchronous condenser transient performance. Exciter ceiling given in per unit of rated full load exciter voltage.

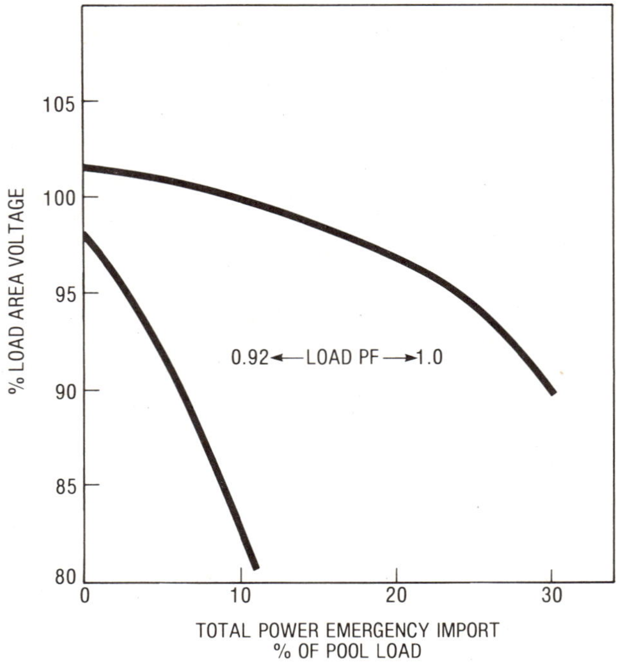

FIGURE 11. Pool energy transfer.

8.4.3. Minimizing Transient Swings

The use of shunt reactive compensation at intermediate switching stations is one means of improving the transient stability of remote generation (see Chapters 2 and 3 and Reference 5). Synchronous condensers and controlled static compensators can contribute to improve transient stability in these cases. Aside from conditions of stability a problem of abnormal voltages at intermediate load areas accompanying transient swings may also be a problem. The situation illustrated in Figure 12 shows the application of a condenser for reducing voltage swings following a fault. A synchronous condenser has a large field time-constant, however, so that in order to be effective in improving the transient voltage swings, the excitation system should be equipped with a supplementary control providing an error signal proportional to the rate of change of voltage.

This lead function offsets the inherent phase lags in the condenser, so that the reactive output is in phase with the transient requirements. Figure 13 shows a stabilizer function added to the excitation computer representation. Figure 14 shows the effectiveness of the supplementary control in improving the condenser performance for a simulated cyclic swing in system voltage.

The necessity for voltage regulator supplementary control to obtain the maximum condenser benefits during transient swings is more specifically

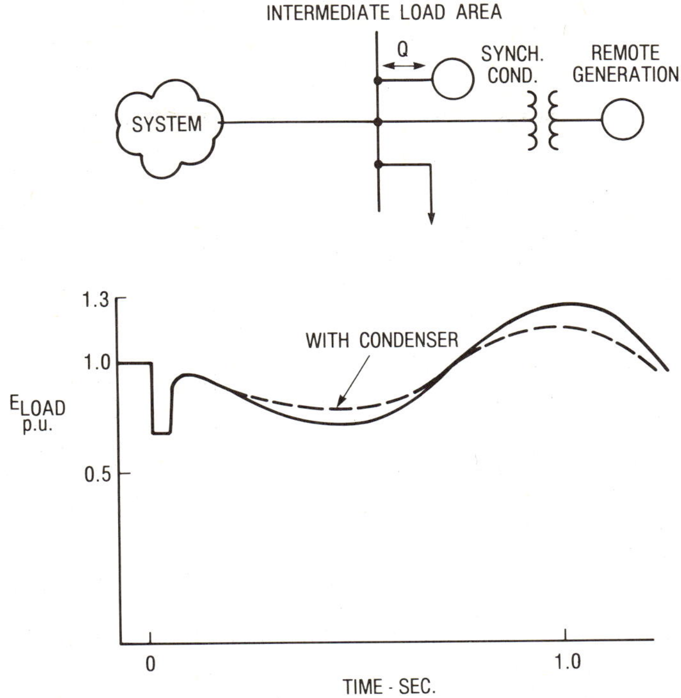

FIGURE 12. Load area voltage swings following transient disturbance.

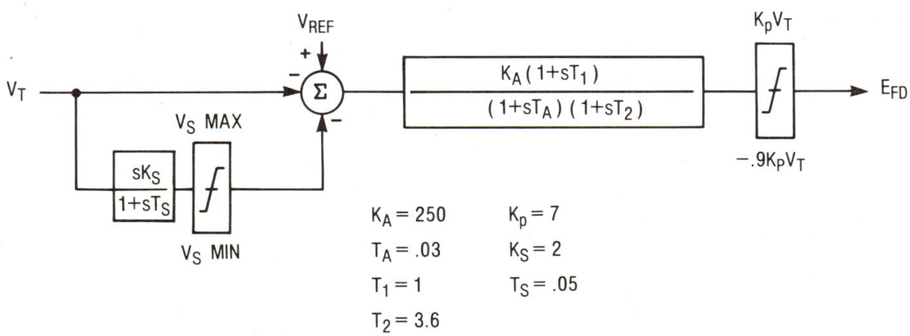

FIGURE 13. Computer representation of excitation system. Bus-fed static excitation system.

demonstrated by stability studies of a large interconnected system. An example of simulation results is shown in Figure 15. The utility system is characterized by major generating units remote from the load area. In occupying a location between widely separated generators the load areas are subject to wide voltage variations during transient swings. A 140-MVAr synchronous condenser is used to provide voltage control and helps to maintain transient stability during disturbances. The particular

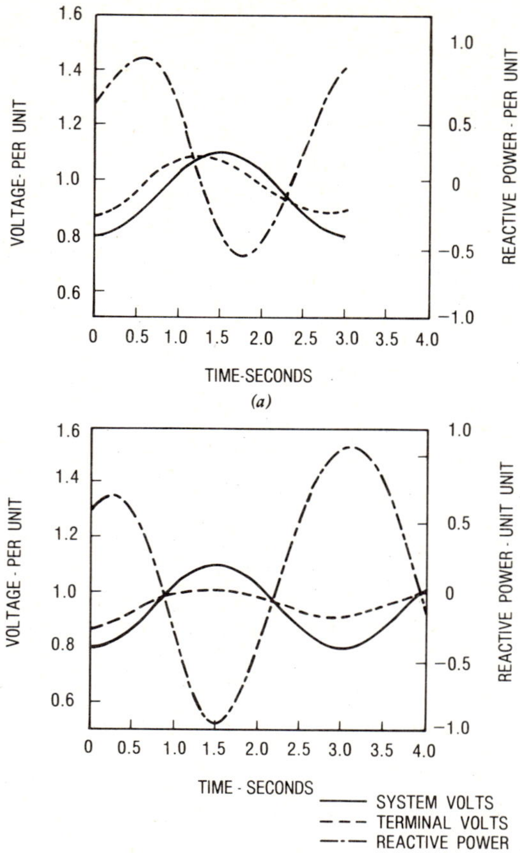

(b)

FIGURE 14. Synchronous condenser performance during system voltage swings. (*a*) No supplementary control. (*b*) With supplementary control.

case shown is for a 3.5-cycle three-phase fault on a major EHV circuit, which resulted in transient instability of the system. Despite the addition of the condenser with a high-ceiling static excitation system, this case is shown to be unstable without the supplementary control. One reason for this is delay in the voltage regulator control. Figure 15*c* shows the phase lead and MVAr output increase provided by the supplementary control, which results in maintaining system stability.

Emphasizing a point made earlier in regard to the low-voltage capabilities inherent to the condenser, it is interesting to note in this practical example that despite a voltage reduction to about 80% at the condenser location the MVAr output is substantially above its nominal rating.

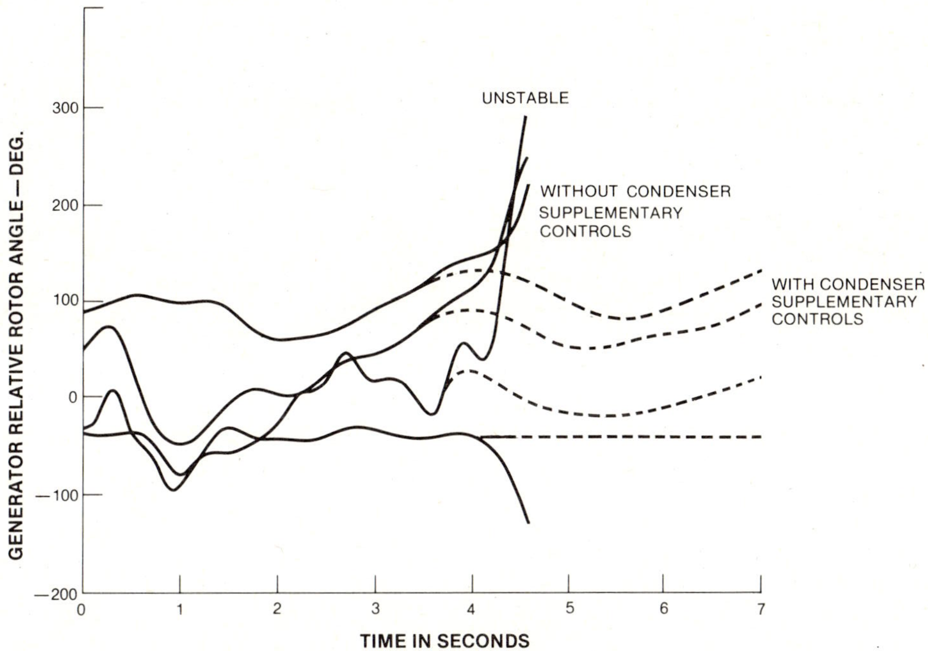

(*a*) GENERATOR ROTOR ANGLES

FIGURE 15. System transient stability improvement using synchronous condenser with voltage regulator supplementary controls.

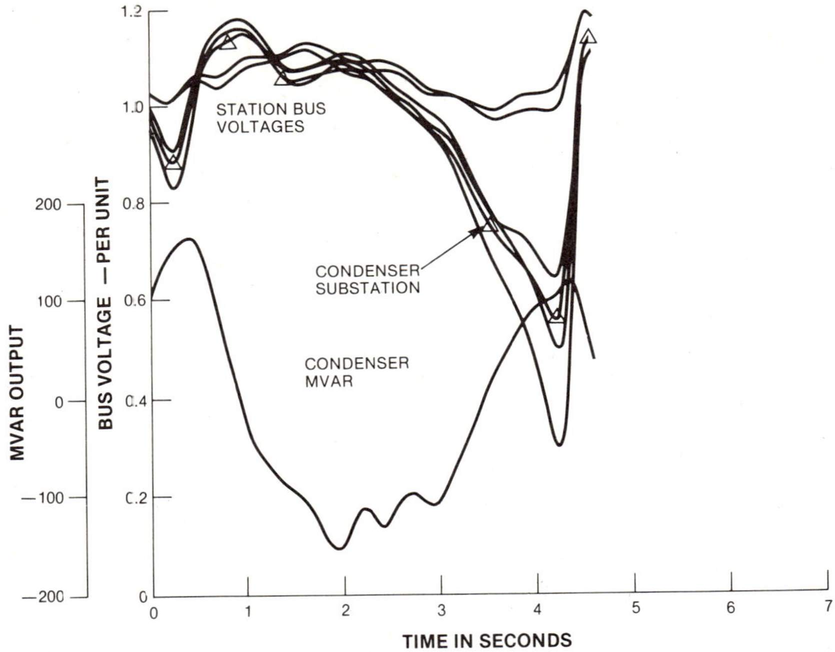

(*b*) PERFORMANCE WITHOUT CONDENSER SUPPLEMENTARY CONTROLS

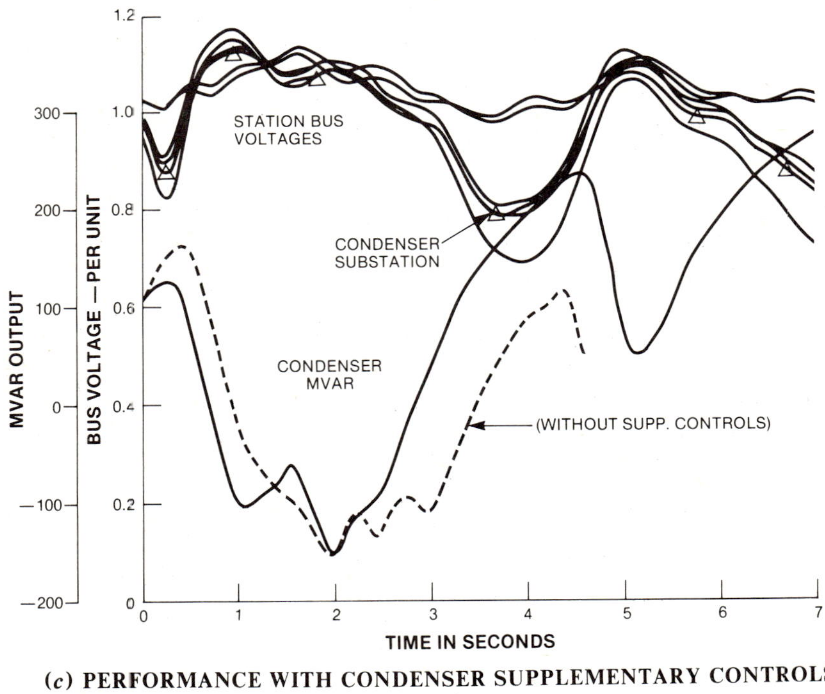

(*c*) PERFORMANCE WITH CONDENSER SUPPLEMENTARY CONTROLS

FIGURE 15. (*continued*).

8.4.4. HVDC Applications

Figure 16 shows the use of synchronous condensers at the receiving end of a high-voltage dc line. In this application the condenser fulfills several needs:

1. It provides a portion of the reactive power requirements of the converter.
2. It permits the dc converter control to maintain acceptable control of the ac voltage where the receiving system short-circuit capacity is low.

When the receiving system does not in itself meet short-circuit requirements, additional short-circuit capacity from synchronous condensers can fill this need, as well as providing a portion of the reactive requirements of the inverting process. Of the available reactive power supplies, the condenser has the unique ability to provide short-term voltage stability in the subtransient range. One demanding condition is the sudden blocking of the dc link for fault clearing which results in a sudden release of reactive power loading of the converter and corresponding instantaneous rise in the ac system voltages. These considerations establish the condenser capacity required for acceptable operation, with the remaining normal and post-subtransient reactive requirements being provided in the most economic manner. A fixed block of reactive power is available from the capacitor banks associated with filter banks. Additional controlled reactive power supply will generally be provided by a controlled compensator. With a strong ac receiving system, that is, a high short-circuit level, all of the variable reactive requirements may be supplied by the system or from switched shunt capacitors.

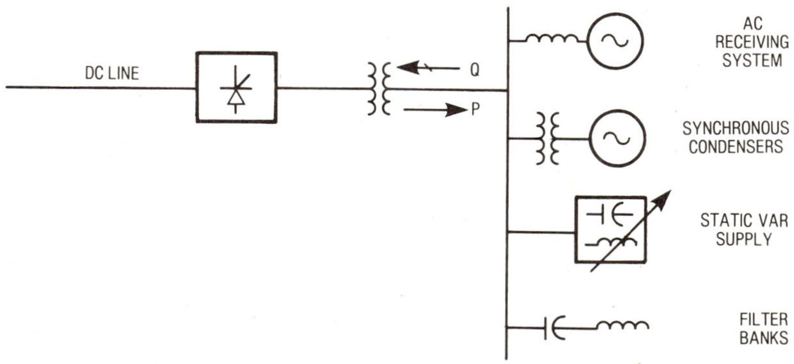

FIGURE 16. HVDC application of synchronous condenser.

8.5. STARTING METHODS

Practical starting methods for large condensers include reduced voltage, starting motor, and static starter. Full voltage or across-the-line starting is not generally used for large condensers because of the magnitude of system voltage dips and also the severity of condenser duty. Generally, starting will be infrequent, the units being shut down only when required for maintenance. Likewise, starting time is generally not critical, 15–20 min being acceptable.

8.5.1. Starting Motor

This method uses a wound-rotor motor with one less pair of poles than the main condenser to accelerate the unit to rated speed and synchronize to the line. It has the advantages of eliminating any voltage dip on the system, as well as stator or amortisseur winding stresses during the startup. There is also a considerable background of experience with this starting method which has been used extensively for startup of both synchronous condensers and pumped hydro units. A motor rating of about 0.5% of the condenser rating is used.

Torque control of the motor during startup is by means of a liquid rheostat controlled as shown schematically in Figure 17. During the acceleration period the control positions the electrodes to maintain constant torque. At approximately 98% speed, the control responds to the pulsed output of speed matching relays to bring the slip to a very small value and allow the automatic synchronizing relay to initiate breaker closing.

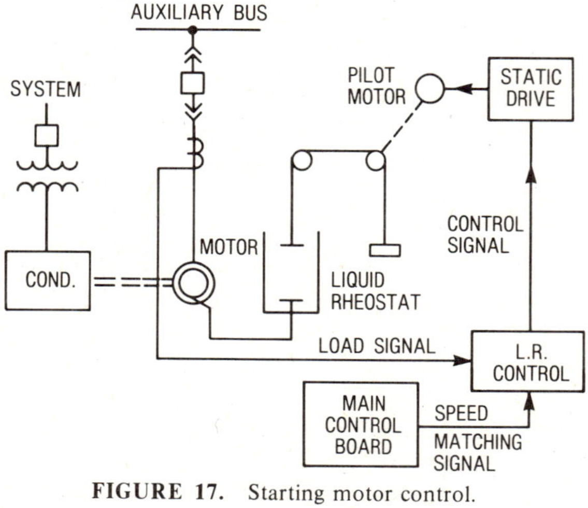

FIGURE 17. Starting motor control.

Where the condenser has a direct-connected dc exciter, this may be used as a driving motor to bring the condenser up to speed and synchronize to the line. Two 250-MVAr condensers presently in service use dc exciter starting, in an automated system. Torque control during starting in this case is through a controlled dc voltage to the exciter/motor.

8.5.2. Reduced Voltage Starting

Two arrangements for reduced voltage starting are shown in Figures 18*a* and *b*. Various switching arrangements, particularly with respect to autotransformers, have been used. The choice is a matter of economics rather than any practical difference in performance. This includes autotransformer connections which utilize tripping of a neutral breaker with the series winding serving as a reactor during the start-run transition, sometimes referred to as the Korndorfer method. The magnetizing impedance of the usual starting autotransformer under these circumstances is so high that there is practically no difference in performance from the open transfers of the arrangements shown in Figure 18.

The reduced voltage tap of the transformer delta winding of Figure 18*a* may be a symmetrical center tap for half voltage or a corner-delta connection for a lower voltage tap.[6] The corner-delta tap introduces unbalanced impedances which result in unbalanced current flow during starting. However, the unbalanced current flows are generally small and of primary interest to the relaying designer with regard to sensitive ground relay setting. The transformer design should recognize the need to avoid excessive unbalance of the tap voltages.

The short-circuit current on a tap tends to be higher in amperes than that of the full winding. Where this may exceed the interrupting or momentary rating of the planned breaker a current-limiting reactor (CLR) as shown in Figure 18*a* may be required.

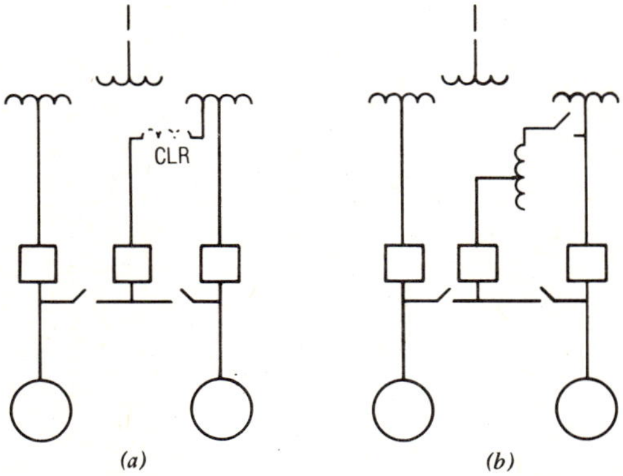

FIGURE 18. Reduced voltage starting. (*a*) Transformer tap. (*b*) Autotransformer.

The unit is started as an induction motor on the reduced voltage tap. Excitation is applied when full speed is reached and the unit pulls into step. Because of the reluctance torque, the unit may pull in on the "wrong pole" initially. As part of the automatic starting sequence, sufficient field current is applied to insure that if necessary an intentional pole-slip will be made. A check is made using a reactive power relay to be sure that the condenser has the correct pole orientation before proceeding further. Following this check, the transfer to full voltage is made by tripping the start breaker and closing the running breaker. At the same time that the starting breaker is tripped, full field forcing by the excitation system is used to minimize the transient current and resultant voltage dip. Typically the starting tap voltage is 35–50%. Where a particularly low tap voltage is used, the transfer transient can be much larger than the initial transient unless control means are used to reduce this. Figure 19 illustrates a case where a slight delay in closing the running breaker reduced the second voltage dip to a small value.[7]

Heavy amortisseurs of the type illustrated in Figure 3 are required for this starting method. As unit sizes increase, it becomes more difficult to provide amortisseur bars of adequate thermal capability and a different starting method is required. Hydrogen-cooled condenser ratings above about 170 MVA will generally require a non-amortisseur type start such as a starting motor.

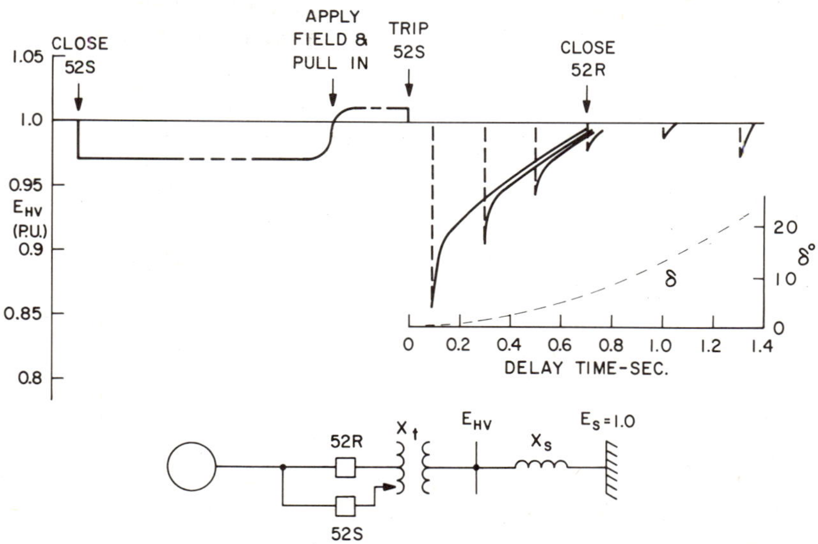

FIGURE 19. Delayed start-run transfer. © 1976 IEEE.

8.5.3. Static Starting

Static starting shown schematically in Figure 20 is basically a synchronous or "back-to-back" type of start in which the condenser is accelerated to rated speed in synchronism with the static equivalent of a starting generator. The static starting equipment is a self-contained static equivalent of not only a starting generator but also its excitation, governor, valves, and synchronizing controls. The major elements of the static starting system are the following:

1. *Static Starter Cubicles* which contain the power thyristor elements and associated control.
2. *Commutating reactors* and the *smoothing reactor* located adjacent to the starter cubicles.
3. *Power switchgear* for supply to the starter and for connection to the unit to be started.

Basically the starter is like the converter equipment of a HVDC terminal except that the receiving "system" into which the power is sent is at changing frequency. During acceleration the line-side converter operates as a rectifier while the machine side converter operates as an inverter. Above a certain minimum speed (5–10%) the condenser can supply the reactive commutation required for operation of the inverter. Below this speed it is necessary to establish the rotating stator flux by successively switching the inverter output from phase to phase.

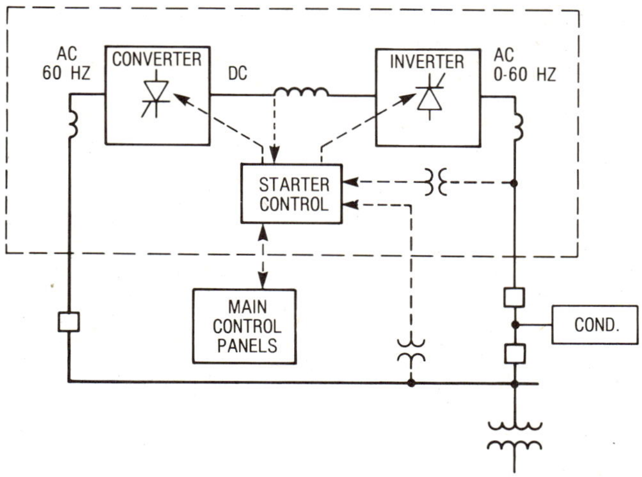

FIGURE 20. Schematic arrangement of static starting system.

8.6. STATION DESIGN CONSIDERATIONS

While detailed discussion of station equipment is beyond the scope of this chapter, several considerations unique to the condenser will be mentioned.

8.6.1. Basic One-Line Arrangement

The one-line diagram of condenser installation varies considerably with differences in the existing substation arrangement to be accommodated and the starting method employed. Figure 21 shows a typical arrangement using a wound-rotor motor for starting.

Where there are tie autotransformers interconnecting two transmission voltage levels at the substation, an economic connection to the power sys-

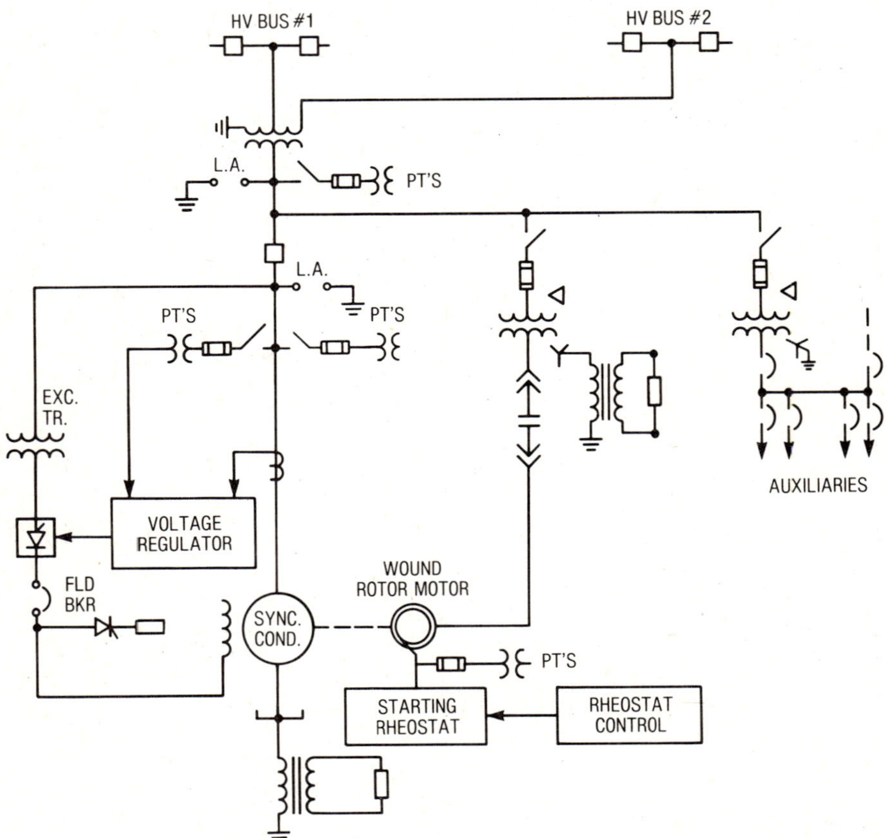

FIGURE 21. Typical one-line arrangement of condenser installation.

tem may be as shown, through the delta tertiary winding of a tie autotransformer. Alternately, the condenser may have its own step-down transformer. Two important and related considerations in either case are the impedance and turns ratio of the transformer. In general, the lowest impedance consistent with circuit-breaker interrupting duties is desirable, to permit maximum utilization of the condenser capability. A low impedance also eases the problem of selecting a transformer turns ratio to match a wide range of conditions, allowing the desired condenser reactive power over the expected range of transmission voltage level variation without exceeding the condenser normal voltage tolerance limits of $\pm 5\%$.

8.6.2. Control and Protection

In nearly all cases condensers are completely automated with remote start-stop and voltage control provisions.

The protective relaying requirements are essentially the same as for a turbine-generator unit; however, there are several special requirements. A different type of loss-of-field relaying is applied, since near-zero field current is normal for a condenser. This function is provided by the combined logic of field current and terminal voltage so that low field current together with low terminal voltage would be considered abnormal.

Under-frequency relaying is also provided on condensers to protect for cases where they become isolated from the system by a remote trip-out.

An important protective function is the condenser stator and rotor winding overload control. Since the only means of "unloading" a condenser is by reducing the field current, this function is normally provided as part of the excitation equipment and consists of automatic runback controls supplementing the normal maximum excitation limit functions.

8.6.3. Auxiliary Systems

The basic auxiliary systems of a condenser in addition to the starting means are for lubrication, cooling, and hydrogen control.

The total auxiliary power requirement for a condenser operating at rated output is of the order of one-tenth of one percent, most of which is associated with the cooling water system.

The ventilation of the machine of Figure 2 is totally enclosed, with hydrogen gas circulated by means of shaft-mounted fans, through gas-to-water heat exchangers. Since cooling water is frequently at a premium, either cooling towers or water-air heat exchangers are generally employed. A closed system containing a glycol mixture for freeze protection is commonly used in low-temperature areas. In other cases freeze control is accomplished by cooling system design allowing automatic draining.

A hydrogen control system maintains normal pressure by means of pressure regulators, providing makeup for leakage from the condenser shell.

The lubricating oil system is usually located in a pit directly beneath the condenser. Lubricating oil is supplied to each bearing at the proper pressure and flow. Also, high pressure oil is supplied to the bearings during start-up in order to lift the journals above the bearing surface, to reduce starting friction and to reduce babbitt wear. A dc backup lube oil pump provides emergency backup to the normal ac lube oil pump.

8.7. SUMMARY

Large hydrogen-cooled condensers have played a major role in the reactive supply requirements of utility systems. Their characteristic of providing a stepless controlled reactive supply for both overexcited and underexcited requirements makes them effective for normal control of transmission voltage levels. More important however is their emergency voltage support capabilities where major disturbances and/or loss of transmission or generation facilities threaten a system collapse. Condensers automatically provide capability in excess of nameplate ratings during an interval required for transmission line reclosing and prime-mover governor controls to act. System disturbances may dictate a need for either capacitive or inductive reactive power to maintain reasonable voltage levels. The MVAr which can be made available during these critical emergencies is an important consideration in the comparison of alternate forms of reactive supply. In general, detailed system studies with a suitable transient stability computer program are required to determine specific requirements.

The basic condenser starting control and protection requirements have been well developed over many years. More recently improvements have been made in two areas to permit more effective transient performance. One is the use of voltage regulator supplementary controls to increase the condenser effectiveness during transient swings. Another is the improvement in overload controls to permit maximum advantage to be taken of the condenser short-time capabilities.

REFERENCES

1. C. R. Canady and J. H. Drake, "Operation of Synchronous Condensers on the Southern California Edison Company System," *Trans. AIEE, Power Appar. Syst.* **71**, 1051–1058 (1952).

2. J. A. Oliver et al., "345 MVA Fully Water-Cooled Synchronous Condenser for Dumont Station," *Trans. IEEE, Power Appar. Syst.* **90,** 2758–2777 (Nov/Dec 1971).

3. F. H. Ryder, J. L. Fink, C. M. Stairs, and R. V. Pohl, "System Design Considerations of the Eel River HVDC Converter Station," *CIGRE 10-05,* 1972.

4. P. G. Brown, H. R. Propst, and J. B. Tice, "Unity Power Factor Essential to Emergency kW Transportation," *Gen. Electr. Co., Electr. Forum,* **1**(2), (Fall 1975).

5. E. W. Kimbark, "How to Improve System Stability Without Risking Subsynchronous Resonance," *Trans. IEEE, Power Appar. Syst.* **96**(5), 1608–1619 (Sept/Oct 1977).

6. P. G. Brown and G. W. Otte, "Electrical Design Considerations in Pumped Storage Hydro Plants," *Trans. IEEE 63-1227 Power Appar. Syst., Special Supplement,* 625–641 (1963).

7. P. G. Brown, "Reduced Voltage Starting Performance of Synchronous Condensers and Pumped Hydro Units," *Trans. IEEE Power Appar. Syst.* 505–511, 643–647 (March/April 1976).

Chapter 9

REACTIVE COMPENSATION AND THE ELECTRIC ARC FURNACE

T. J. E. MILLER and A. R. OLTROGGE

9.1. INTRODUCTION

For the efficient use of electrical power in the steelmaker's arc furnace, a unique combination of problems must be solved, particularly in voltage stabilization, but also in power factor correction and harmonic filtering.

Voltage stabilization can markedly improve furnace operation: either by increasing the maximum power and therefore the rate of steel production; or by enabling the furnace to work at the same maximum power but with a shorter arc, which reduces refractory wear and helps to limit voltage "flicker."

While voltage stabilization is always a benefit to the steelmaker, to the supply utility it may be *necessary* as a remedy for voltage disturbances (flicker) caused by the rapid, large, and erratic variations in furnace current. Such disturbances might otherwise be a nuisance to neighboring customers.

Methods that have been used for voltage stabilization include connection of the furnace at a higher network voltage, synchronous condensers with buffer reactors, and the modern high-speed thyristor-controlled and saturated-reactor compensators. The rapid response of high-speed compensators helps to resolve the conflict between furnace performance and flicker reduction which is met with the synchronous condenser/buffer reactor combination. Connecting the furnace at a higher network voltage is often expensive and sometimes impractical, and in many such cases the thyristor-controlled or saturated-reactor compensator proves to be economic and technically advantageous.

The order in which the subject matter is treated is intended to reflect the importance of voltage stabilization to the steelmaking process. This is in contrast to many publications which only treat the static compensator as a means for limiting flicker on the external system. Flicker correction is considered in Sections 9.3 to 9.5.

9.2. THE ARC FURNACE AS AN ELECTRICAL LOAD

9.2.1. The Arc Furnace in Steelmaking

Worldwide electric arc furnace capacity is today continuing to increase at a significant rate in spite of an overall slowing of the growth in steel production. In the United States alone, between 1975 and 1981, some 15 million tons of newly installed arc furnace capacity brought the arc furnace's share up to one-third of total steel production.[1] There are at present about 850 electric furnaces with electrodes more than 6 in. in diameter, and their capacities range from a few tons to over 400 tons. The largest furnaces now are more than 30 ft in diameter and require up to 110 MW of electric power. Figure 1 shows a typical large furnace.

These statistics underline the importance of the arc furnace both as a steel producer and as a load on the electric supply system.

9.2.2. Electrical Supply Requirements of Arc Furnaces

Furnace Operating Characteristics. In order to appreciate the benefits of voltage stabilization to the furnace, it is necessary to look at the furnace operating characteristics. These can be derived as follows.

The arc can be represented as a variable resistance in a simple single-phase equivalent circuit of the furnace and its supply system (see Figure 2). Although this model is a sweeping simplification of a real furnace, it gives a surprisingly accurate account of furnace operation in terms of averaged quantities.

It is convenient to ignore all resistances except that of the arc. The reactance X is the sum of all reactances in series with the arc: it includes the reactances of the flexible leads and the electrode-arm busbar, the electrodes, the furnace transformer and the main steelworks transformer, and the short-circuit reactance of the supply network. All reactances are referred to the secondary winding of the furnace transformer. The short-circuit reactance of the supply network may be of the order of 15–20% of the total, while the leads from the secondary terminals of the furnace transformer to the graphite electrodes may account for as much as 70%. The emf E is the open-circuit voltage at the furnace transformer second-

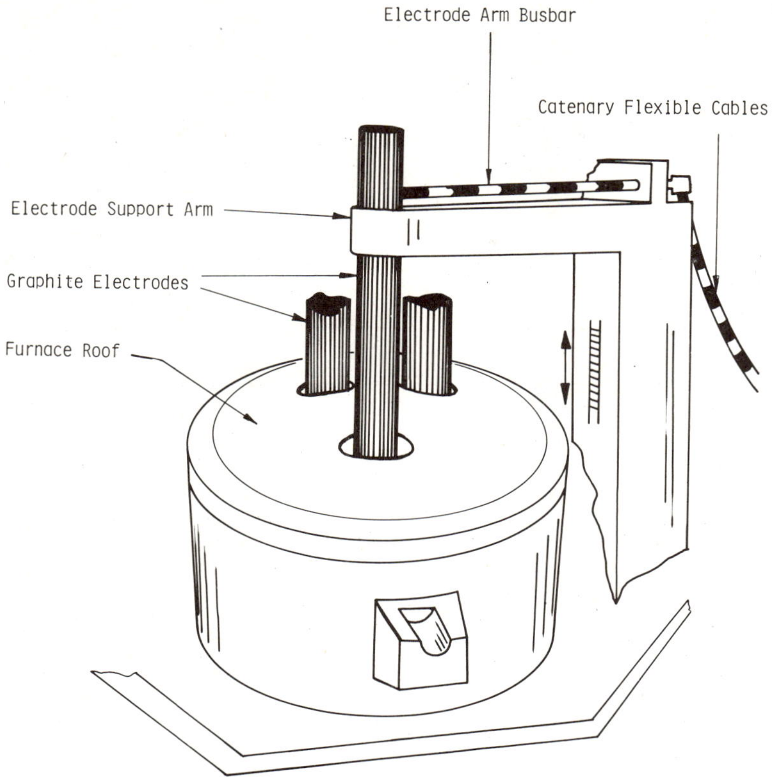

FIGURE 1. General layout of a large modern electric arc furnace (electrical connections are shown for only one phase).

ary terminals, and together with X defines the Thévenin equivalent circuit of the supply to the arc.† The operation of the furnace will be described as though the furnace currents were balanced and sinusoidal.

Control of the circuit of Figure 2 is by vertical movement of the graphite electrodes, which controls the arc length and therefore its voltage (discussed later); and by tapchanging on the furnace transformer, which varies E. Tapchanging also varies X slightly; but this effect will be ignored.

It is well known that in a circuit of the form of Figure 2 the power which can be delivered to the load (i.e., the arc) as R varies is limited to the maximum value

$$P_{max} = E^2/2X \text{ per phase.} \tag{1}$$

† In practice X may be increased by a factor of the order of 1.1 to account for the harmonics, while E is increased by about 5% to allow for uncertainties in reactance values and variation in the supply voltage.

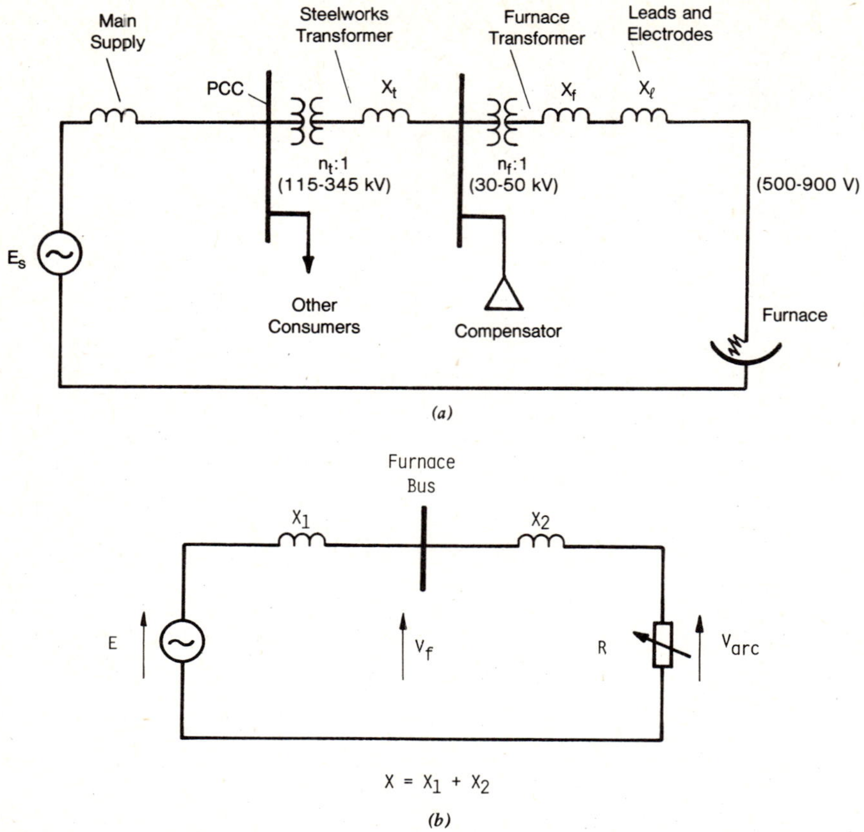

FIGURE 2. Derivation of single-phase equivalent circuit for estimating furnace operating characteristics. © 1981 IEEE.

The value of R at the maximum-power condition is

$$R_{\text{Pmax}} = X \tag{2}$$

and the current corresponding to maximum power is $I_{\text{Pmax}} = E/\sqrt{2}X$. *The voltage drop across the arc is then equal to the voltage drop across X, both being equal to* $E/\sqrt{2}$.

Figure 3 shows the variation of furnace power and other quantities as R varies. Plotted to a base of arc current, these curves are called the "operating characteristics" of the furnace. The tapsetting is assumed fixed in Figure 3. The furnace rating is approximately 60 MW. E is 700 V rms (line–line), and X is 4.04 mOhm, of which 80% is "downstream" of the furnace transformer primary (the "furnace bus" in Figure 2), and 20% is "upstream," giving $X_1 =$ 0.81 mOhm and $X_2 =$ 3.23 mOhm. The primary of the furnace transformer is the most usual

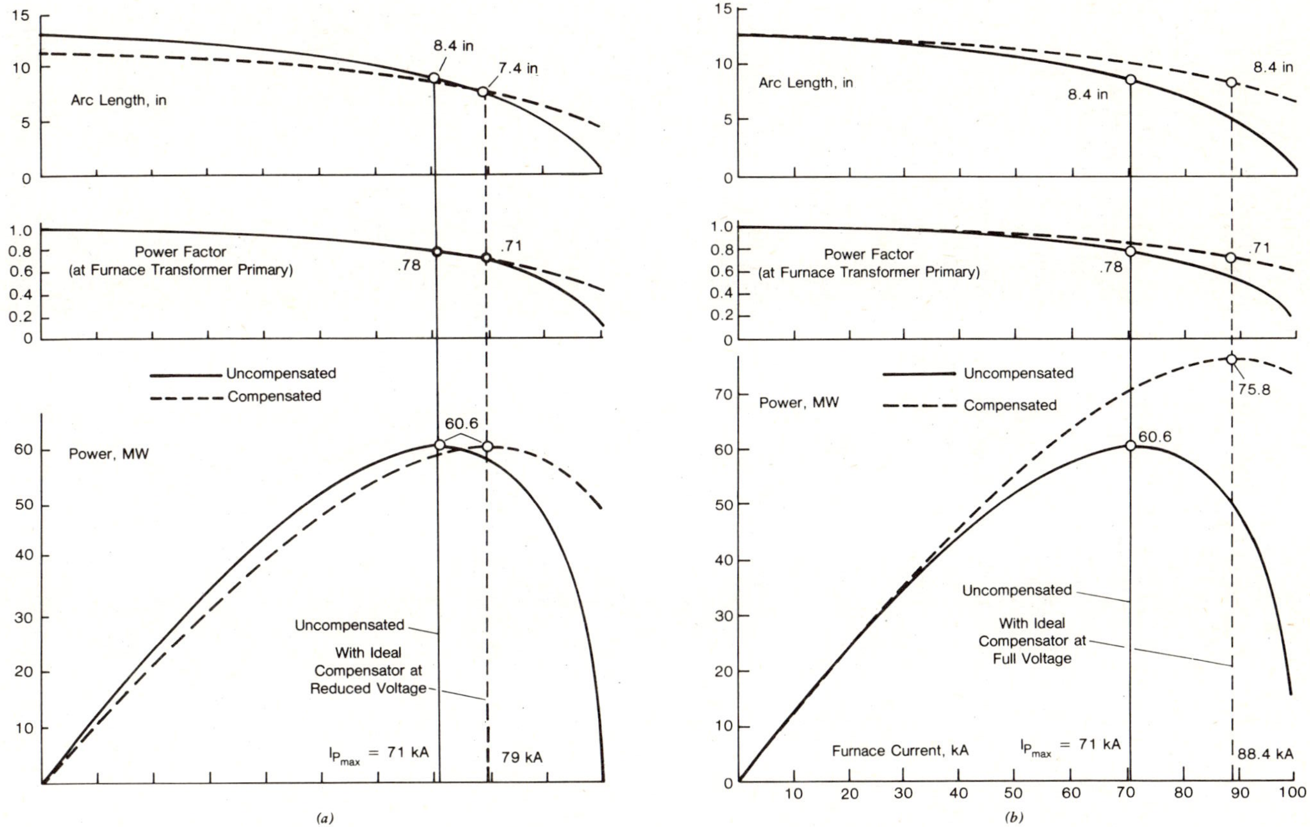

FIGURE 3. (*a*) Furnace operating characteristics showing effect of compensation at reduced voltage. (*b*) Furnace operating characteristics showing effect of compensation at full voltage. © 1981 IEEE.

metering point, and the power factor shown in Figure 3 is at this point. It is here that a compensator would normally be connected.

Choice of Operating Point and the Importance of Supply Reactance. It is known that satisfactory arc stability is achieved at the maximum-power point. A current lower than I_{Pmax} gives a higher power-factor, but a less stable arc and lower power. A higher current, on the other hand, gives a lower power-factor and a lower power, implying uneconomic utilization of the furnace. Taken by itself, this consideration suggests the desirability of operating the furnace near to the maximum-power point. The freedom to operate at this point is, however, limited by the wear rate of the refractory sidewall lining. Refractory maintenance cost is a significant fraction of steel production cost, and is becoming more important with the trend towards higher power inputs in a given furnace diameter. It is therefore important to understand how the wear rate is related to the operating point and the operating power.

The refractory wear rate at a given power level is approximately proportional to the arc length.† Now the arc length depends almost exclusively on the arc voltage, being almost independent of current. Approximately,

$$V_{arc} = kl_{arc} + 40 \text{ Volts rms}, \tag{3}$$

where $k = 29$ V/in. (11.5 V/cm). At the maximum power point $V_{arc} = E/\sqrt{2}$, which from Equation 1 becomes $V_{arc} = \sqrt{P_{max}X}$. Thus for a given maximum arc length (i.e., voltage) and a desired maximum arc power P_{max}, the reactance X must not exceed $V^2_{arc(max)} / P_{max}$. If X exceeds this value, the furnace must be operated at a higher current, that is, to the right of the maximum-power point, in order to limit the arc length. This resort is by no means uncommon in practice, but of course the power and the power-factor are both reduced in consequence. This is illustrated in Figure 3*a* by the solid curves of arc power, length, and power-factor.

The dotted curves in Figure 3*a* show the effect of eliminating X_1, that is, reducing X to 80% of its original value. E is reduced by the square root of this factor in order to keep P_{max} constant. It can be seen that operation at the same maximum power is now possible with a reduced arc length, that is, with less refractory wear. The reduction in X reduces the arc length at *all* levels, not just at maximum power.

† W.E. Schwabe, whose choice of constants appears in Equation 3, has suggested a Refractory Wear Index[3] (RWI) defined by:

$$\frac{\text{Arc Power} \times \text{Arc Length}}{(\text{Arc-sidewall Clearance})^2}$$

In summary, the key to making the desired power input equal to the maximum power with an acceptably short arc is to have the *total reactance X as low as possible.*

Modern furnaces have special arrangements of their flexible leads and electrode-arm busbars to minimize the reactance.[3] The extent to which X can be reduced is limited by practical and economic factors, and by the degree to which the furnace builder can control the reactance upstream of the steelworks. In many cases lower reactance has been achieved by reinforcing the steelworks supply, with direct connection to a higher network voltage having a higher fault level and a lower effective short-circuit reactance. Even when the primary reason for this has been to reduce flicker, the benefits to the furnace are nevertheless the same.

Benefits of Compensation. It is now clear how a compensator can provide the same benefits as a reduction in X. A compensator may be connected at the furnace busbar (see Figure 2a). If it is capable of maintaining constant voltage, that is, $V_f = E$, then X_1 is effectively eliminated. If, as in Figure 3a, the furnace busbar voltage V_f is stabilized to $E\sqrt{X_2/(X_1 + X_2)}$, then the maximum power is unchanged while the arc length and the refractory wear are both reduced.

An alternative compensation strategy is shown in Figure 3b, where the furnace busbar voltage is stabilized to E. Again X_1 is effectively eliminated, and according to Equation 1, this increases the maximum power by the ratio $(X_1 + X_2)/X_2$, which may be as much as 1.3. The arc voltage, and therefore the arc length, at the maximum-power point are unchanged. This is illustrated by the dotted lines in Figure 3b. In the example, $(X_1 + X_2)/X_2 = 1.25$, and the maximum power is increased by this factor. The power factor at the maximum power point is slightly reduced, while the current is increased by the factor 1.25. In this case the power-factor correction kVAr requirement is increased along with the maximum power. At the maximum-power point, for example, the reactive power is increased by the ratio $(X_1 + X_2)/X_2$.

The only types of compensator capable of realizing the benefits of voltage stabilization are active compensators with very rapid response. These include the saturated-reactor compensator, the thyristor-controlled reactor, and the thyristor-switched capacitor. The rapid response is necessary because the compensator works essentially by providing all or most of the reactive current required by the furnace, and as we have seen this can vary rapidly. The synchronous condenser was applied in a few installations before the development of these high-speed compensators. Because of the need for a buffer reactor in series with the furnace busbar and upstream of the condenser, the flicker improvement was limited to the point of common coupling and was obtained at the expense of an *increase* of flicker at the furnace busbar. This had a deleterious effect on furnace performance if two furnaces were operating in parallel.

9.3. FLICKER AND PRINCIPLES OF ITS COMPENSATION

9.3.1. General Nature of the Flicker Problem

Rapid fluctuations in the currents supplied to arc furnaces (and other large loads, e.g., rolling mills, mine hoists) are sometimes large enough to cause voltage fluctuations in the supply system, resulting in lamp flicker. Television receivers and sensitive electronic equipment can also be adversely affected. To minimize these effects, the voltage fluctuations must be kept below some agreed "threshold of irritation." The term "voltage flicker" is often used to embrace all adverse effects of rapid voltage fluctuations.

Characterization of the Flicker Problem. Electric utilities differ in the methods they use to characterize the "threshold of irritation" due to flicker, but broadly speaking there are two main approaches, the SCVD method and the flickermeter method. There are fundamental differences between the two. SCVD = Short-circuit voltage depression, discussed below. Note that the SCVD will be worse at the PCC (point of common coupling) than at other, more remote, points in the supply system where other loads are connected.

(a) The SCVD Method. The SCVD method is based on a correlation between the *incidence of flicker complaints* from other consumers near the steelworks, and the *short-circuit voltage depression* at the point of common coupling. The SCVD is usually expressed as the percentage decrease in voltage at the PCC when the furnace goes from open-circuit to short-circuit on all three phases. The correlation between the SCVD and the incidence of flicker complaints is determined from past surveys. Three of the best known are described in References 18, 30, and 31, and a simplified (though widely used) representation of the UIE/UNIPEDE[31] and AIEE[30] survey results is reproduced in Figure 4 (see also Reference 2). Typical of the standards and recommendations that are applied is the one used in the United Kingdom,[18] where the SCVD is limited to 2.0% at voltages at and below 132 kV, and 1.6% at higher voltages. These figures apply to the SCVD calculated with minimum short-circuit level in the supply system.

The SCVD method of dealing with the flicker problem has important practical advantages. It is simple, and it defines the "threshold of irritation" in a way which should be automatically acceptable to the utility's customers as a whole. With this method it is possible to predict *at the design stage* whether a planned new furnace installation will cause objectionable flicker, because the SCVD can be calculated in advance from the

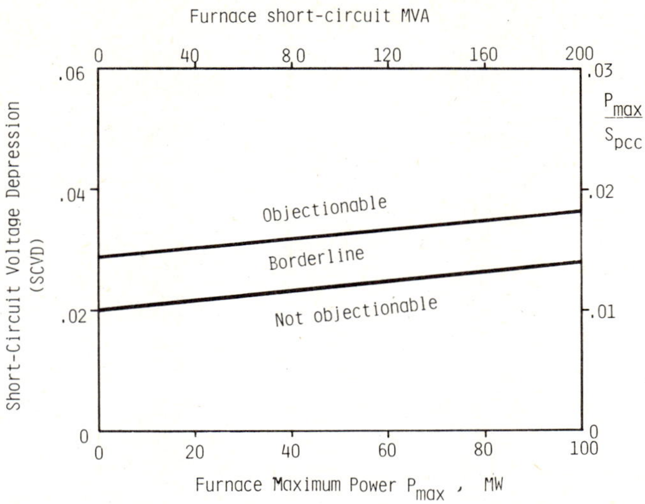

FIGURE 4. Typical relationship between SCVD and objectionability of flicker.

power and reactive power ratings of the furnace and the impedance of the supply system. All of these parameters are usually available at the design stage.

(b) The Flickermeter Method. A difficulty with the SCVD method is that it does not *define* flicker, and therefore it does not indicate how actual voltage flicker can be quantified for the purpose of measurement or prediction. Measurement is important in evaluating the flicker in both compensated and uncompensated installations, while prediction is important in the design of control circuitry for certain types of compensator.

The difficulty of precisely defining flicker is reflected in the large variety of flickermeters which have been tried,[18,22,32,38] using various principles of measurement. Some use a photocell responding to the light from a tungsten filament lamp, while others are directly sensitive to input *voltage.* Most types essentially measure the rms value of the fluctuation (or modulation) of the voltage, because this is the most important single common factor in the flicker perceived by a large number of consumers. There are, however, wide differences in the frequency-responses and in the interpretation of different flickermeters. Some attempt to reproduce the "visual sensitivity" curve of the human eye,[22] while others employ no frequency-response weighting at all. Some types incorporate harmonic filters, on the grounds that harmonics can affect the meter reading without contributing much to the rms value of the fluctuation. Flickermeters also differ widely in the averaging period they use to determine the root-mean-square value. Some give a continuous reading; some give a sampled reading at intervals of a few seconds or minutes; others integrate continuously to give a cumulative "flicker dose."

It appears to be relatively uncommon for electric utilities to specify the "threshold of irritation" in terms of a reading on a particular flickermeter, although the increasing use of these instruments may change this, and already some purchasers of compensation equipment have used such a criterion in their specifications.

The main value of the flickermeter so far has been in making "before and after" measurements of flicker when a compensator is added to an existing furnace installation. However, it is of little help in planning a new furnace installation or in designing a compensator, because it is not at present possible to predict from design data what the flickermeter reading will be, either with or without the compensator.

An example of a "visual sensitivity" curve is reproduced in Figure 5, not for the purpose of describing any particular flickermeter, but merely to show that the eye is sensitive, via electric lighting, to the spectral character of the flicker, and can detect very small voltage variations. The "threshold of perception" of continuous sinusoidal voltage modulation at 7–8 Hz has been put at around 0.3% rms. Figure 5 shows a peak sensitivity to voltage fluctuation frequencies around 6–10 Hz, but perception of sharp variations lasting less than one half-cycle has been recorded. This serves to show that any compensator intended to reduce flicker must be both fast and accurate in response.

The SCVD recommendations of one utility (the U.K. Electricity Council) have been correlated with a particular value of the rms value of the

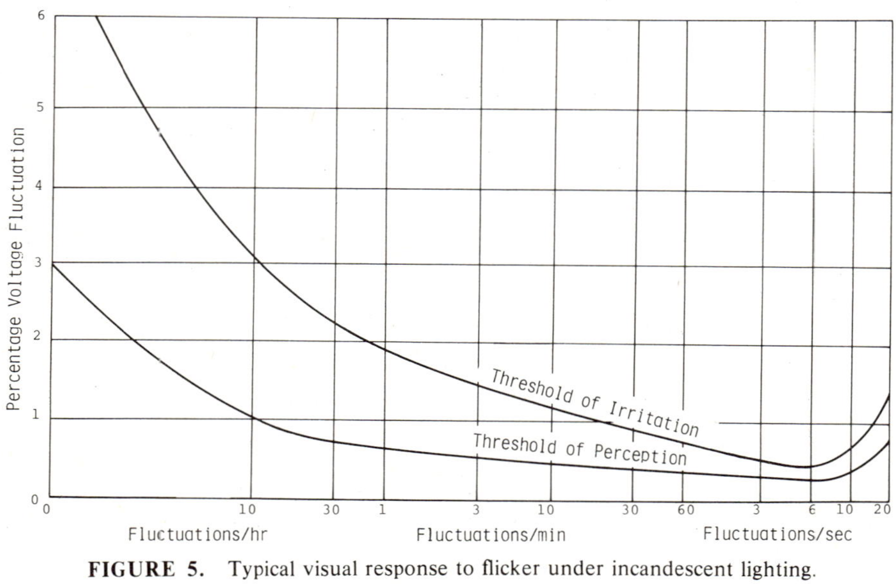

FIGURE 5. Typical visual response to flicker under incandescent lighting.

voltage fluctuation. The survey results reported in Reference 18 showed that, based on complaints, the "threshold of irritation" could be defined by an rms fluctuation exceeding 0.25% for 1% of the time. This is the so-called "gauge-point" fluctuation.

Nature of Arc-Furnace Current Variations. The voltage fluctuations result from the large and sometimes erratic variations in current flowing through the common supply impedance upstream of the point of measurement. A recording of typical furnace power and reactive power is shown in Figure 6. The current tends to be irregular because of ignition delay and the nonlinear resistance of the arc, and because the arc moves about, sometimes erratically, under the combined influence of electromagnetic forces, convection currents, the movement of the electrodes, and the caving-in and sliding of the melting "charge" (which consists mainly of scrap metal in most cases). The currents are unbalanced, distorted, and fluctuate by large amounts even between consecutive half-cycles. The variability and the distortion tend to be more severe during the first few minutes of a melting cycle, as the graphite electrodes are being driven down into the charge ("bore-down"). As the pool of molten metal grows, the arc becomes shorter and more stable and the subsequent "refining" period is characterized by much steadier currents with relatively little distortion. The furnace power may be reduced in this period to a small fraction of its "bore-down" value, by tapchanging on the furnace transformer.

An idea of the size of the current variations in a large furnace can be gained from Figure 6 and the example in Section 9.1, in which a 60-MW furnace operating at an electrode voltage of 700 V has arc currents over 70,000 A per phase. The average reactive power is of the order of 60 MVAr, requiring a very large capacitor bank for complete power factor correction.

Spectral analysis of the distorted furnace current does not result in discrete integral-order harmonic components, but in a continuous

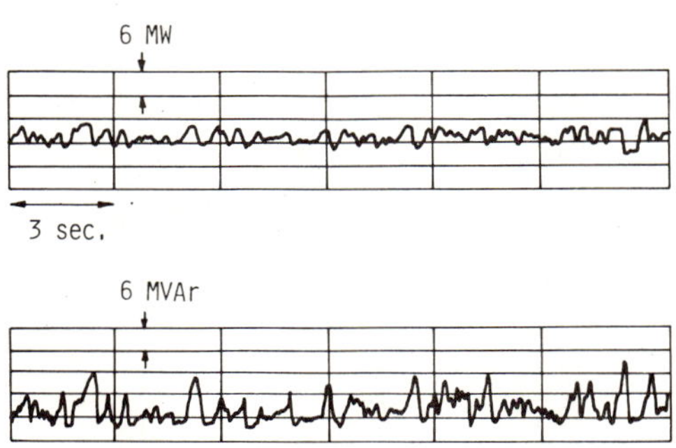

FIGURE 6. Power and reactive power variations of a typical electric arc furnace. © 1981 IEEE.

(though uneven) spectrum whose amplitude is broadly inversely related to frequency.[15,24] It is nevertheless important to extract the spectral amplitudes at the low integral multiples of fundamental frequency for the purpose of rating the harmonic filters associated with the furnace. These are usually tuned to harmonics of such orders (typically 2, 3, 4, 5, and 7). Table 1 summarizes the results of a digital Fourier analysis of furnace currents, used in the determination of filter ratings.[6]

TABLE 1
Illustration of Harmonic Content of Arc Furnace Current at Two Stages of the Melting Cycle

		Harmonic Current % of Fundamental				
Furnace Condition	Harmonic Order:	2	3	4	5	7
Initial melting (active arc)		7.7	5.8	2.5	4.2	3.1
Refining (stable arc)			2.0		2.1	

9.3.2. Flicker Compensation Strategies

If the calculated SCVD of a planned furnace installation falls in the Borderline or Objectionable regions of Figure 4, or if flicker complaints are already being provoked by an existing installation, then the choice must be made between reducing the furnace load, stiffening the supply, or installing compensating equipment.

The supply can be stiffened by tapping it at a higher voltage level, or by installing additional lines, or both. This option may be expensive, but it is often adopted when future expansion of the steelworks or of neighboring loads is anticipated.

When compensation is to be employed, the choice can be made between several different types of compensator. Some of the practical merits and demerits of the different types are summarized in Table 2, and the principles of operation of the most important modern types are discussed in Sections 9.3 and 9.4.

TABLE 2
Some Practical Advantages and Disadvantages of Different Types of Flicker Compensator

Flicker Compensating Equipment or Technique	Advantages	Disadvantages
Thyristor-controlled reactor	Rapid response Independent operation of phases	Requires shunt capacitor for p.f. correction Generates harmonics
Thyristor-switched capacitor	No harmonics generated No reactors required Independent operation of phases	Limited speed of response
Tapped reactor/ saturated reactor	Rapid response gives large flicker suppression Independent operation of phases Transformer-type construction	Requires large shunt capacitor bank for p.f. correction Generates harmonics Applicable to only one furnace
Harmonic-compensated saturated reactor	Rapid response Negligible harmonics generated Transformer-type construction	Requires shunt capacitor for p.f. correction Energizing transients Phases not independently controlled
Synchronous condenser	—	Requires regular maintenance Limited flicker suppression capability, even with buffer reactor
Increase supply short-circuit level	High reliability Future expansion of furnace installation	High cost

Basic Notions of Compensator Performance. The ability of a compensator to reduce voltage flicker depends, in broad terms, on the *compensation ratio* and the *speed of response,* which are defined as follows.

(a) Compensation Ratio. This is defined as

$$C = \frac{\text{Reactive power rating of controllable part of compensator}}{\text{Variation in Reactive Power of Furnace}} . \tag{4}$$

The controllable part of a compensator is distinct from the fixed shunt capacitors which may be included for overall power-factor improvement. It is not uncommon to have $C > 1$, especially where the compensator is required to correct unbalance. This is because the compensating reactive power swing needed in one phase when the other two are short-circuited exceeds the per-phase reactive power swing needed under balanced conditions. Sometimes the compensator is "over-rated" (i.e., $C > 1$) in order to compensate for the voltage drop in the main supply transformer and improve the voltage stability at the PCC.

(b) Speed of Response. The reactive power fluctuation $\Delta Q_\ell(t)$ of the furnace and the compensator's response to it are not instantaneous events, but are both extended in time. The compensator's response $Q_\gamma(t)$ does not in general have the same form as $\Delta Q_\ell(t)$, and in any case both $\Delta Q_\ell(t)$ and $Q_\gamma(t)$ are not unique but depend on the definition of reactive power and the method by which it is measured. $Q_\gamma(t)$ in some cases includes a contribution from the harmonic filters. These factors make it difficult to define the response of the compensator in terms of a single number, and great care should be exercised in interpreting claimed figures for response time or even frequency-response.

Notwithstanding these difficulties it is helpful to examine in a general way the effect of delay in the response of the compensator, under idealized conditions in which the furnace reactive power is modulated sinusoidally at a single frequency:

$$\Delta Q_\ell(t) = \Delta \hat{Q}_\ell \sin \omega_m t . \tag{5}$$

Suppose that the compensator's reactive power response $Q_\gamma(t)$ is also sinusoidal, of amplitude $C\Delta\hat{Q}_\ell$, but retarded in phase by an angle γ. Thus

$$Q_\gamma(t) = C\Delta\hat{Q}_\ell \sin(\omega_m t - \gamma). \tag{6}$$

In general C and γ are both functions of the modulating frequency ω_m and of the type of compensator.† The residual reactive power fluctuation after compensation is

$$Q_\epsilon = \Delta\hat{Q}_\ell(t) - Q_\gamma(t) . \tag{7}$$

† This treatment pre-supposes that reactive power can be defined in such a way as to have an instantaneous value as expressed in Equations 5 and 6.

The relationship between $\Delta Q_\ell(t)$, $Q_\gamma(t)$, and $Q_\epsilon(t)$ is shown in phasor terms in Figure 7. From this diagram it is possible to define a *flicker suppression ratio, F,* as

$$F = \left| \frac{\Delta Q_\ell}{Q_\epsilon} \right| = \frac{1}{\sqrt{1 + C^2 - 2C\cos\gamma}} . \tag{8}$$

The flicker suppression ratio F is, of course, a function of the fluctuation frequency ω_m. F is plotted in Figure 7*b* as a function of the retard angle γ for various values of C. These curves apply for fluctuations at one frequency taken in isolation. It can be seen that with $C = 100\%$, for $\gamma = \pi/3$ there is no improvement in flicker, and for $\gamma > \pi/3$, $F < 1$, that is, the flicker is actually *amplified* by the compensator. A flicker suppression ratio of $F = 4$ at a fluctuation frequency of 10 Hz is representative of a typical compensation requirement, and with $C = 100\%$ this requires $\gamma < .08$ radians (14°). If, as is sometimes done, the compensator delay is represented by a fixed time delay T, then

$$\gamma = \omega_m T = 2\pi f_m T , \tag{9}$$

and to achieve $F = 4$ at 10 Hz requires $T \leqslant 4$ msec (with $C = 100\%$).

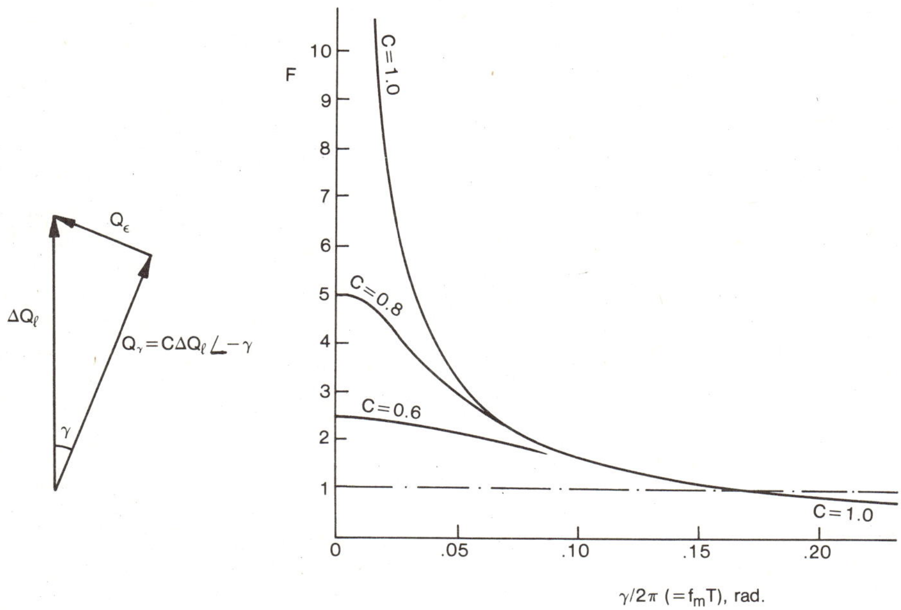

(*a*) (*b*)

FIGURE 7. (*a*) Effect of compensation ratio C and delay γ on compensation of reactive power fluctuations at a single modulating frequency. (*b*) Flicker suppression ratio F as a function of the retard angle γ with $C = 1.0$.

The other curves in Figure 7*b* show that for a given flicker suppression factor, there is a tradeoff between the size of the compensator (C) and its speed of response. A slower compensator may need to be oversized in order to achieve comparable performance with a faster one.

Reactive power fluctuations from a real furnace can never be characterized by a single modulation frequency, and the response of a particular type of compensator to a real furnace cannot be completely defined in terms of a single response delay *T*. Nevertheless, the flicker suppression factor is a useful figure of merit for a compensator's performance. It is often quoted in the literature, although several alternative definitions are used, some corresponding to the SCVD criterion and some to field measurements using a particular type of flickermeter. This is in contrast with Equation 8 which represents the simplest *theoretical* definition. More sophisticated theoretical treatments are possible.

In an attempt to avoid the problem of reproducing the furnace flicker exactly, the flicker with and without the compensator has been measured simultaneously by means of the mimic circuit shown in Figure 8 (see Reference 22). However, this method assumes that the furnace currents are just as distorted and erratic *with* the compensator stabilizing the voltage, as they are without it.

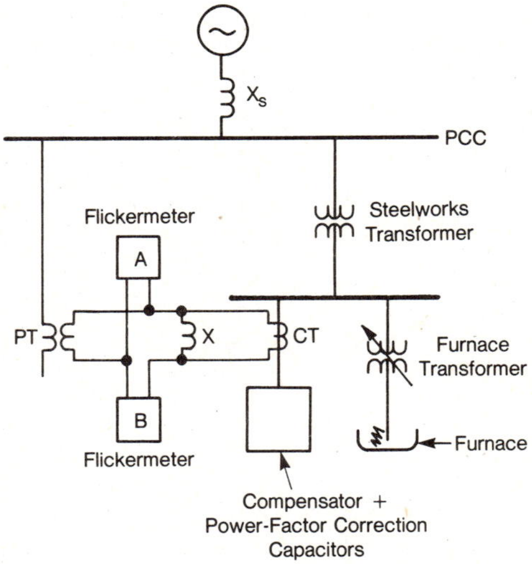

FIGURE 8. Mimic circuit for measuring flicker with and without the compensator. Reactance *X* in the secondary of the CT develops a voltage that represents the compensator's effect on the PCC voltage. This voltage is subtracted from the input to flickermeter B, which reads the flicker without the compensator.

9.3.3. Types of Compensator

Two approaches to the compensation of flicker will be described. Both of these are superior in overall performance to the older synchronous condenser approach, which is not described. (See References 19, 20, and Table 2).

The *saturated-reactor compensator* was the first of the modern fast-response compensators to be applied widely. First reported in service by Friedlander in 1964,[8] this compensator appears in two main variants: the earlier *tapped-reactor* arrangement and the *busbar* compensator,[14] which employs the twin- or treble-tripler type of polyphase harmonic-compensated self-saturating reactor.[9,39,40] The tapped-reactor scheme is essentially a single-phase device, and although it is rapid in response it requires a series element, which makes it unsuitable for multiple-furnace installations. Its rapid response is achieved at the expense of a large harmonic filtering requirement.

The second approach to flicker compensation is that employed in the thyristor-controlled reactor (TCR) or thyristor-switched capacitor (TSC) compensators.[4,5,22,27,38,41–46] Like the synchronous condenser and the busbar variant of the saturated-reactor compensator, these compensators are connected in parallel with the furnace (or furnaces) at the furnace busbar. In simplified terms, the principle is to measure the reactive power (or current) of the furnace as fast as possible, and to control the compensator in such a way that the sum of the furnace reactive power and the compensator reactive power is as nearly constant as possible.

Harmonic filters are often necessary to absorb the furnace harmonics. Conceptually the compensator is only intended to stabilize and compensate the fundamental-frequency reactive power. The TCR and the tapped-reactor types themselves generate harmonic currents in addition to those of the furnace. The duty for the harmonic filters can therefore vary according to the type of compensator used. It also depends strongly on the frequency-response characteristics of the supply system.

A third approach to flicker compensation is a direct attempt to achieve *instantaneous* compensation of reactive current including harmonic currents.[33] One device suggested for this is a tolerance-band current controller,[33-35] which is controlled to provide all components of the furnace current except that in-phase fundamental-frequency component which is required to produce the useful power of the furnace. Although this approach is in a sense the most direct it requires sophisticated power-electronics technology with a large number of thyristor switches. It has so far not been widely applied, and is not described here.

9.4. THYRISTOR-CONTROLLED COMPENSATORS

The basic element of all thyristor-controlled compensators is a capacitor or reactor in series with an ac thyristor switch, as shown in Figures 9 and 10. The effective fundamental-frequency susceptance is controlled by switching, the control signals being applied to the gates of the thyristors. Connected in parallel with the arc furnace, the controlled susceptance supplies or absorbs reactive power in such a way that the combined reactive power of the furnace and the compensator is as nearly constant as possible. In TCR-compensated systems this constant reactive power is usually equal to the maximum frequently repeated reactive power peaks of the furnace, and the overall power factor is corrected by means of shunt capacitors. This is called "indirect" compensation. In TSC-compensated systems the constant reactive power is near zero, the TSC generating what the furnace absorbs. This is called "direct" compensation. This strategy automatically reduces the voltage fluctuations at and upstream of the compensator connection point. Fixed capacitors in parallel with the compensator can be used to adjust the overall average power factor.

9.4.1. Relationship between Compensator Reactive Power and Thyristor Gating Angle

It is important to understand how the reactive power of the compensator is determined by the *gating angles* of the thyristor switches.

The *thyristor-switched capacitor* (TSC) is comprised in each phase of several smaller capacitor units in parallel, each one having its own thyris-

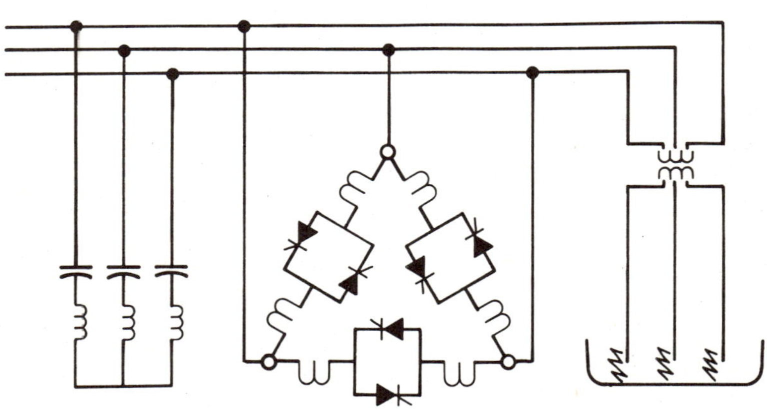

FIGURE 9. General arrangement of a thyristor-controlled reactor compensator applied to an arc furnace. Note the reactors which tune the shunt power-factor correction capacitors to low-order integral harmonic frequencies. © 1981 IEEE.

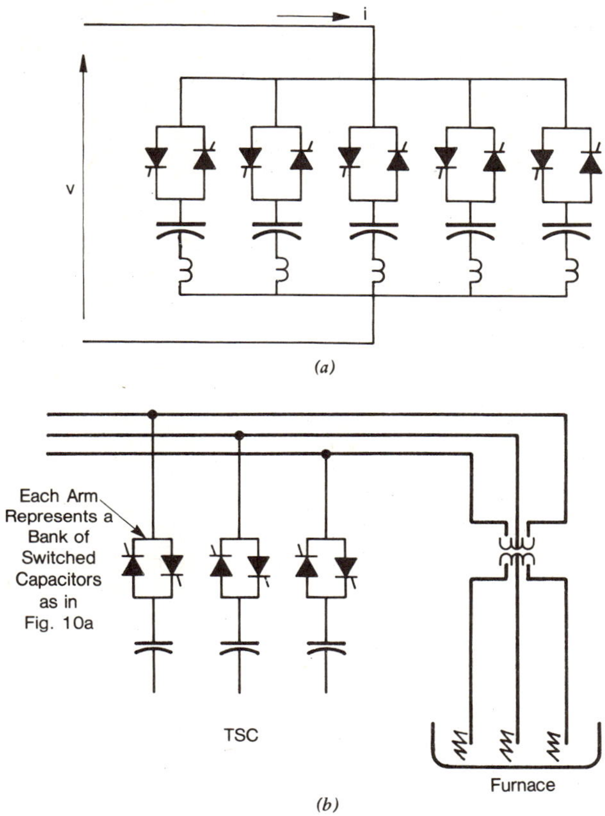

FIGURE 10. (*a*) Principle of TSC compensator. (*b*) General arrangement of TSC compensator applied to an arc furnace. Note the tuning reactors.

tor switch (Figure 10). The demand for reactive power is met by switching into (or out of) conduction the appropriate *number* of capacitor units according to the relationship shown in Figure 26 of Chapter 4. The ideal switching waveforms of voltage and current for one capacitor unit are shown in Figure 11. In the "off" state the capacitor has a precharge voltage equal to the positive or negative peak of the ac voltage, which it has retained from the current-zero when it was last switched off. In order to minimize transient currents the capacitor is switched on only when the ac voltage is equal to (or very close to) the precharge voltage. This happens only once per period so the control circuit always has at least one full period in which to "decide" whether or not to switch on a particular waiting capacitor unit. Similarly the control circuit always has at least one *half*-period in which to "decide" whether or not to switch *off* a particular conducting capacitor (by withholding the gating pulse to the conducting thyristor at the next natural current-zero). In the interest of rapid

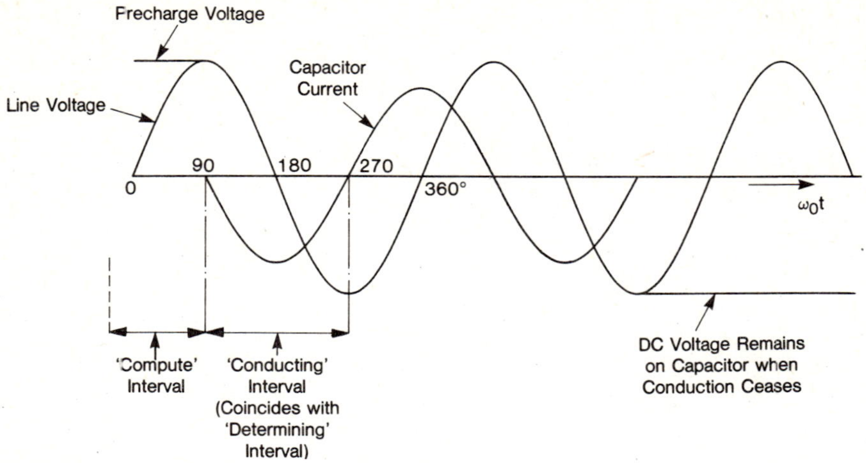

FIGURE 11. Switching voltage and current waveforms in the thyristor-switched capacitor (ideal case).

response these control decisions (which include measurement of the reactive power requirement and coordination of the states of all thyristors and capacitors) are usually left until the last few milliseconds of the available wait time.

Once a thyristor has been gated, it continues to conduct until the next natural (i.e., circuit-determined) current-zero. In the absence of harmonics originating outside the TSC, conduction periods will always be integral multiples of one exact half-period, and the TSC itself generates no harmonics. The issuance of a gating pulse to a thyristor switch commits the associated capacitor unit to conduct for the next half-period and so deliver compensating reactive power according to the definition

$$q = \frac{2}{T} \int_{T/2} v(t)\, i_c(t)\, \mathrm{d}t\,. \tag{10}$$

This emphasizes the fact the reactive power response is not an instantaneous event but is defined in terms of an average taken through an interval of time. Because the integration interval is not one exact period, harmonics in v or i_c can and do distort this result. The overall effect of the harmonics must be small enough to allow the control system to switch in the correct number of capacitors to provide the required "fundamental" reactive power.

The integration interval in Equation 10 can be called the *determining interval* for the reactive power of the compensator. The gating pulse decision cannot be made any later than the start of the *conducting* interval. In the TSC the conducting interval coincides with the determining interval. Therefore the "compute" interval (during which the reactive power demand is measured and the decision whether or not to gate the thyristor is made) must always be completed by the start of the determining interval.

The *thyristor-controlled reactor* (TCR) comprises in each phase only one reactor, controlled by a thyristor switch (Figure 9). The demand for reactive power is met by controlling the *duration* of the conducting interval in each half-period, as shown in Figure 12, by issuing gating pulses to the thyristors either later or earlier (relative to a voltage peak), according as the reactive power demand is less or greater. This process is called *phase control.*† As in the TSC, conduction ceases at a natural current-zero. Since the reactor is left with zero energy storage at a current-zero, it can subsequently be switched on again without transients at any point in the voltage cycle. However, in order to maintain balance between the oppositely poled thyristors in the switch, and avoid dc and even-harmonic currents, the gating pulses are restricted to the 90° period immediately following a voltage peak; that is, $90° < \alpha < 180°$.

The phase-controlled currents contain harmonics which have to be absorbed by parallel-connected filters. The *fundamental* current component I_{L1} is related to the gating angle α by the equation

$$I_{L1} = \frac{V}{\pi \omega L}(2\pi - 2\alpha + \sin 2\alpha), \tag{11}$$

where L is the inductance of the reactor and ω is the fundamental radian frequency. This relationship is plotted in Figure 7 of Chapter 4. It is

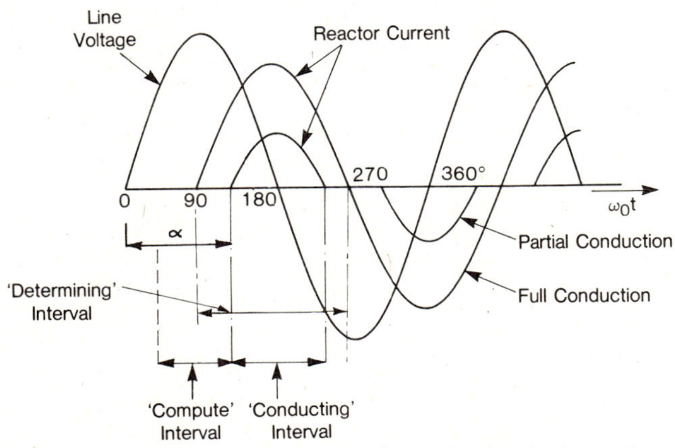

FIGURE 12. Voltage and current switching waveforms in the thyristor phase-controlled reactor.

† The switching of capacitors in the TSC is sometimes called "integral cycle control" or, more strictly, "integral half-cycle control."

from I_{L1} that the reactive power delivered by the compensator is determined, according to the definition

$$q = VI_{L1} \,. \tag{12}$$

This again must be interpreted in terms of an average taken over one half-period, because I_{L1} in Equation 11 was determined by Fourier analysis as

$$I_{L1} = \frac{2}{\pi} \int_{T/2} i_L(\alpha) \sin \omega t \, dt \,. \tag{13}$$

The issuance of a gating pulse to a thyristor switch at an angle α commits the associated reactor to conduct until the next current-zero and so deliver compensating reactive power according to the definition in Equations 12 and 13. The integration interval (that is, the *determining interval*) in Equation 13 is from 90° to 270° on the voltage waveform, and unless $\alpha = 90°$ it starts *before* the gating pulse is issued. In other words, the conducting interval is generally shorter than the determining interval, and since these two intervals are both symmetrical about the 180° point on the voltage waveform, the gating pulse is generally issued *after* the start of the determining interval. This gives rise to the possibility of extending the "compute" interval beyond the start of the determining interval, right up to the start of the conducting interval, which improves the speed with which the TCR responds to a given demand for "fundamental" reactive power.

The gating angle α may be determined by an algorithmic circuit operating according to Figure 13. At each voltage peak the trajectory labeled "prospective compensating reactive power" is initiated. This trajectory is defined by Equation 11, with α increasing linearly through time. The curve labeled $q_d(t)$ is a continuous signal representing the computed reactive power which is needed for compensating the load (discussed later). A comparator detects the instant when the two curves cross, and causes a gating pulse to be issued to the appropriate thyristor. Logic circuitry determines which of the two oppositely poled thyristors in each switch should be gated.

9.4.2. Determination of Reactive Power Demand

Although the TCR and the TSC differ as to the time and manner in which gating pulse decisions are made, they both require a signal which accurately represents the most up-to-date value of the "fundamental" or "displacement" reactive power of the furnace. This reactive power can be defined per phase by

$$q_\ell(t) = \frac{1}{T} \int_{t-T}^{t} v(t) i(t - T/4) \, dt \,, \tag{14}$$

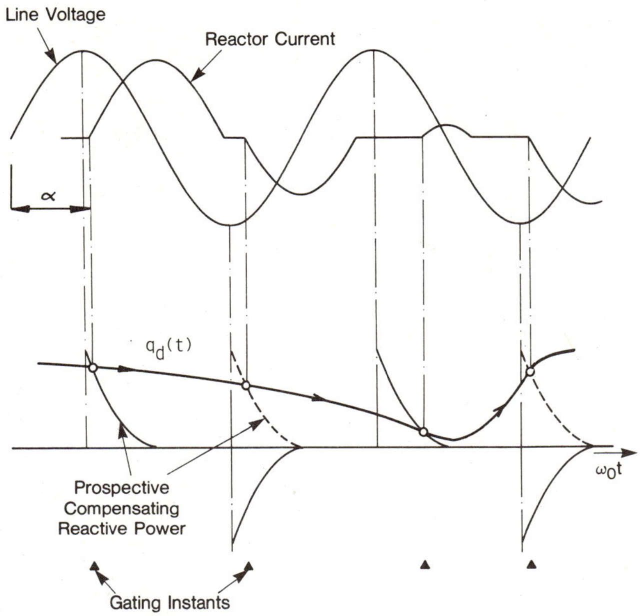

FIGURE 13. Principles of algorithmic circuit for determining thyristor gating angles.

that is, as a "running average." This relationship can be realized in an algorithmic circuit, which gives a continuous signal representing the load reactive power in the period which has just finished at time t. (In practice the integrand may be more complicated than shown here, if the compensator is expected to correct phase unbalance; see Chapter 1.) The integration interval is shown here as exactly one period. With any other value of integration interval, $q_\ell(t)$ is corrupted by harmonic components of the current i (voltage harmonics being negligible), and $q_\ell(t)$ no longer means the "fundamental" reactive power. This is in principle undesirable because the compensator proper is intended to compensate only the "fundamental" reactive power while the harmonics or "harmonic reactive power" are absorbed in parallel-connected filters. In practice it is found that an integration interval of one period is too long for acceptable flicker suppression, and compensators which use the "running average" method may employ a shorter integration interval, sacrificing accuracy for speed in an attempt to optimize the flicker suppression capability.

Another method used to determine $q_\ell(t)$ is based on the relationship

$$\hat{v}\hat{i}_\ell \sin\phi = -\hat{v}\sin\omega t \cdot \hat{i}_\ell \sin(\omega t - \phi - \pi/2) + \hat{v}\sin(\omega t - \pi/2) \cdot \hat{i}_\ell \sin(\omega t - \phi)\ . \tag{15}$$

The left-hand side is equal to twice the load reactive power per phase under steady-state, fundamental-frequency conditions. This relationship can be realized by the algorithmic circuit shown in Figure 14. Such a circuit can be made extremely rapid in response. For example, if the current $\hat{i}_\ell \sin(\omega t - \phi)$ is suddenly switched on at time $t = \phi/\omega$, the output can reach a steady value $\hat{v}\hat{i} \sin \phi$ within 2 msec. However, the result is corrupted by harmonic components in the current. These can be removed by filtering, as shown in Figure 14, but filtering introduces a further response delay. Some of this can be recovered by means of phase-lead compensation techniques. A realistic step-response time of the type described above is then of the order of 4–5 msec.

An additional problem may arise as a result of taking products of v and i_ℓ. If i_ℓ has the form of a fundamental-frequency sinewave modulated at a frequency f_{m}, then the output of the measuring circuit may have components at the frequencies f_{m}, f, $2f - f_{\mathrm{m}}$, $2f$, and $2f + f_{\mathrm{m}}$. The compensator is, of course, intended to respond only to the f_{m} component, but there may be some residual response at higher frequencies. Should there be a supply system resonance at or near the frequency $2f - f_{\mathrm{m}}$ there can result undesirable interference between this resonance and the compensator. This can be avoided by means of adequate roll-off of the compensator's frequency-response at higher frequencies.

Several other algorithmic circuits for determining the load reactive power have been described in the literature. The requirements for reactive-power compensation and phase balancing are discussed in more detail in Chapter 1.

9.4.3. Example of Flicker Compensation Results with a TCR Compensator

Figure 15 shows an actual recording of the compensated and uncompensated voltage fluctuations at the 230-kV point of common coupling. The TCR compensator in this example is rated 60 MVAr and connected at 34.5 kV.

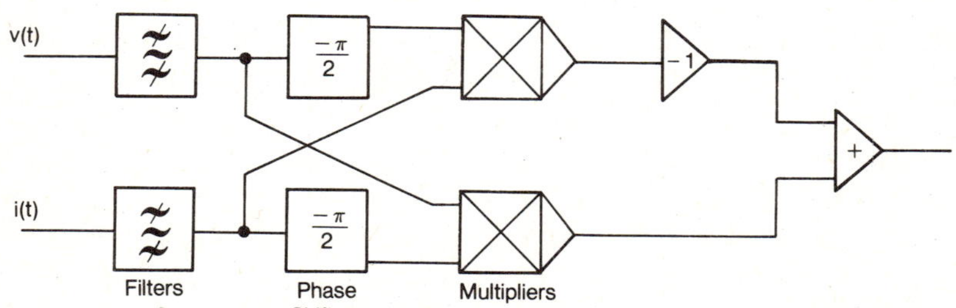

FIGURE 14. Algorithmic circuit for determining instantaneous reactive power $q_\ell(t)$.

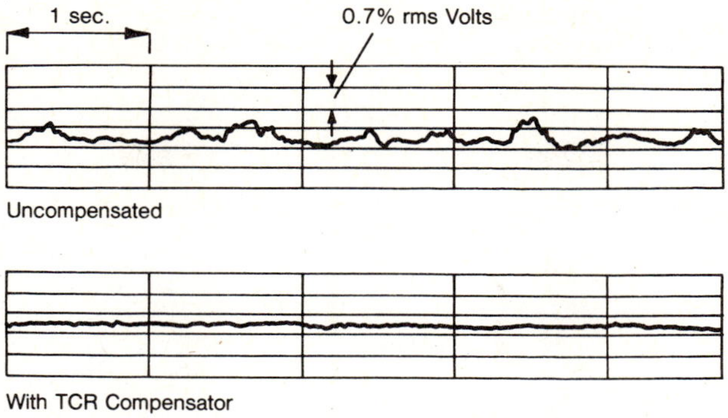

FIGURE 15. Typical improvement in the rms voltage at the PCC with a thyristor-controlled reactor compensator.

9.5. SATURATED-REACTOR COMPENSATORS

9.5.1. The Tapped-Reactor/Saturated-Reactor Compensator

The ideal device for eliminating flicker would be a *constant-voltage* reactive compensator. An approximation to such a device is the self-saturating reactor. Figure 16 shows the relationship between terminal voltage and *fundamental* current for a single-phase saturated reactor with just one winding on a closed iron core. The *V-I* curve is related to the *B-H* curve of the core steel, but differs from it in shape because of harmonic currents which are inevitable with such nonlinearity. For operation above the ac knee-point in Figure 16, the fundamental *V-I* characteristic can be approximately represented by the equation

$$\mathbf{V} = \mathbf{V}_k + j X_r \mathbf{I}_r . \tag{16}$$

Since the average power in the reactor is zero, $\mathbf{V}$ and $\mathbf{V}_k$ are in phase and the phasor notation can be dropped. Equation 16 is then represented by the straight line in Figure 16. The characteristic is not quite constant-voltage, owing to the finite slope reactance X_r, which is typically 0.1–0.15 pu. The furnace impedance can vary from full open-circuit to full short-circuit. With such a widely varying load the finite slope reactance X_r prevents the reactor from drawing enough current to match the short-circuit current of the furnace, and only partial compensation is achieved with the plain saturated reactor. To overcome this, a *tapped reactor* is introduced (Figure 17) which redistributes the "leverage" that the saturated reactor and the furnace currents have on the furnace-busbar voltage. It is even possible to overcompensate so that approximately constant voltage is achieved at some point upstream of the compensator connection (for example, at the PCC).

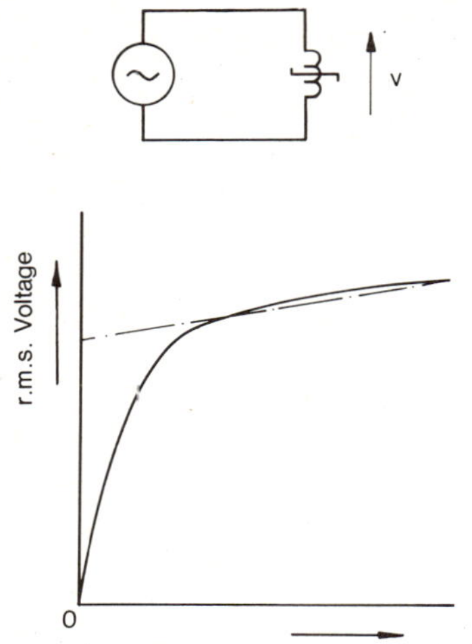

FIGURE 16. Relationship between voltage and fundamental current for a simple single-phase closed-core saturating reactor.

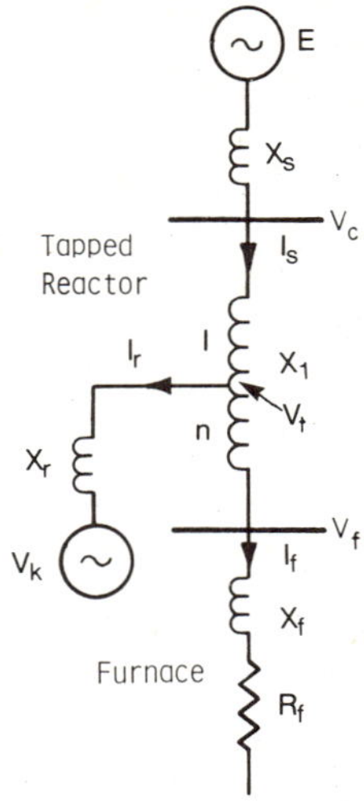

FIGURE 17. Principle of the tapped-reactor/saturated-reactor compensator.

The principle of the scheme is shown by the following simplified analysis, in which all resistances (except that of the arc) are ignored (see references 7, 8, and 36). The voltage divides across the two sections of the tapped reactor according to the relations

$$\mathbf{V}_c - \mathbf{V}_t = j\,(\mathbf{I}_s + n\mathbf{I}_f)X_1 \tag{17}$$

$$\mathbf{V}_t - \mathbf{V}_f = j\,n\,(\mathbf{I}_s + n\mathbf{I}_f)X_1\,, \tag{18}$$

and the furnace voltage is given by

$$\mathbf{V}_f = \mathbf{V}_c - j(\mathbf{I}_s + n\mathbf{I}_f)(1+n)X_1\,. \tag{19}$$

The supply current $\mathbf{I}_s$ must satisfy the equation

$$\mathbf{E} - \mathbf{V}_k = jX_s\mathbf{I}_s + (\mathbf{V}_c - \mathbf{V}_t) + jX_r\mathbf{I}_r \tag{20}$$

$$= j\,(X_s + X_1 + X_r)\mathbf{I}_s + j\,(nX_1 - X_r)\mathbf{I}_f\,.$$

If X_1, the tapped-reactor total reactance, and n, its tapping ratio, are chosen to satisfy

$$nX_1 = X_r\,, \tag{21}$$

then the supply current $\mathbf{I}_s$ *is independent of the furnace current* $\mathbf{I}_f$. $\mathbf{I}_s$ is then given by Equations 20 and 21:

$$\mathbf{E} - \mathbf{V}_k = j(X_s + \frac{1+n}{n}\,X_r)\mathbf{I}_s \tag{22}$$

Neglecting any phase change between $\mathbf{E}$ and $\mathbf{V}_k$ (which in practice is small anyway), the supply current is constant. The voltage at all points upstream of the tapped reactor is therefore constant and free from fluctuations. In particular, since $\mathbf{E} - j X_s \mathbf{I}_s = \mathbf{V}_c$, the voltage at the supply end of the tapped reactor is constant and is given by

$$\mathbf{V}_c = \mathbf{V}_k + j \frac{1+n}{n} X_r \mathbf{I}_s \tag{23}$$

The voltage supplied to the primary of the furnace transformer is now given by Equations 19, 21, and 23 as

$$\mathbf{V}_f = \mathbf{V}_k - j(1+n) X_r \mathbf{I}_f , \tag{24}$$

which is *independent of the supply voltage* $\mathbf{E}$ (or $\mathbf{V}_c$) and is determined exclusively by the compensator and the furnace current.

A disadvantage of the plain tapped-reactor/saturated-reactor arrangement described so far is that the furnace open-circuit voltage is reduced to V_k (see Equation 24 with $\mathbf{I}_f=0$) from its original level which was intermediate between E and V_c (depending on the other loads connected at the PCC and elsewhere). At the same time the short-circuit reactance is increased from $(X_s + X_f)$ to $X_s+X_f+X_r(1+n)$. Both of these effects reduce the furnace power. In order to restore the furnace supply to the condition it had without the compensator, the voltage boosting winding shown in Reference 8 has been successfully used. The alternative of reducing k by means of a capacitor in series with the saturated reactor is not used because it delays the response of the compensator.[7] The power factor is also reduced by the tapped-reactor/saturated-reactor compensator, and capacitors are generally connected at the supply end of the tapped reactor. These are usually divided into banks and combined with linear air-core reactors to make harmonic filters to absorb harmonic currents from the furnace as well as from the compensator, and to avoid system resonances. Filters for frequencies up to the 15th have been employed.[8]

9.5.2. The Polyphase Harmonic-Compensated Self-Saturating Reactor Compensator

A second type of saturated-reactor compensator, quite distinct from the tapped-reactor arrangement described in Section 9.4.1, is the polyphase harmonic-compensated reactor. Sometimes called a busbar compensator (because it can be connected directly to the furnace busbar without the need for a tapped series reactor), this compensator was developed by Friedlander (see Figure 38 of Chapter 5). Two main variants are commonly used, the twin-tripler and treble-tripler reactors.

The basic principle is the same as that of the tapped-reactor/saturated-reactor scheme, that is, to use the flat V-I characteristic of a saturating

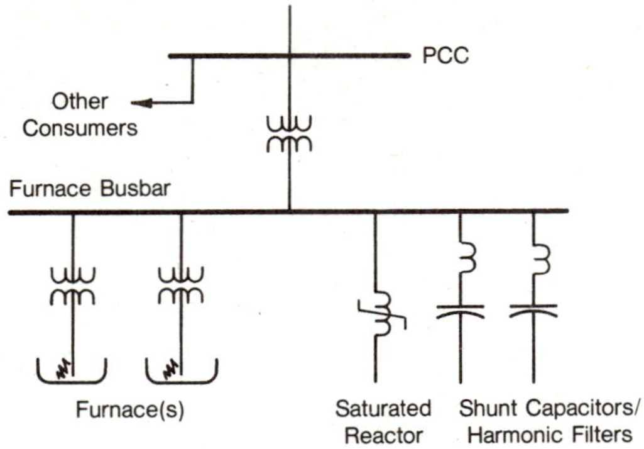

FIGURE 18. General arrangement for flicker compensation with busbar-type of saturated-reactor compensator.

iron core to approximate an ideal constant-voltage reactive characteristic. In the twin- and treble-tripler reactors, however, the three phases are mutually coupled on a multiple-limb core in a phase-and-frequency-multiplying arrangement which results in the elimination of harmonic currents up to orders $2n \pm 1$, where n is the number of "phases" produced (cf. the pulse-multiplication techniques used in rectifiers). In the twin-tripler $n=6$, and the lowest-order harmonic currents are the 11th and 13th. In the treble-tripler $n=9$, and the lowest-order harmonic currents are are the 17th and 19th under balanced conditions. By loading the main reactor with a "mesh tuning reactor," which itself is loaded with a short-circuited delta-connected winding, the lowest-order harmonics can be further suppressed, leaving negligible residual harmonics below the 23rd and 25th in the twin-tripler and the 35th and 37th in the treble-tripler under balanced conditions (1–1.5% and 2–4% respectively).

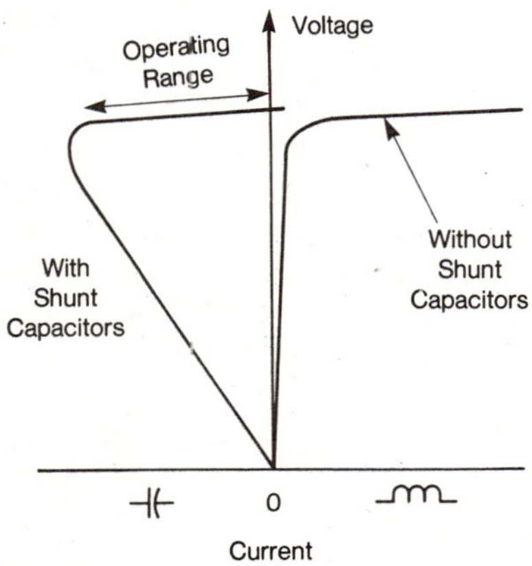

FIGURE 19. V-I characteristic of saturated-reactor compensator of the busbar-type with and without shunt capacitors.

Both of these harmonic suppression techniques have the additional effect of improving the linearity and flatness of the *V-I* characteristic to within ±0.3% over the normal range of currents (i.e., from about 10% to 100% of rated current).[11] However, the linearity as well as the size of the iron core can be unfavorably affected unless the slope is kept above about 7%. In arc furnace applications this residual slope reduces the effective compensation ratio. The obtainable flicker suppression ratio is somewhat less than what is obtainable with the tapped-reactor/saturated-reactor compensator.

The internal harmonic cancelation in the saturated reactor is imperfect under unbalanced conditions, but the harmonic currents which appear are usually not excessive in the presence of current harmonics from the arc furnace.

Shunt capacitors are usually provided for overall power-factor correction, and these may be divided up into harmonic filters, typically for orders 2, 3, 5, 7 and sometimes also for 4 and 6 and a "high-pass" branch (Figure 18). The shunt capacitors modify the resulting *V-I* characteristic as shown in Figure 19. Compensators rated up to 117 MVAr have been applied at voltages up to 70 kV, for rolling mills and other highly variable large loads as well as for arc furnaces.

REFERENCES

1. G. J. McManus, "Electric Furnace Succeeds in Technology and Profit," *Iron Age*, MP-7 to MP-18, February 4, 1980.
2. A. R. Oltrogge, "Arc Furnace Voltage Can be Critical," *Electrical World*, December 1, 1974.
3. E. R. Freeman and J. E. Medley, "Efficient Use of Power in Electric Arc Furnaces," *Proc. IEE, Electr. Power Appl.* **1** (1), 17–24, February 1978.
4. H. Frank and K. Pettersson, "Raising the Production of Arc Furnaces by Stabilizing the Voltage with Thyristor-Switched Capacitors," *ASEA J.* **50** (1), 9–16 (1977).
5. Y. Sundberg, "The Arc Furnace as a Load on the Network," *ASEA J.* **49** (4), 75–87 (1976).
6. L. E. Bock and A. H. Moore, "Application of Capacitors in Arc Furnace Power Supply Systems," *Industr. Power Syst.* **19** (3), 18–23, September 1976.
7. G. F. L. Dixon, E. Friedlander, R. Seddon, and D. J. Young, "Static Shunt Compensation for Voltage-Flicker Suppression," Paper 7, IEE Symposium on Transient, Fluctuating, and Distorting Loads and their Effect on Power Systems and Communications, February 1963. IEE Conference Report Series No. 8 (1964), "Abnormal Loads on Power Systems."
8. E. Friedlander, A. Telahun, and D. J. Young, "Arc-Furnace Flicker Compensation in Ethiopia," *GEC J. Sci. Tech.* **32** (1), 2–10 (1965).

9. D. H. Berry et al., "Saturated Reactor Compensator Achieves Major Reduction of Flicker Caused by Arc Furnace Installation," presented at CEA Spring Meeting, Toronto, March 22–24, 1976.

10. M. W. Kennedy, J. Loughran, and D. J. Young, "Applications of a Static Suppressor to Reduce Voltage Fluctuations Caused by a Multiple Arc Furnace Installation," *IEE Conf. Publ.* 110, 130–137, April 22–24, 1974.

11. E. Clegg, A. J. Heath, and D. J. Young, "The Static Compensator for the BSC Anchor Project," *IEE Conf. Publ.* 110, 110–116, April 22–24, 1974.

12. Y. H. George and R. Gosselin, "Static Compensators of Simplified Construction as a Remedy to Power System Disturbances," *IEE Conf. Publ.* 110, 117–123, April 22–24, 1974.

13. C. G. Cooksley et al., "The Compensation of Voltage Dips and Harmonics at British Steel Corporation's Anchor Development," *IEE Conf. Publ.* 110, 124–129, April 22–24, 1974.

14. C. B. Cooper, E. Friedlander and D. J. Young, "Requirements and Compensation Methods for Scrap Melting Arc Furnaces," *IEE Conf. Publ.* 110, 146–150, April 22–24, 1974.

15. R. Coates and G. L. Brewer, "The Measurement and Analysis of Waveform Distortion Caused by a Large Multi-Furnace Arc Furnace Installation," *IEE Conf. Publ.* 110, 135–145, April 22–24, 1974.

16. H. S. Brown, Hunter Memorial Lecture, November 1978, *Proc. IEE, Electr. Power Appl.*, **2** (3), 99–107, June 1979.

17. R. Armstrong, "Predicting the Electrical Performance of Arc Furnaces," *Proc. IEE, Electr. Power Appl.* **1** (3), 86–90, August 1978.

18. G. F. L. Dixon and P. G. Kendall, "Supply to Arc Furnaces: Measurement and Prediction of Supply-Voltage Fluctuation," *Proc. IEE* **119** (4), 456–465 (April 1972).

19. C. Concordia, L. G. Levoy, and C. H. Thomas, "Selection of Buffer Reactors and Synchronous Condensers on Power Systems Supplying Arc-Furnace Loads," *AIEE Trans.* **76,** 123–135, July 1957.

20. C. Concordia, "Voltage Dip and Synchronous-Condenser Swings Caused by Arc-Furnace Loads," *AIEE Trans.* **74,** 951–956, October 1955.

21. P. M. Black and L. F. Lischer, "The Application of a Series Capacitor to a Synchronous Condenser for Reducing Voltage Flicker," *AIEE Trans.* **70,** 145–150 (1951).

22. A. Seki, J. Nishidai, and K. Murotani, "Suppression of Flicker Due to Arc Furnaces by a Thyristor-Controlled VAr Compensator," *Paper A78 590-2, IEEE* Summer Meeting, Los Angeles, July 1978.

23. M. Chanas, "Perturbations on Industrial and Distribution Systems — Their Compensation by Static Means," Pt. 1, *Rev. Gén. Electr.* **87** (12), 925–943, December 1978, (in French).

24. M. Lemoine, *Op. cit., Pt. 2,* 944–962.

25. P. Charles, *Op cit., Pt. 3, Rev. Gén. Electr.* **88** (1), 49–57, January 1979.

26. M. Boidin and G. Drouin, *Op cit., Pt. 4,* 58–72.

27. E. Wanner and W. Herbst, "Static Power Factor Compensators for use with Arc Furnaces," *Brown Boveri Publ. CH-IW 511 790 E.*

28. W. E. Schwabe and R. Kasper, "Flicker Caused by UHP Arc Furnaces Using Scrap and Directly Reduced Materials," paper presented to CNBE-Brazilian National Committee–UIE.

29. W. E. Schwabe and C. G. Robinson, "Report on Ultra-high Power Operation of Electric Steel Furnaces," *J. Metals,* 67–75, April 1967.

30. "Survey of Arc Furnace Installations on Power Systems and Resulting Lamp Flicker," *AIEE Trans.,* 1957.

31. J. Lemmenmeier et al., UIE/UNIPEDE Survey, International Symposium on Electric Arc Furnaces, Liège 1973.

32. A. Dejoue and P. Gaussens, "Dispositifs Experimentaux pur l'étude du Flicker," *Bull. Soc. Francaise Electr.* **7** (77), 263–272, May 1957.

33. M. Depenbrock, "Compensation of Rapidly Varying Reactive Currents," *ETZ(A)* **98,** 408–411, June 1977, (in German).

34. G. Klinger, "Control and Regulation Device of a Self-Guiding Compensation Commutator Connection," *ETZ(A)* **98,** 411–414 (1977).

35. G. Klinger, "Tolerance Band Controlled Single-Phase Inverter Circuits with Optimum Selection of Control Values," *ETZ(A)* **97,** 87–90, February 1976.

36. E. Friedlander, "Voltage-Flicker Compensation with AC Saturated Reactors," *GEC J. Sci. Tech* **29** (2), 107–114 (1962).

37. E. Friedlander, "Flicker Suppression for Single-Phase Resistance-Welding Machines," *GEC J. Sci. Tech.* **32** (2), 79–84 (1965).

38. T. E. Harras and R. W. Lye, "A Static VAr Compensator for Flicker Control on the Saskatchewan Power Corporation System," Presented to Canadian Electrical Association, Montreal, March 27, 1974.

39. E. Friedlander, "Static Network Stabilization: Recent Progress in Reactive Power Control," *GEC J. Sci. Tech.* **33** (2), 58–65 (1966).

40. E. Friedlander, "The Development of Saturated Reactors for Network Stabilization as Applicable to Magnet Power Supplies," Second International Conference on Magnet Technology, 1964.

41. R. J. Moran, "Static VAr Control for the Steel Industry," *Industr. Power Syst.* **21** (1), 4–7 (1978).

42. T. Sjokvist, "Thyristor Switched Capacitors for Reactive Power Compensation," *Iron and Steel Engr.* **54** (4), 40–47, April 1977.

43. R. J. Moran et al., "Solid-State Converters for Power System Voltage Control," *Proc. Amer. Power Conf.* 1041–1047 (1975).

44. L. Gyugyi, "Static Shunt Compensation for Voltage Flicker Reduction and Power Factor Correction," *Proc. Amer. Power Conf.,* **37** 1271–1285 (1976).

45. H. Frank and S. Ivner, "TYCAP, Power-Factor Correction Equipment Using Thyristor-Controlled Capacitors for Arc Furnaces," *ASEA J.* **46,** 147–152 (1973).

46. J. W. Motto et al., "The Static VAr Generator and Alternative Approaches to Power Systems VAr Compensation," *Proc. Amer. Power Conf.* **37,** 1032–1040 (1975).

47. O. K. Hill and C. G. Robinson, "Large Arc Furnaces and the Effect of Key Dimensions on the Performance of UHP Furnaces," *Iron and Steel Engr.,* 33–36, July 1979.

48. W. E. Schwabe, "Ultra-High Power Arc Furnances," *Iron and Steel Engr.,* 132–137, September 1969.

49. B. Bowman, G. R. Jordan, and F. Fitzgerald, "The Physics of High-Current Arcs," *J. Iron and Steel Inst.,* 748–805, June 1969.

50. W. E. Schwabe and C. G. Robinson, "UHP: The Big Power Play in Arc Furnace Steelmaking," *Magazine Metals Prod.* **33,** 61–73, September 1968.

51. W. E. Schwabe and C. G. Robinson, "New Developments in UHP: Theory and Practice," Congrés Internationale Sur le Four Electrique à Arc en Acière, Cannes, France, June 7–9, 1971.

52. D. A. McCallum and C. B. Cooper, "AC System Network Control Methods—Compensation," IEE Symposium on Power System Dynamics—Control and Compensation, University of Newcastle Upon Tyne, Henderson Hall, September 12, 1974.

53. I. Hosono et al., "Suppression and Measurement of Arc Furnace Flicker with a Large Static VAr Compensator," *Trans. IEEE, Power Appar. Syst.* **98,** 2276–2284 (1979).

54. P. Duveau, "Flickermétre Numérique," *Rev. Gén. Electr.* **80,** 851–853 (1971).

55. D. J. Young, "Static Shunt Compensators—Saturated Reactor Types," *IEE Digest* 1979/52 (1979).

56. J. J. Trageser, "Power Usage and Electrical Circuit Analysis for Electric Arc Furnaces," *Trans. IEEE* **IA-16,** 277–284 (1980).

Chapter 10

HARMONICS

A. H. MOORE

10.1. INTRODUCTION

Certain loads present a nonlinear impedance to ac current. Their current waveforms are distorted from pure sinewaves at fundamental frequency. In the language of Fourier analysis they contain harmonic components or "harmonics."[(1)] Examples of large loads that generate harmonics are static power converters (rectifiers and inverters), power supplies, electric arc furnaces, and thyristor ac power controllers. The thyristor-controlled reactor (TCR) is another example of a power system component that generates harmonics.

In this chapter we first examine the generation of harmonics by certain of these loads, and the adverse effects of the harmonics are summarized. After a brief introduction to the topic of harmonic resonances, filters are discussed. Finally, an account is given of the effect of harmonics on telephone communications; this includes the definition of the widely used distortion and telephone influence factors.

10.2. HARMONIC SOURCES

Static power converters (rectifiers, inverters, and other switching devices) are a common source of harmonic currents. The distortion of the current waveform (i.e., the generation of harmonics) results from the switching. A typical high-power rectifier is shown in Figure 1, with voltage and current waveforms in Figure 2. These are theoretical waveforms based on the following assumptions:

1. There is a sufficient dc load inductance to produce a nonpulsating (constant) dc current.
2. There is zero commutating (ac source) inductance.

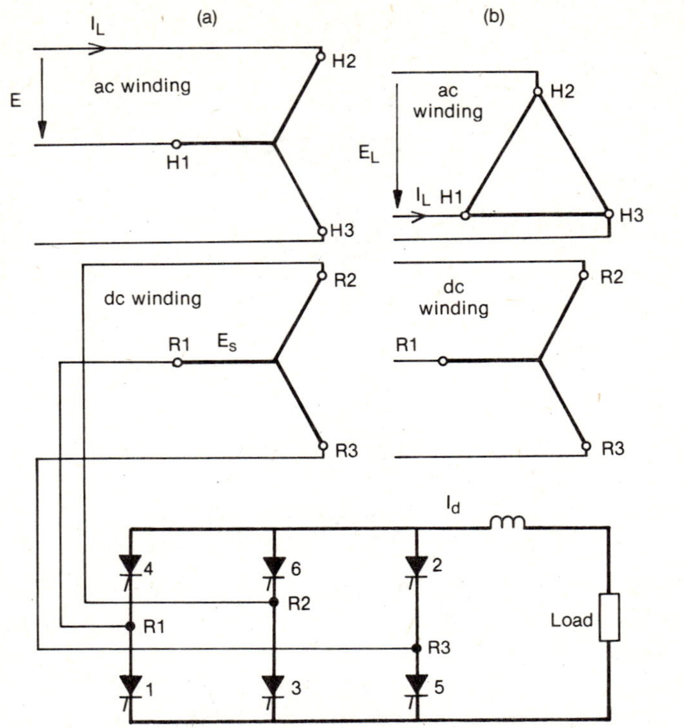

FIGURE 1. (*a*) ANSI-24 and (*b*) ANSI-23 three-phase rectifier circuits.

By Fourier analysis the ac waveform of Figure 2*c* can be decomposed into the following series of harmonic components:

$$i = \sqrt{2}\, I_1 \left[\sin \omega t - \frac{\sin 5\omega t}{5} - \frac{\sin 7\omega t}{7} + \frac{\sin 11\omega t}{11} + \frac{\sin 13\omega t}{13} + \cdots + \frac{2}{\sqrt{3}} \cos \frac{n\pi}{6} \frac{\sin n\omega t}{n} + \cdots \right], \tag{1}$$

where I_1 is the rms value of the fundamental component of the ac line current (equal to $6/\pi\ I_d$), and n is the order of the harmonic. An alternative transformer connection is shown in Figure 1*b*, with the primary winding in delta. The current in the secondary or "dc" winding is the same as in Figure 2*c*, but the 30° phase shift introduced by the delta/wye connection produces the stepped line current waveform of Figure 2*d*. Fourier analysis of this primary current wave yields the following series of harmonics:

$$i = \sqrt{2}\, I_1 \left[\sin \omega t + \frac{\sin 5\omega t}{5} + \frac{\sin 7\omega t}{7} + \frac{\sin 11\omega t}{11} + \frac{\sin 13\omega t}{13} \cdots \right]. \tag{2}$$

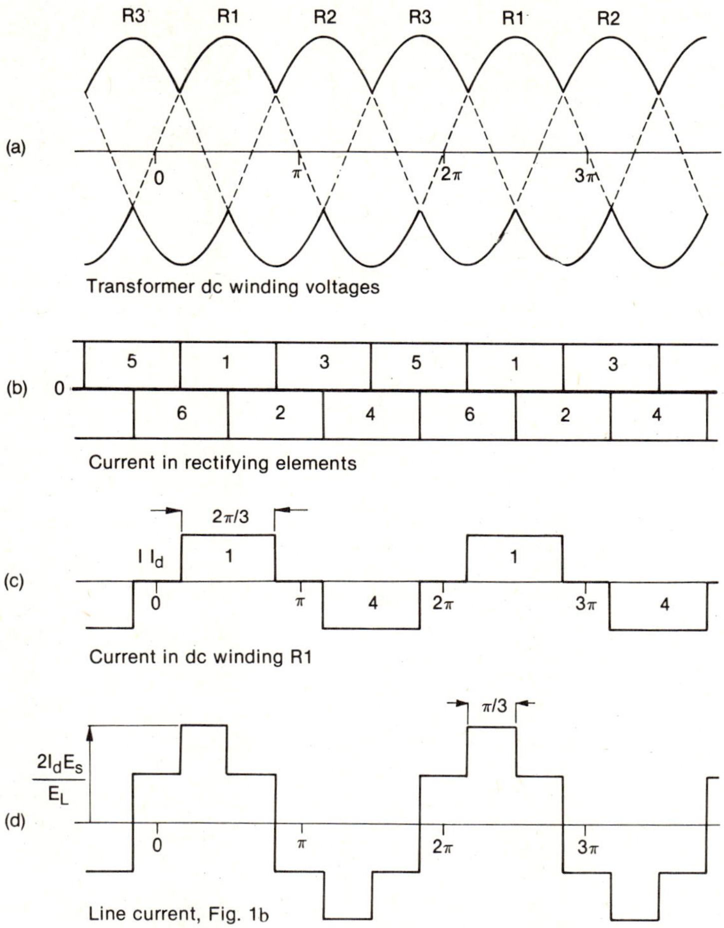

FIGURE 2. Voltage and current waveforms for Figure 1*a*.

The harmonic "spectra" represented by Equations 1 and 2 are identical except that alternate harmonic pairs (5th and 7th, 17th and 19th, etc.) are negative in Equation 1 while all are positive in Equation 2. This suggests that if the two types of rectifier are paralleled, with equal line currents, these alternate pairs of harmonics will cancel in the common source path.

When two equally rated rectifiers are combined, one with a transformer phase shift 30° from the other, the resulting system is said to operate "12-pulse," or with a pulse number of 12. Additional rectifier units equally phase shifted from one another create other pulse numbers. Three units phase shifted from one another by 20° constitute an 18-pulse system; four units separated by 15° result in 24-pulse operation. When the pulse number q is used in Equation 4, the orders of the harmonics may be determined. Table 1 gives the harmonics for $q = 6$, 12, 18, and 24. Note also that the magnitude of the harmonic current is

TABLE 1
Characteristic AC Supply Line Harmonic Currents in 6-, 12-, 18-, and 24-Pulse Rectifiers

Harmonic Order n	Rectifier System Pulse Number, q 6	12	18	24	Harmonic Freq. Base=60	6-Pulse Rectifier Harmonic Current In Percent of Fundamental: Theoretical	$X_c = 0.15$ $\alpha = 0$	$X_c = 0.15$[a] $\alpha = 32°$
5	×				300	20	16.2	18.9
7	×				420	14.2	9.3	12.7
11	×	×			660	9.09	3.04	6.70
13	×	×			780	7.69	1.86	4.96
17	×		×		1020	5.88	1.261	2.60
19	×		×		1140	5.26	1.08	1.78
23	×	×		×	1380	4.35	0.66	0.63
25	×	×		×	1500	4.00	0.53	0.28
29	×				1740	3.45	0.43	0.36
31	×				1860	3.23	0.39	0.48
35	×	×	×		2100	2.86	0.28	0.59
37	×	×	×		2220	2.70	0.24	0.59

[a] X_c is per-unit commutating reactance.
[b] Phase retard $\alpha = 32°$ produces 15% reduction in dc voltage.

$$I_n = \frac{I_1}{n}, \tag{3}$$

and that only the following "characteristic" harmonics appear under ideal conditions:

$$n = kq \pm 1 \tag{4}$$

Here, k is any integer, I_1 is the amplitude of the fundamental component, and q is the pulse number.

The square-cornered wave shapes of Figure 2 are possible only with idealized zero commutating reactance. The current waveforms will be modified by finite amounts of commutating reactance and phase retard. These two factors will affect the magnitude and phase angle of each harmonic but the same harmonic orders will prevail. For circuits such as those in Figure 1, Figure 3*a* shows how finite commutating reactance results in a gradual rise time and fall time represented by the angle μ. Controlling the dc voltage by thyristor phase retard delays the time of commutation by the angle α and reduces the commutation angle μ (Figure 3*b*). The reduction in μ further modifies the harmonic amplitudes.

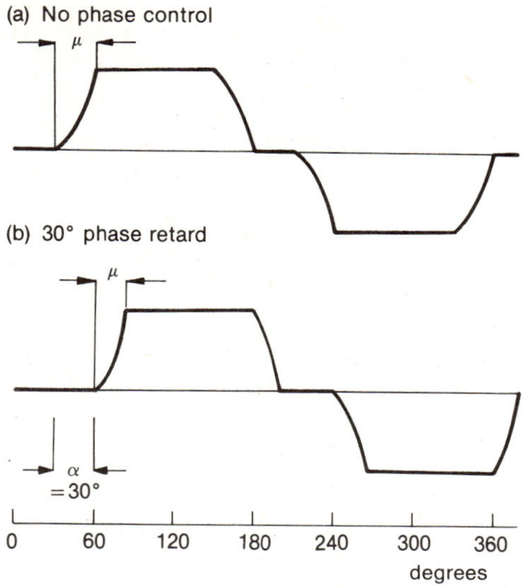

FIGURE 3. Effect of phase control and overlap.

Harmonic current magnitudes in percent of the fundamental are tabulated in Table 1. The column headed "Theoretical" follows Equation 3. The other two columns list calculated magnitudes for an assumed commutating reactance X_c of 0.15 per unit — one for phase retard angle of $\alpha = 0°$ and the other for $\alpha = 32°$. Magnitudes for other values of X_c and α can be calculated using the classical equations of Fourier analysis.

With finite "real world" commutating reactance X_c, harmonic currents are reduced. Read[(2)] presents curves of harmonic current amplitude versus X_c for various angles of phase retard for harmonics up to the 25th. The last two columns of Table 1 were taken from the printout of a computer calculation based on the classical equations relating these quantities.

It is not to be inferred from Table 1 that complete harmonic cancelation occurs for the harmonics "eliminated" by higher than 6-pulse operation. There will always be residual harmonics due to imperfect transformer phase shift, impedance unbalances, or unequal loadings or phase retard angles in the rectifiers. Under reasonably balanced conditions, residual harmonics are often assumed to be 10 to 20% of the value present in a 6-pulse rectifier of the same total rating. Under unbalanced conditions, greater amounts of residual harmonics are produced. For example, if one unit of a 4-unit balanced 24-pulse rectifier system is removed for service, the harmonics will be those of a 12-pulse rectifier plus a 6-pulse rectifier. Aluminum potline rectifier systems are always rated to deliver full potline power with one unit out of service. The harmonic duty of capacitor banks should always be calculated for projected conditions with the highest degree of unbalance.

As well as solid state rectifiers, there are other examples of switching devices that control the power in different types of loads. The ac thyristor controller described in Chapters 5 and 6 is used for heating control in furnaces and other types of resistive loads. We have already seen in Chapter 5 an example of the harmonics generated by thyristor-controlled reactors. Cycloconverters, frequency converters, and inverters are among the less common solid-state power converters found in power systems in large sizes, and these all contribute harmonics in varying degrees.

Unlike the harmonics of static power converters (such as rectifiers), which can be calculated from periodic waveforms, the harmonics generated by electric arc furnaces are unpredictable because of the cycle-by-cycle variation of the arc, particularly when boring into new scrap. The arc current is nonperiodic, and analysis reveals a continuous spectrum of harmonic frequencies of both integer and noninteger orders. Even so, harmonic measurements have shown that integer-order harmonic frequencies, particularly the 3rd, 5th, and 7th, predominate over the noninteger ones, and that the amplitude of the harmonics decreases with order. Fourier analysis of typical magnetic tape recordings taken of arc furnace currents yielded the harmonic percentages in Table 1 of Chapter 9. It will be noted from that table, which was for selected maximum arc activity periods of five consecutive 5-cycle periods, that low order harmonics prevail, and that even harmonics are present. Later, when the arc is steady, and the surface of the molten steel is flat, the harmonic magnitudes decrease considerably and the even harmonics virtually disappear. Figure 4 is a trace of the spectral response measured by means of a FFT-computing spectrum analyzer in the primary of a furnace transformer while the arc is boring into new scrap (active arcing). Distinct lines for integer orders, as well as a high threshold of noninteger orders, are seen. The frequency range is 0–500 Hz and the magnitudes, displayed on a logarithmic scale, are peaks averaged over several seconds. In most cases the harmonic magnitudes just reported would not by themselves be troublesome to power systems were it not for the possibility of their amplification by resonance of the power capacitors almost always associated with the inherently low power-factor arc furnaces.

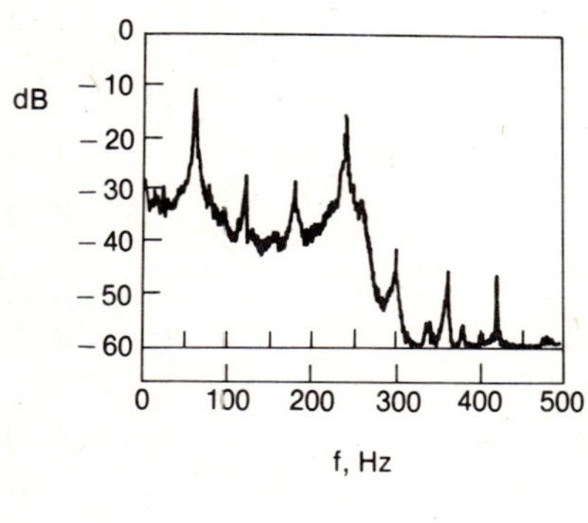

FIGURE 4. Spectrum of arc furnace line current during active arcing.

Another potentially troublesome source of harmonics is the magnetizing inrush current of transformers. The harmonic magnitudes can be very large, although they persist only for a short time (up to a second or two for large transformers). However, unlike the harmonics from rectifiers, even harmonics are included because of asymmetry. Second and fourth order harmonics can be particularly high and need to be considered when designing filters.

10.3. EFFECT OF HARMONICS ON ELECTRICAL EQUIPMENT

Blown capacitor fuses or failed capacitors in power capacitor banks are often the first evidence of excessive ac harmonic levels. ANSI Standard C55.1-1980 and NEMA CP1-1973 cover the characteristics of shunt power capacitors. In these standards, considerable attention is given to harmonics, and allowances are made for increases in both the effective voltage and current due to harmonics. Continuous operation with excessive harmonic current can lead to increased voltage stress and overtemperature, and can shorten the life of capacitors. Typically, a 10% increase in voltage stress will result in a 7% increase in temperature, reducing the life expectancy to 30%. This "life" analysis does not allow for capacitor failure initiated by dielectric corona. The damage done by corona produced by excessive peak voltages depends on both the intensity and duration of the corona. There have been numerous cases of premature failure — in the order of months rather than years — as a result of inadequate provision for harmonic voltages.

Harmonic currents can cause overheating of rotating machinery, particularly solid-rotor (nonsalient-pole) synchronous generators. Harmonic currents produce a magnetomotive force that causes currents to flow in the solid rotor surface, adding to the heating. Positive-sequence rectifier harmonics [following the equation $n = kq + 1$ (7th, 13th, etc.)] rotate forwards and cause harmonic orders 6, 12, and so on, in the rotor. Those harmonics following the equation $n = kq - 1$ (5th, 11th, etc.) are negative sequence, rotate against the rotation of the rotor, and again produce harmonic orders 6, 12, and so on, in the rotor. The resulting pulsating magnetic field caused by the oppositely rotating pairs of magnetomotive forces sets up localized heating in the rotor which may require a derating of the machine. The derating for 6-pulse operation, where the 5th and 7th harmonics dominate, can be considerable, depending on the particular machine design. Derating for balanced 12-pulse rectifier operation is generally minimal. The presence of rotor amortisseur windings greatly alleviates the rotor heating problem.

Induction motors are much less affected by harmonics than are solid-rotor synchronous generators. However, excessive harmonic currents can overheat induction motors, especially when they are connected into systems where capacitors in resonance with the system are aggravating one or more harmonics.

Harmonic currents carried by transformers will increase the load (I^2R) loss by a factor greater than the mere increase in rms current. The amount of the increase depends on the proportion of I^2R loss proportional to frequency squared (eddy current loss), and the amount proportional to the first power of frequency (stray load loss). The same is true of current-limiting or tuning reactors. As a result, designers of reactors need to know the amount and order of each significant harmonic so that they can apply the proper factor for I^2R loss contributed by the fundamental and each contributing harmonic.

10.4. RESONANCE, SHUNT CAPACITORS, AND FILTERS

When static power capacitors are added to a power system for reactive power compensation, there will be one or more frequencies at which the capacitors will be in parallel resonance with the inductance of the system. Harmonics "injected" into the system at coincident frequencies will be amplified.

To illustrate the principles of harmonic current flow and resonance, reference is made to the network in Figure 5*a*. The typical harmonic source may be assumed to be a generator of constant harmonic currents $\mathbf{i}_n$, each of which can be calculated by Fourier analysis or determined by test.

This being the case, the one-line diagram of Figure 5*a* may be represented on a per-phase basis by Figure 5*b*. The harmonic current $\mathbf{i}_n$ divides between the capacitor and the supply according to the equation

$$\mathbf{i}_n = \mathbf{i}_{sn} + \mathbf{i}_{fn} \tag{5}$$

The impedance of the capacitor branch at any frequency is given by

$$\mathbf{Z}_f = \mathbf{Z}_{fc} + \mathbf{Z}_{f1} \tag{6}$$

where $\mathbf{Z}_{fc}$ is the impedance of the capacitor and $\mathbf{Z}_{f1}$ is the impedance of the tuning reactor, if fitted (discussed later). The subscript "f" alludes to the filtering action of the capacitor branch. The harmonic current $\mathbf{i}_n$ divides between the capacitor and the supply in proportion to the admittances of these parallel branches. If $\mathbf{Z}_s$ is the equivalent impedance of the supply (including the supply transformer shown in Figure 5*a*), then

$$\mathbf{i}_{fn} = \frac{\mathbf{Z}_s}{\mathbf{Z}_f + \mathbf{Z}_s}\,\mathbf{i}_n = \rho_f\,\mathbf{i}_n \tag{7}$$

and

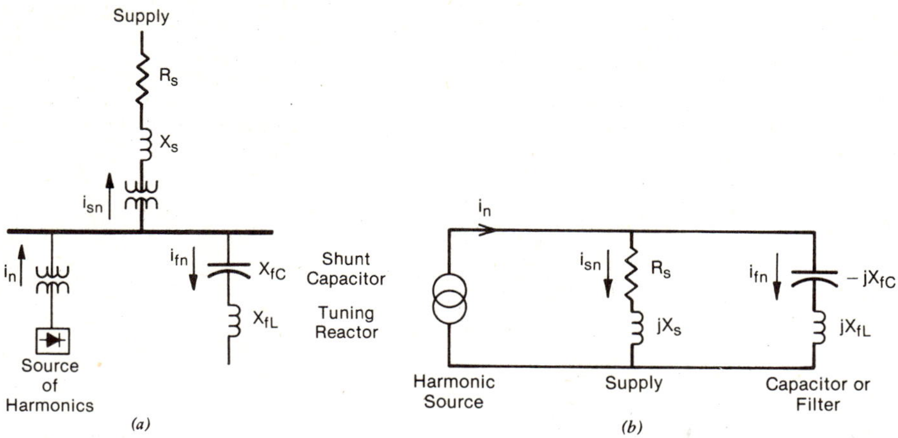

FIGURE 5. (*a*) One-line diagram of system with harmonic source. (*b*) Per-phase equivalent circuit.

$$\mathbf{i}_{sn} = \frac{\mathbf{Z}_f}{\mathbf{Z}_f + \mathbf{Z}_s}\,\mathbf{i}_n = \rho_s \mathbf{i}_n \tag{8}$$

If the *distribution factor* ρ_s is large at a particular harmonic frequency coincident with one of the harmonics generated by the harmonic sources, then amplification of the harmonic current will occur and the currents in the capacitor and the supply may be excessive. In particular if $\mathbf{Z}_f + \mathbf{Z}_s \to 0$ at some harmonic frequency, the system is resonant at that frequency. It is clearly important to avoid this condition at harmonic frequencies near those excited by the harmonic source. In other words, ρ_s must be kept low at these frequencies.

The function of the tuning reactor shown in series with the capacitor in Figure 5*a* is to form a series-resonant branch or filter, for which $\mathbf{Z}_f \to 0$ at the resonant frequency (e.g., the 5th harmonic). As a result, $\rho_s \to 0$ so that the harmonic current "escaping" into the supply is reduced or eliminated; while $\rho_f \to 1$ so that $\mathbf{i}_{fn} = \mathbf{i}_n$, that is, all the harmonic current generated enters the filter.

A numerical example shows how these relationships can be used to explore the performance of an untuned capacitor and a tuned capacitor filter. Assume that the step-down transformer impedance is much greater than the source impedance so that for a narrow range of frequencies the approximation X_s/R_s = constant may be used. Parameters assumed are:

Bus voltage = 13.8 kV
Short circuit MVA = 476
Capacitor reactive power = 19.04 MVAr

$$\frac{X_s}{R_s} = 10$$

Since inductive reactance is proportional to frequency, and capacitive reactance is inversely proportional to frequency, we have

$$X_s = \frac{n(13.8)^2}{476} = 0.4n$$

$$X_{fc} = \frac{(13.8)^2}{19.04n} = \frac{10}{n}$$

$$R_s = \frac{X_s}{10} = 0.04n$$

and

$$\rho_f = \frac{0.04 + j0.4}{0.04 + j(0.4 - 10/n^2)} \tag{9}$$

In Figure 6*a* ρ_f and ρ_s are plotted against the harmonic order n. There is a parallel resonance between the capacitor and the supply at the 5th harmonic; note $\rho_s \simeq \rho_f$.

If a tuning reactor X_{f1} is added to the capacitor to form a series 5th-harmonic filter, and if the resistance of the reactor is considered negligible in comparison with the system resistance, then at 60 Hz,

$$X_{f1} = \frac{X_{fc}}{(5)^2} = \frac{10}{25} = 0.4 \text{ Ohm} .$$

The equation for ρ_f now becomes

$$\rho_f = \frac{R_s + jX_s}{R_s + j(X_s + X_{f1} - X_{fc})} . \tag{10}$$

Figure 6*b* shows the response in terms of ρ_f and ρ_s with the tuning reactor. Note that $\rho_f = 1$ at the 5th harmonic ($\rho_s = 0$); and ρ_f is a maximum (parallel resonance) at a lower harmonic order, about 3.54.

It can be seen in Figure 6 that at high frequencies the tuning reactor reduces the ratio ρ_f/ρ_s as compared with the value obtained without it. For this reason, a high-pass filter may be required in parallel with the 5th harmonic filter (see Section 10.5).

More complex systems can be solved by the use of "harmonic load-flow" programs on the digital computer. Figure 7 shows a hypothetical

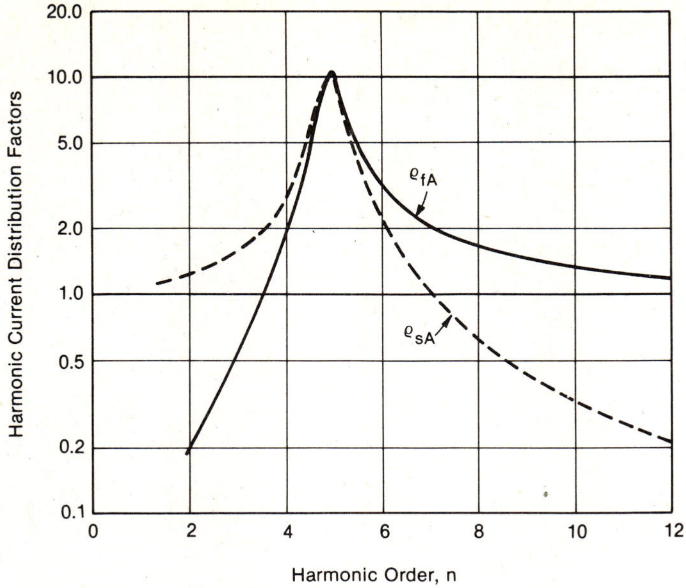

(a)

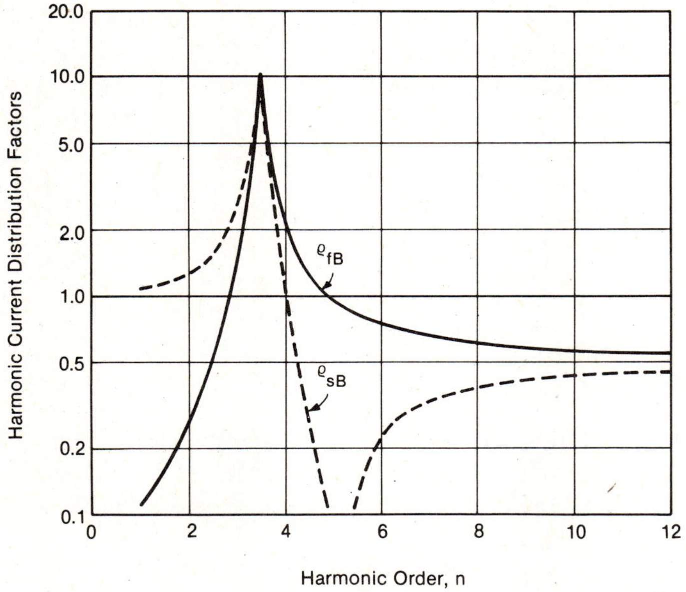

(b)

FIGURE 6. Harmonic current distribution factors versus harmonic order. (*a*) Capacitor bank without tuning reactor. (*b*) Capacitor bank with tuning reactor.

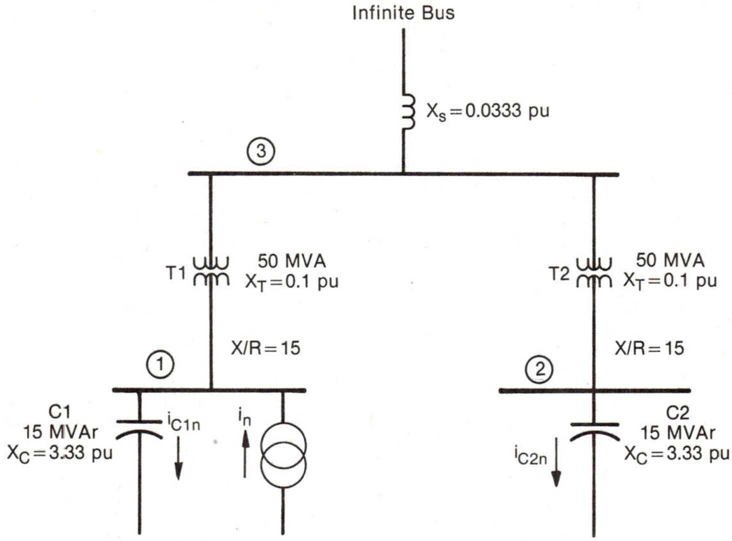

FIGURE 7. Hypothetical electrochemical plant illustrating the use of the harmonic load-flow program. Rectifier at bus 1 produces harmonics i_n.

industrial plant with two main busses, 1 and 2. A source of harmonic currents is shown on bus 1. For each harmonic frequency the proportion of the total harmonic current that flows into each capacitor can be determined in terms of distribution factors ρ_{f1} and ρ_{f2}. Likewise the proportion entering the supply is determined by ρ_s. Figure 8 shows ρ_{f1} and ρ_{f2} calculated by a harmonic load-flow program. The dual resonant peaks are caused by the presence of two L-C circuits. For capacitor C1 the two peaks are at 268 Hz and 350 Hz. If the circuit breaker to transformer T2 is opened, there will be a single resonant peak at

$$f = 60\sqrt{\frac{3.333}{0.1 + 0.0333}} = 300 \text{ Hz}.$$

This demonstrates how a change in the system configuration — even on another bus — can affect parallel resonance. In effect, the presence of capacitor C2 on bus 2 alters the harmonic impedance of the system upstream of transformer T1.

The frequency responses that are met in practice are often more complicated than the examples just described, because of interconnections. Without going into the details of an extensive network, we can explore an example of a more complex frequency response in terms of a long transmission line. Because of the distributed nature of its parameters, the long line behaves like a ladder network containing a large number of inductive and capacitive reactances, but the basic equations are simple enough to be developed in closed form. Figure 9 shows an example of a

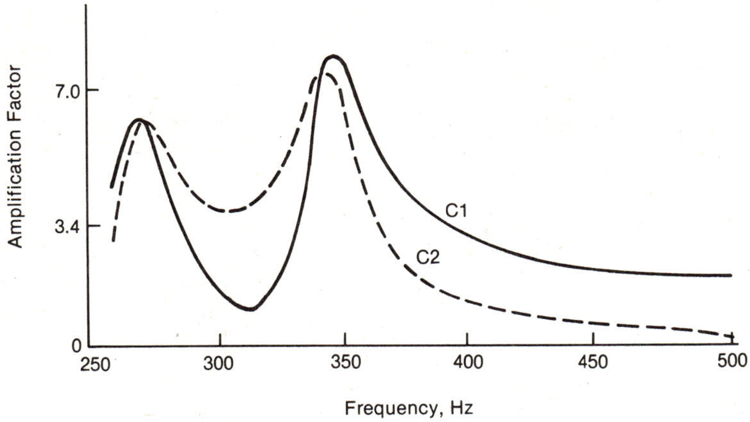

FIGURE 8. Harmonic load-flow for circuit of Figure 7.

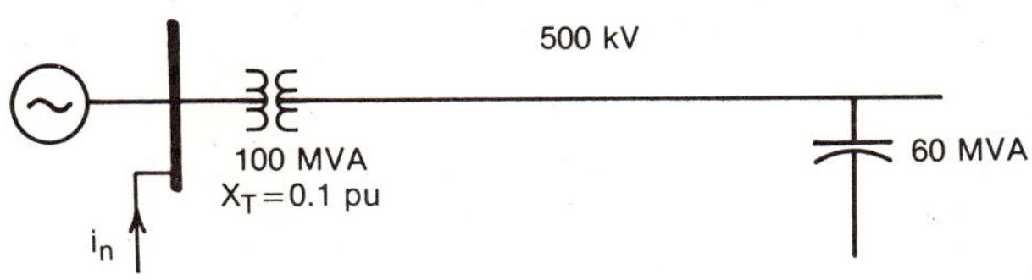

FIGURE 9. Effect of harmonics on transmission line.

100-mi section of 500-kV transmission line similar to those studied earlier (see Table 3 in Chapter 2). From Equation 2 of Chapter 2, if the line resistance is included, it can be shown that the driving-point impedance per phase at the sending end is given by

$$\mathbf{Z}_s = \frac{\mathbf{V}_s}{\mathbf{I}_s} = \frac{\mathbf{Z}_r \cosh \Gamma a + \mathbf{Z}_0 \sinh \Gamma a}{(\mathbf{Z}_r/\mathbf{Z}_0) \sinh \Gamma a + \cosh \Gamma a} + \mathbf{Z}_{ts} \qquad (11)$$

where $\mathbf{Z}_0 = [(r + j\omega l)(g + j\omega c)]^{1/2}$ is the characteristic impedance at the harmonic frequency ω; Γ is given by Equation 1 in Chapter 2; and $\mathbf{Z}_r$ is the impedance per phase at the receiving end. $\mathbf{Z}_{ts}$ is the impedance of the transformer at the sending end. The remaining symbols are the same as in Chapter 2.

Figures 10*a* and *b* show the magnitude Z_s as a function of frequency with and without the sending-end transformer. As the frequency is increased, a series of alternate poles and zeroes are met. The poles in particular represent parallel-type resonances in which large circulating currents form standing waves along the line. This is a condition to be avoided, and it is important to study the harmonic characteristics of such systems whenever reactive compensation is applied.

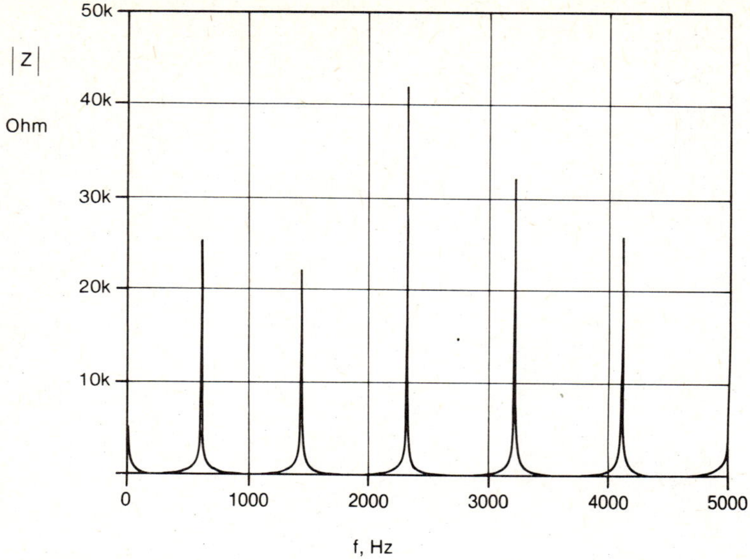

(a)

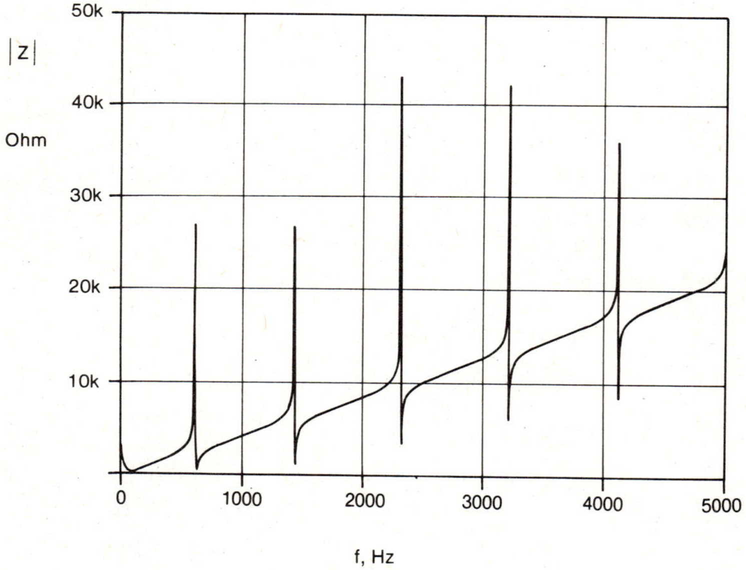

(b)

FIGURE 10. Driving-point impedance of transmission line versus frequency. (*a*) Without transformer. (*b*) With transformer.

10.5. FILTER SYSTEMS

Where shunt capacitors are applied for power-factor improvement in systems with large harmonic-producing loads, it is often difficult to avoid resonance or near-resonance at a dominant harmonic. The solution often is to create a filter system consisting of one or more capacitor banks equipped with series tuning reactors. Filters may also be required if the harmonic outflow, even without the amplifying effect of resonance, is excessive. Figure 11*a* shows a simple example of a single-frequency tuned filter having a very high admittance at its tuned frequency. Figure 11*b* shows a system of multiple filters tuned to different frequencies.

An example of the response of a typical filter system is shown in Figure 12. This is for a parallel combination of two filters tuned to the 5th and 7th harmonics. The complex impedance locus passes essentially through zero at the tuned frequencies of 300 and 420 Hz, the small offset being mainly due to the resistance of the tuning reactors. The filter response can be represented in other ways; for example, in terms of the magnitude and phase of the admittance or impedance, plotted or tabulated against frequency. The distribution factors ρ_f and ρ_s may also be used. "Harmonic load flow" computer programs are often used to model the response of the entire network including filters and harmonic sources. These calculations take into account the variation of the impedances in the system with frequency.

The single tuned filter has been used successfully in many industrial plants, notably in electrochemical installations such as aluminum smelters and chlorine-caustic plants where large rectifier systems are employed for electrolysis of the product. These plants usually take advantage of multi-

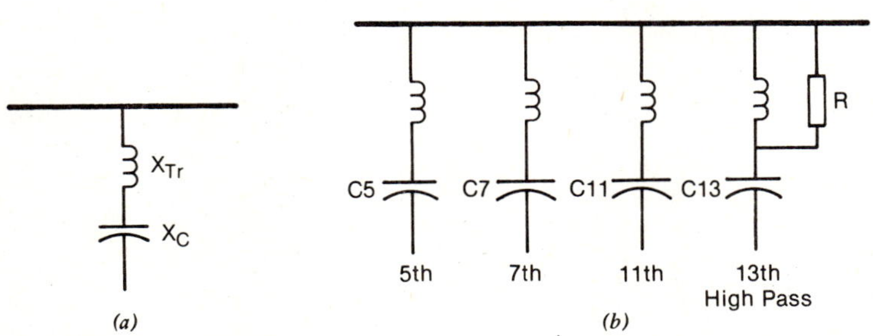

FIGURE 11. Typical shunt filter. (*a*) Tuned filter. (*b*) Bank of tuned filters with high-pass filter.

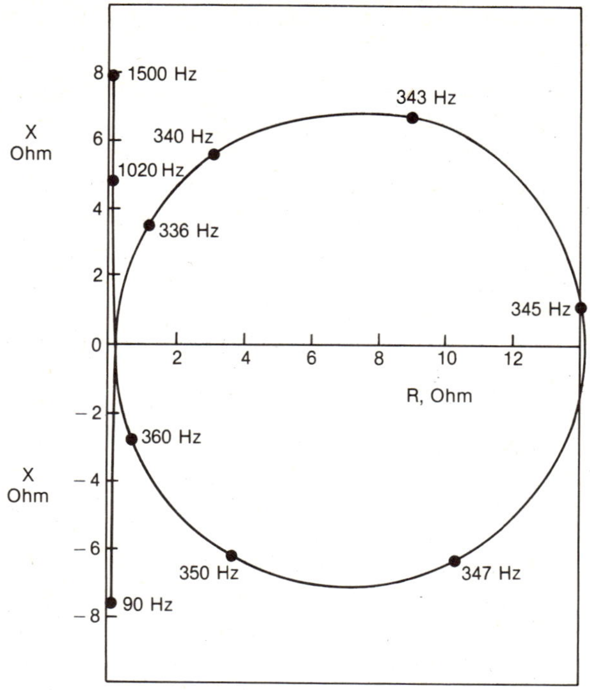

FIGURE 12. Filter impedance as a function of frequency.

ple phasing, in which overall pulse numbers of 24 or greater are achieved by phase-shifting building blocks of 6-pulse rectifiers. By tuning the capacitor bank to a harmonic at or near the 5th (the lowest characteristic harmonic of a 6-pulse rectifier), parallel resonance with the system at any rectifier harmonic is avoided since ρ_s will be less than unity for all frequencies above the filter tuned frequency. It is important that the resultant parallel resonance of the filter with the system reactance not be at a harmonic which will be encountered during operation of the system. For example, if resonance should be at or very close to the 4th harmonic, the strong 4th harmonic in the magnetizing inrush of a large transformer on the bus can excite high peak 4th harmonic voltages.

It has already been pointed out in connection with Figure 6 that where the single tuned filter provides insufficient filtering above its tuned frequency, multiple filters such as that in Figure 9*b* may be required. To demonstrate the performance of this type of filter, the following assumptions were made:

1. The 13th harmonic filter was eliminated and the 11th harmonic filter was converted into a high-pass filter by means of a resistor shunting the reactor.

2. The fundamental-frequency reactive power of the 5th harmonic capacitor bank was 40% greater than for either the 7th or 11th harmonic filter.

Four different values of resistance were investigated:

A. $R = X_0$
B. $R = 2X_0$
C. $R = \dfrac{X_0}{2}$
D. $R = \infty$ (shunting resistor absent)

where $X_0 = \sqrt{L_{11}/C_{11}}$.

A harmonic load-flow digital computer program was used to calculate harmonic current into the system in terms of the magnitude ρ_s for a range of harmonics from the third to the 34th. The result is plotted in Figure 13. The sharpest filtering of the 11th harmonic was obtained in case D (with no shunt resistor), but at higher harmonics ρ_s approaches the rather high value of 0.25. The best filtering of higher-order harmonics is obtained in case C, with $R = X_0/2$, but at the expense of poorer filtering of harmonic orders 11 through 23. Cases A and B, with $R = X_0$

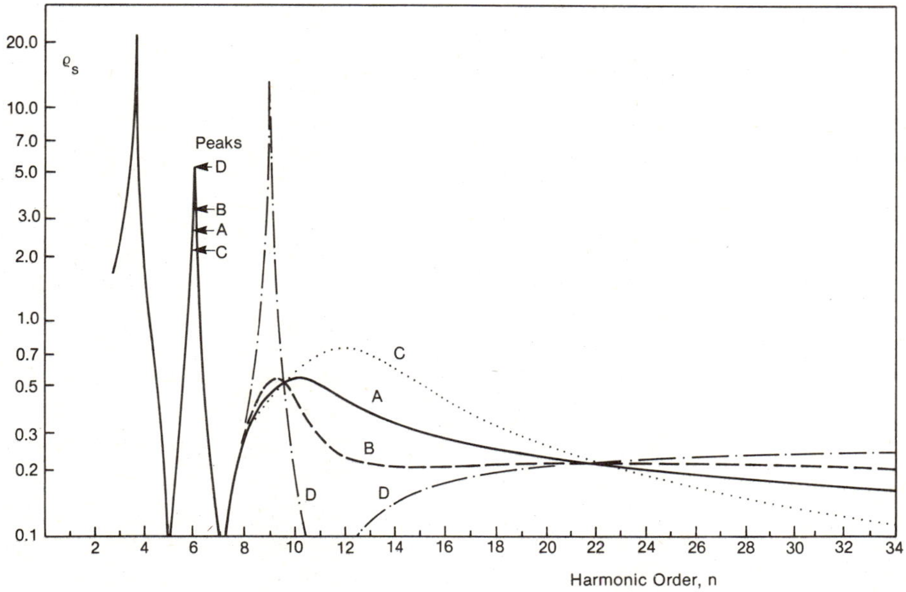

FIGURE 13. Harmonic distribution factor ρ_s for 5th/7th/11th/high-pass filter.

and $R = 2X_0$ respectively, give filtering characteristics between the preceding two limits. In a practical case the resistance shunting the tuning reactor would typically be chosen in this range, as giving the best compromise between the higher- and lower-order harmonics.

Harmonic Voltage Distortion. Harmonic currents flowing into a system will produce harmonic voltages, with each harmonic voltage following the equation

$$\mathbf{V}_n = \mathbf{Z}_n \mathbf{I}_n , \tag{12}$$

where $\mathbf{I}_n$ is the harmonic current into the system and $\mathbf{Z}_n$ is the harmonic driving-point impedance of the system. A measure of the deviation from a sinusoidal voltage wave may be made by calculating the so-called voltage distortion factor according to the following equation:

$$DF = \left[\frac{\sum_{2}^{N} V_n^2}{V_1^2} \right]^{1/2} \times 100\% . \tag{13}$$

IEEE Standard 519-1981 suggests permissible upper limits for the voltage distortion factor. It must be pointed out, however, that these limits are not absolute and that a particular power system might be especially sensitive to a certain harmonic or range of harmonics. In some systems, telephone interference may be detectable even though the voltage distortion factor is less than the upper limit.

10.6. TELEPHONE INTERFERENCE

Noise originating from harmonic currents and voltages in power systems can be coupled into wire communication circuits through magnetic and electrostatic fields. Weighting factors have been developed to compare the relative effectiveness of different frequencies in interfering with telephone conversations. The response of the telephone receiver, the coupling between power and telephone circuits, as well as the sensitivity of the human ear to varying frequencies enter into the weighting factors.

In the United States, the Joint Subcommittee for Development and Research, Edison Electric Institute, and Bell Telephone System, has established a system of weighting characteristics called *TIF* factors. The contribution of each individual harmonic current or voltage of a disturbing power circuit is the product of the current or voltage and the weighting value for that frequency. For current, it is designated *I.T*; for voltage it is *kV.T*. The influence of a complete current wave is the square root of the sum of squares (RSS) of all the individual frequency $I.T$ products including the fundamental. Similarly the influence of a complete voltage wave is the RSS of all the individual $kV.T$ products.

The *TIF* factors take into account:

1. The relative subjective effect of frequency f in the message circuit as heard in the telephone

2. Coupling between the power and telephone circuits which is assumed to be directly proportional to frequency f.

The first item, called C-message weighting for the latest BTS telephone sets (500 type), is the interfering effect at frequency f relative to 1000 Hz as determined by subjective tests.

$$TIF_f = 5\ P_f\ f \tag{14}$$

where P_f is the relative interfering effect at f with P_f equal to 1 at 1000 Hz. Thus, $TIF_{1000} = 5 \times 1 \times 1000 = 5000$.

Figure 14 displays in graphical form the latest (1960) *TIF* factors. Because the C-message weighting is largely controlled by the characteristics of the listeners, it is unlikely that any major re-evaluation of *TIF* will be required in the future.

Measurement of current or voltage *TIF* can be made in several ways. Certain instruments can be used to measure *TIF* directly. For example, a Western Electric 106A current coupling unit connected to a Western Electric 3A Noise Measuring Set will yield the current *TIF* directly. The 3A set includes a C-message weighting network while the coupler introduces the multiplier for f. Similarly voltage *TIF* can be measured by substituting a voltage *TIF* coupling unit for the current coupler.

Another method is to measure harmonic current or voltage with one of the various harmonic analyzers now available and multiply each harmonic magnitude so derived by the appropriate single frequency *TIF* factor (see

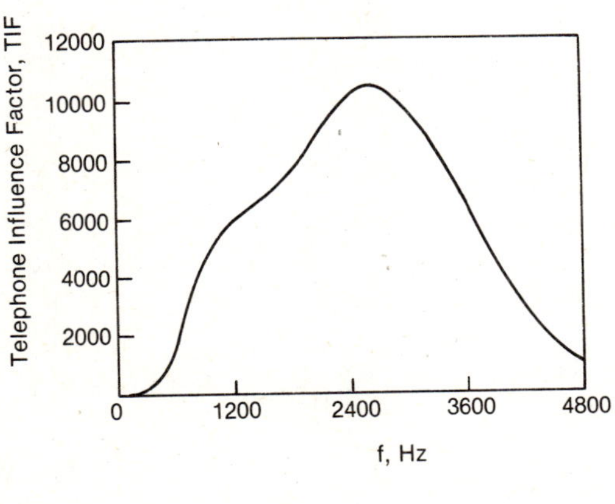

FIGURE 14. Telephone influence factor weighting curve.

Figure 14). From these values the $I.T$ or $kV.T$ of the complete current or voltage wave can be calculated as

$$I.T = \left[\sum_{f=60 \text{ Hz}}^{5000} (I_f \cdot T_f)^2 \right]^{1/2} \tag{15}$$

or

$$kV.T = \left[\sum_{f=60 \text{ Hz}}^{5000} (kV_f \cdot T_f)^2 \right]^{1/2} \tag{16}$$

where $I_f \cdot T_f$ = single frequency $I.T$ product
$kV_f \cdot T_f$ = single frequency $kV.T$ product
T_f = corresponding single frequency *TIF* from Figure 14 or a table of such factors.

Whether a certain value of $I.T$ product or $kV.T$ product will actually interfere with a voice communication system is dependent on a number of factors relating to the physical interaction of the power system and the communication system.

The complete story of "Inductive Coordination" between the power system and the voice communication circuit is beyond the scope of this chapter, but is adequately covered in the references. However, a few comments are worthy of note.

1. Except for very close separations of these circuits such as joint use of poles, induced voltages by current *TIF* are more important than electric field induction by voltage *TIF*.

2. Induced voltage is essentially longitudinal, and is due chiefly to residual (zero-sequence) current in the power line. If the harmonic currents were either positive sequence or negative sequence as from 6-pulse rectifier modules, and the power circuit was perfectly balanced, essentially no residual harmonic current would flow and *TIF* would be minimal.

3. The unbalanced impedances of ac lines convert a part of the balanced harmonics to zero sequence which are residual as seen by the voice communication circuit.

4. A high-voltage power transmission network may be converted to a lower-voltage subtransmission network via a power transformer. Noise coupled to the lower-voltage network may be troublesome to a voice communication circuit sharing the same poles as the subtransmission network.

5. It is difficult to assign definite limits on $I.T$ *or* $kV.T$ products because of these and other variables.

REFERENCES

1. E. W. Kimbark, *Direct Current Transmission.* Vol. I, Chap. 8, Wiley, New York, 1971.
2. J. C. Read, "The Calculation of Rectifier and Inverter Performance Characteristics," *J. IEE,* **92,** pt. 2, 495–509, (1945).
3. J. H. Galloway, "Harmonic Line Currents in Large Thyristor Six-Pulse Converters," *IEEE Industry App. Soc. 8th Annu. Mtg. Record,* 753–759 (Oct.1973).
4. H. A. Gauper, "Regulating and Industrial Aspects of Power Conversion Interference," *IEEE Industr. Static Power Conversion Conf. Rec.,* Cat. No. 34C20, pp. 109–111 (1965).
5. W. C. Ball and C. K. Poarch, "Telephone Influence Factor (TIF) and its Measurement," *AIEE Trans.* **79,** pt. II, 659–664 (Jan. 1961).
6. *Limits for Harmonics in the United Kingdom Electricity Supply System,* System Design and Development Committee, The Electricity Council (Britain), Recommendation G5/3, Sept. 1976.
7. W. G. Sherman, "Summation of Harmonics with Random Phase Angles," *Proc. IEE* **119,** 1643–1648 (Nov. 1972).
8. N. B. Rowe, "The Summations of Randomly-Varying Phasors or Vectors with Particular Reference to Harmonic Levels," *IEE Conf. Publ.* 110, 177–181 (April 1974).
9. R. D. Evans, "Harmonics and Load Balance of Multiphase Rectifiers," *Trans. AIEE* **62,** 182–187 (1943).
10. G. L. Brewer, "A Simple Method for Estimating Harmonic Filter Performance," *IEE Conf. Publ.* 110, 162–167 (April 1974).
11. D. E. Steeper and R. P. Stratford, "Reactive Compensation and Harmonic Suppression for Industrial Power Systems Using Thyristor Converters," *IEEE Trans.* **IA-12,** 232–255 (May 1976).
12. A. H. Moore, "Application of Power Capacitors to Electrochemical Rectifier Systems," *IEEE Trans.* **IA-13**(5), 399–406 (Sept./Oct. 1977).
13. *IEEE Guide for Harmonic Control and Reactive Compensation of Power Converters,* IEEE Standard 519-1981.
14. Engineering Reports of the Joint Subcommittee on Development and Research of the Edison Electric Institute and the Bell Telephone System, 5 Volumes, July 1926 to January 1943.
15. *Report on Harmonic Distortion caused by Converter Equipment,* British Electricity Boards A.C.E. Report No. 15, 1970.
16. W. L. Kidd and K. J. Duke, "Harmonic Voltage Distortion and Harmonic Currents in the British Distribution Network, Their Effects and Limitation," *IEE Conf. Publ.* 110, 228–234 (April 1974).
17. G. Deloux, "International Standardization and Electric Power Supply Network Disturbances," *IEE Conf. Publ.* 110, 223–227 (April 1974).
18. C. G. Amato, "A Simple and Speedy Method for Determining the Fourier Coefficients of Power Converter Waveforms," *IEEE Industry Gen. Appl. 4th Annu. Mtg. Rec.,* 477–483, (Oct. 1969). Reprinted in *Power Semiconductor Applications* **II,** 44–50, IEEE Press, New York, 1972.
19. J. H. Galloway, "Line Current Waveforms and Harmonics for a Large Multiphase Converter System," *IEEE Trans.* **IA-13,** 394–399 (Sept. 1977).
20. R. P. Stratford, "Analysis and Control of Harmonic Currents in Systems with Static Power Converters," *IEEE Trans.* **IA-17,** 71–81 (Jan. 1981).

Chapter 11

REACTIVE POWER COORDINATION

K. A. WIRGAU

11.1. INTRODUCTION

In recent years, increased attention has been given to improving power system operation by reduction of fuel consumption, and by better utilization of existing equipment to defer new equipment purchases.[1-13] Other related factors influencing utilities are inflation, fuel shortages and price increases, and increased pressures to borrow less money. One approach addressing these concerns is *reactive power management.*

Normally, there are two basic types of reactive power flows of concern in a power system:

1. Reactive power consumed by loads.
2. Reactive power consumed within the network.

The components which absorb reactive power are generators and synchronous condensers operated with a leading angle, shunt reactors, line and transformer inductances, static reactive power compensators, and loads. Reactive power is generated by generators and synchronous condensers operated with a lagging angle, static capacitors, static compensators, and the capacitance of lines and cables. With all the interaction between system components, a coordinated procedure is needed to control voltage and reactive power flow in such a way as to minimize transmission losses. Optimization computer programs assist in achieving this goal.

Reactive power management can be defined as the control of generator voltages, variable transformer tap settings, compensation, and switchable shunt capacitor and reactor banks plus the allocation of new shunt capacitor and reactor banks in a manner that best achieves a reduction in system losses and/or voltage control. Reactive power management by electric utilities, under steady-state and dynamic system conditions, can be subdivided into the following categories:

1. Reactive power planning.
2. System operations planning.
3. Reactive power dispatch and control.

Reactive power planning is concerned with the installation or removal of reactive power equipment in a power system. Typically, this effort is directed at system conditions from several months to several years in the future.

System operations planning is concerned with the improvement of operating practices utilizing existing reactive power equipment. This planning is performed for system conditions anticipated to occur a few days to a year into the future.

Reactive power dispatch and control determines actual equipment operations. The associated analysis is performed seconds to hours prior to its implementation.

The term *equipment* refers to the reactive power compensating devices, as well as the monitoring, control, and communication equipment required to carry out the real-time dispatching function.

Reactive power compensating equipment that may be installed, removed, or controlled, includes: switched shunt capacitors, shunt reactors, series capacitors, static compensators, synchronous condensers, generators, and load tap-changing transformers. The ancillary equipment includes: measuring devices for reactive power, relays, automatic controls (e.g., substation automation), switches and circuit breakers, and communication equipment (e.g., power line carrier).

11.2. REACTIVE POWER MANAGEMENT

Reactive power *planning* is an integral function of reactive power management. The planning problem is mathematically large (considering economics and security) and has not been solved satisfactorily to date for large systems (e.g., 500 buses, 100 contingencies). Research and development is presently underway to solve the size problem and is explained in this chapter.

The objective in reactive power planning is to minimize the cost of necessary reactive power equipment to enable the power system to operate in an acceptable manner in the event of any one of a collection of contingencies occurring. This problem involves determining both the optimal installation of reactive support which satisfies every contingency in a simultaneous manner and the individual dispatches of the reactive support. The purpose of adding new reactive power devices is to control the post-contingency voltage variation.

By observing the relationships between the power system network equations before and after an equipment problem (e.g., contingency or outage), a methodology has been developed for obtaining the sparse factorization of hundreds of buses and hundreds of contingencies by only considering the sparse factors of the base case power system. The set of sparse factors is a by-product of the power flow solution.

A special version of the linear programming simplex algorithm was developed to accommodate this new formulation. The new algorithm exploits the special block diagonal structure of the problem as well as some special properties of the solution process. The formulation has been successfully applied to a 600 bus system and 285 contingencies.

The following topics present an overview of reactive power management in *operations*.

11.2.1. Utility Objectives

The utility objectives for each management category are twofold:

1. Security.
2. Economics.

In the past, the reactions of utilities to reactive power and voltage problems have usually been prompted by a security problem. A power system is stated to be *secure* if it is able to undergo a disturbance without violating any of its load and operating limits. In practice, the degree of security that can be planned into a power system is limited by economic considerations. Both planning and operations must maintain security while maximizing economy. Economics includes the evaluation of alternatives based on a comparison of total benefit to total cost. In both planning and operations, reactive power management can enhance both the economy and the security of the power system.

11.2.2. Utility Practices

The main simulation programs utilities use in planning (reactive power and system operations) are the power flow and transient stability programs. At the present, considerable engineering judgement and trial-and-error is necessary to plan a power system for satisfactory reactive power flows and voltage profiles.

It has been the practice of utilities to monitor voltage at key locations in their systems. This information, combined with off-line analysis, has provided guidelines for system operators to control voltages and reactive power flows.

In general, reactive power management has suffered due to the lack of the following:

1. Knowledge and understanding by the industry.
2. Computer program availability.
3. Real-time data.
4. Economic incentives.

More attention is now being given to reactive power management, because of higher fuel costs and because of financial and regulatory impediments to the installation of new equipment.

11.2.3. Mathematical Modeling

The major tool used for a reactive power dispatching strategy is an optimal power flow program. Optimal power flow theory for power system work was originally formulated in 1962[14] and has been refined substantially since that time.[15,16] Figure 1 shows a flow chart of an optimal power flow algorithm.

There are many techniques for making the calculations indicated by the blocks in Figure 1. For example, the fast decoupled technique can be used for the power flow routine, and a reduced gradient/steepest descent technique for the optimization. The goal (or objective) is the minimization of some nonlinear objective function (such as transmission system losses) subject to nonlinear equality and inequality constraints or limits. In a specific application, transmission system losses can be minimized by interactively adjusting selected voltages and load tap-changing (LTC) taps. The nonlinear nature of the problem requires an iterative process and

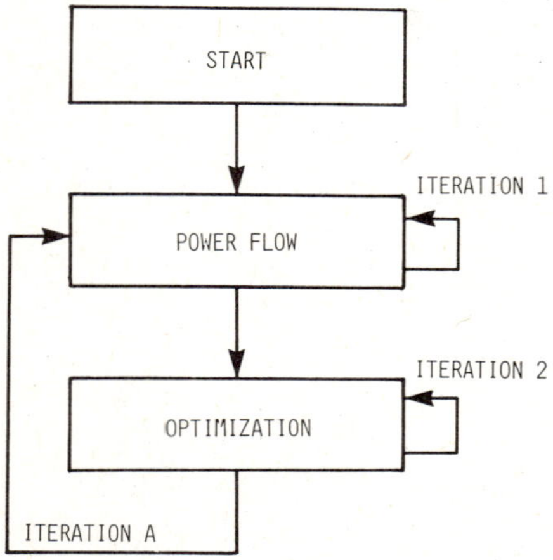

FIGURE 1. Optimal power flow algorithm (reduced gradient/steepest descent).

careful "tuning" may be required to approach the optimum in an efficient manner.

In order to understand Figure 1, a sample three-bus system is used to illustrate the calculation steps for system operations planning (see Figures 2 and 3). Iterations are performed in the power flow routine to balance the power equations (iteration 1 of Figure 1). Next, an optimization routine is added to the end of the power flow routine. In the optimization routine, new values for the control variables (e.g., the terminal voltages of generators) are found and the process is repeated until the objective function f is minimized.

During the optimization, Lagrangian multipliers λ are found to help compute the gradient direction. Depending on the power flow technique used, the Lagrange multipliers are either computed directly (e.g., Newton power flow) or solved iteratively (e.g., fast decoupled power flow). Once the gradient direction ∇f is determined, the step length (or step size) must be found for updating the control variable settings. The first constraint encountered determines the value c_{max}, then a search is made between 0 and c_{max} to determine the minimum value of the objective function along the gradient. The control variables are then modified and convergence is tested. If the objective function has not decreased, the gradient is small, and all limits (constraints) are met, a power flow is calculated and the process terminates. If any of the foregoing items are not met, the process repeats.

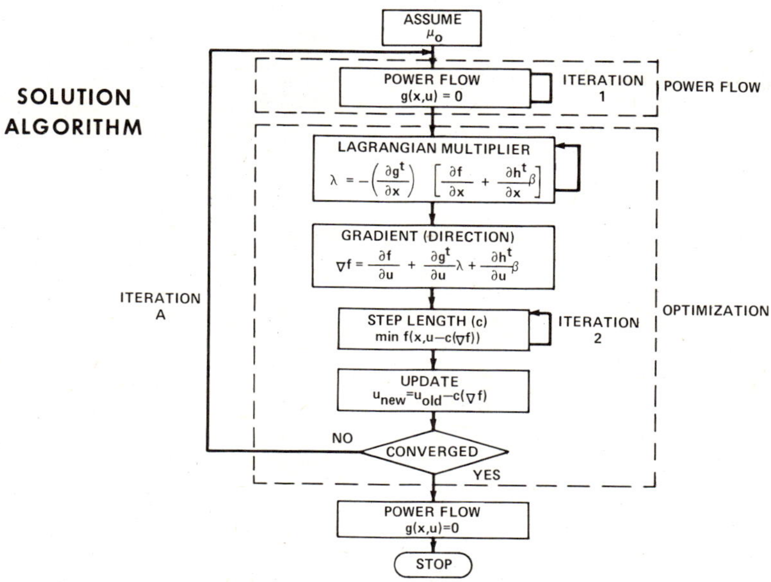

FIGURE 2. System operations planning example.

Reactive power dispatching in system operations planning is the ability to minimize the total power system losses by adjusting parameters in the power system while staying within equipment limits (see Figure 3). A reactive power dispatch assumes that real power has been dispatched and will remain fixed throughout the optimization procedure. Note that the slack bus real power will decrease due to a reduction in total system losses. Therefore system operating costs decrease.

A new algorithm is presently being developed for solving the optimal power flow problem. The algorithm is of the projected Lagrangian type, involving a sequence of sparse, linearly constrained subproblems whose objective functions include a modified Lagrangian term and a modified quadratic penalty function. The program has been successfully applied to a large power system (600 buses).

The application of an optimal power flow program for reactive power *dispatch and control* in real time is presently in its infancy. The development of a security-constrained optimal flow, which computes in less than 5 min, is required. Researchers are presently trying to solve this problem.

References 17–21 are recommended for further reading into the area of optimization in power systems for reactive power management.

11.2.4. Transmission Benefits

The application of a reactive power dispatching strategy to improve power system operation has many identified benefits to an electric utility. Some of these benefits are now discussed.

(a) Cost savings due to reduced system losses. Reduction in total system losses has the benefit of lowering generator fuel cost. Since the real power of generators must supply the system loads and losses, fuel cost will decrease when system losses decrease.

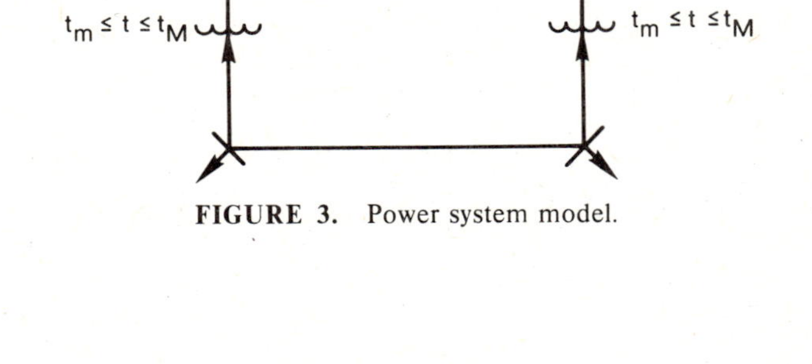

FIGURE 3. Power system model.

(b) Improved voltage profile. The general voltage profile is improved by flattening and increasing the nominal voltage value. This is important because the voltage profile is a measure of the flow of reactive power in a power system. Voltage can vary considerably from location to location. (Although voltage profile is a valuable indicator of system conditions, it does not by itself show the system reactive *reserves* which are available in a disturbance.)

(c) Better voltage control. Better voltage control is possible on a systemwide basis. A global voltage control strategy encompassing the entire system is used, compared with the present local voltage control strategy. Investments in communication equipment and metering equipment are necessary to fully achieve this benefit.

(d) Improved system security. Power system security is improved through better utilization of reactive resources, thus having greater reactive reserves available for system conditions which require sudden reactive demand increases. Although many different conditions can cause a sudden increase in system reactive demand, the substantial increase in loading of already heavily loaded EHV lines, following the tripping of another EHV line, would generally cause the greatest increase in system reactive demand. This condition can be alleviated by reactive power dispatching.

(e) Improved interchange transfer capability. Transmission equipment loading decreases due to reduced reactive power flow. Unloading allows higher real power capability which allows increased interchange transfer capability. Interchange transactions are big business for utilities, so reactive power dispatching is attractive as a means of facilitating such transactions, especially if they can be performed with no major equipment purchases.

(f) Improved system operation. The primary responsibility of system operators is supervision of real power generation and active power and reactive power flows in the system, with due regard to the maintenance of correct voltage levels. Operators sometimes tend to maintain high reactive power reserves, which can result in uneconomical operation. For example, reactive power generating capacity is not fully utilized when operators maintain margins near rotor or stator current limits to avoid overloads caused by system voltage fluctuations. Again, reactive power absorbing capacity may not be fully utilized because of margins maintained for stability reasons or to avoid overheating of the core-ends of generator stators. Also, either too many transformer tap adjustments are

made (decreasing mechanism life) or too few adjustments occur (causing uneconomical reactive power flows).

A reactive power dispatching strategy can give an operator improved guidelines for reactive power flows, reserve level, and voltage control. The process can be automated and integrated with dispatching procedures presently used at dispatching centers.

11.2.5. Experience with Reactive Power Dispatch

The work that has been done to date in applying the optimal power flow theory to power system problems has been relatively limited. General Electric has been working with an electric utility since early 1977 to determine operating savings that can be realized when the control parameters are optimally determined. An initial computer study was executed for a system peak load condition to estimate possible savings. A 2.4% reduction in total system losses (4.4 MW out of 185 MW) was found possible if only the terminal voltages of thermal generators were adjusted, and an estimated 4.5% reduction in total system losses (8.3 MW out of 185 MW) could be achieved by coordinating LTC transformer tap settings with the terminal voltages on thermal generators. The 8.3-MW reduction in total system losses represents a 0.11% reduction in real power generation. The resulting decrease in operating costs can be as high as 0.3%.

11.2.6. Equipment Impact

Reactive power dispatching requires adjustment of control settings on several power system equipments, and monitoring throughout the power system to determine the status of critical parameters. Two questions must be addressed:

1. How does reactive power dispatching affect present equipment?
2. What future impact could reactive power dispatching have on present equipment, or what new equipment may be required?

(a) Impact on present equipment. The controlling parameters on a power system for reactive power dispatch are: the terminal voltage of generators, static compensator control settings, switchable capacitor and reactor banks, and the tap-changing settings of LTC transformers. To make adjustments from a remote control center, a communication scheme is necessary along with equipment that measures the terminal voltages of the generators. The system operator can monitor the present status and make any necessary adjustments.

It may be possible to use present communication equipment to perform monitoring and adjustments. The current tap-changing procedure for transformers involves either a time clock or automatic circuitry to hold constant voltage on a busbar in the local vicinity of the transformer. It tap changing is done by time clock, the procedure remains the same, but a different tap-setting schedule would be needed with monitoring of tap position and, possibly, local voltage.

A scheme could also include the system operator in the control loop. The automatic scheme would have to be deactivated if taps are to be used for optimizing, and a time clock scheme as described above would have to be implemented involving the system operator in the control loop.

Metering power system data (e.g., terminal voltage on various key buses, real and imaginary components of load, etc.) is another aspect to be considered. This is required to build an on-line digital model of the power system. Additional metering points may be needed to simulate the power system for reactive power dispatching. This may require purchasing additional meters and possibly supplementing or adding to the communications system in place.

The major impact on present equipment, therefore, would be additional metering, and either additional or supplemental communications systems for handling increased metering data.

(b) Possible impact on future equipment and new hardware. The excitation system of thermal generators should be investigated for possible improvements to adjust voltage. A faster cycling transformer tap-changing mechanism may be required to assist with voltage control to high voltage levels. If a faster tap-changing mechanism is required, new contact material will probably be developed to handle increased tap-changing duty. Meters for monitoring various parameters (e.g., voltage) may have to be designed with greater accuracy.

11.3. CONCLUSIONS

Electric utilities are increasingly interested in maximizing the utilization of existing transmission equipment. Reactive power dispatching and optimal management provides electric utilities with an opportunity to take a further step in that direction. Initial studies have shown the potential for reducing fuel costs as well as for improving the management of reactive power reserves.

REFERENCES

Reactive Power Planning.

1. H. H. Happ and K. A. Wirgau, "Minimization of VAR Allocation for System Planning," IEEE Milwaukee Symposium on Automatic Computations and Control, 33–36, April 1976.

2. H. H. Happ and K. A. Wirgau, "VAR Allocation in Power System Planning," *IEEE Can. Commun. Power Conf.*, 261–264, October 1976.

3. H. H. Happ and K. A. Wirgau, "Static and Dynamic VAR Compensation in System Planning," *IEEE Trans. Power Appar. Syst.* **97,** 1564–1578 (September/October 1978).

4. H. H. Happ and K. A. Wirgau, "Minimization of VARs in System Planning," *Proc. Amer. Power Conf.* **39,** 1019–1025 (April 1977).

System Operations Planning.

5. R. A. Fernandes, H. H. Happ, and K. A. Wirgau, "System Loss Reduction by Coordinated Tap and Generator Voltage Adjustment," *Proc. Amer. Power Conf.* **40,** 1390–1393 (1978).

6. R. A. Fernandes, H. H. Happ, and K. A. Wirgau, "The Application of Optimal Power Flow for Improved System Operations," *Proc. Sixth Power Syst. Comput. Conf.* (*PSCC*) **1,** 465–472 (August 1978).

7. R. A. Fernandes, H. H. Happ, and K. A. Wirgau, "Optimal Reactive Power Flow: Loss Reduction for Improved System Operation," *IEEE Can. Commun. Power Conf.*, 324–327, October 1978.

8. K. A. Wirgau, "Reactive Power Dispatching," *Electr. Forum* **5** (1), 6–9 (1979).

9. J. F. Aldrich, R. A. Fernandes, L. W. Vicks, H. H. Happ, and K. A. Wirgau, "Benefits of Voltage Scheduling in Power Systems," *IEEE Trans. Power Appar. Syst.* **99,** 1701–1712 (September/October 1980).

10. R. A. Fernandes, H. H. Happ, and K. A. Wirgau, "Optimal Reactive Power Flow for Improved System Operations," *Intern. J. Electr. Power Energy Syst.* **2,** 133–139 (1980).

11. R. A. Fernandes, H. H. Happ, and K. A. Wirgau, "Voltage Scheduling," *IEEE Can. Commun. Power Conf.* 102–105 (October 1980).

12. J. F. Aldrich, R. A. Fernandes, H. H. Happ, and K. A. Wirgau, "The Benefits of Voltage Scheduling in System Operations," (article of a Current Operating Practices Working Group Report on EHV Operating Problems Associated with Reactive Control), *IEEE Trans. Power Appar. Syst.* **100,** 1376–1381 (March 1981).

Reactive Power Dispatch and Control.

13. H. H. Happ and K. A. Wirgau, "Optimal Dispatch and Voltage Scheduling in Power Systems," *IEEE Intern. Symp. Circuits Syst.*, Chicago, Illinois, April 1981.

14. J. Carpentier, Contribution a l'Etude du Dispatching Economique, Bulletin de la Societe Francaise des Electriciens, Serial No. 8, Volume 3, August, 1962.

15. R. C. Burchett, H. H. Happ, D. R. Vierath, and K. A. Wirgau, "Advanced Optimal Power Flow Methodology," *IEEE Can. Commun. Power Conf. Proc.*, 160–172 (October 1980).

16. R. C. Burchett, H. H. Happ, D. R. Vierath, and K. A. Wirgau, "Developments in Optimal Power Dispatch," *IEEE Power Industry Comput. Appl. Conf.*, Philadelphia, Pennsylvania, May 1981.

Optimization Summaries.

17. A. M. Sasson and H. M. Merrill, "Some Applications of Optimization Techniques to Power System Problems," *Proc. IEEE* **62,** 949–972 (July 1974).

18. H. H. Happ, "Optimal Power Dispatch — A Comprehensive Survey," *IEEE Trans. Power Appar. Syst.* **96,** 841–854 (May/June 1977).

19. J. W. Carpentier, "Optimal Power Flows," *Electr. Power Energy Syst.* **1,** 3–15 (April 1979).

20. B. Stott, O. Alsac, and J. L. Marinho, "The Optimal Power Flow Problem," presented at SIAM Int. Conference on Electrical Power Problems: The Mathematical Challenge, Seattle, Washington, March 1980.

21. H. H. Happ, K. A. Wirgau, "A Review of the Optimal Power Flow," *J. Franklin Inst.* **311,** 231-264 (1981).

SELECTED BIBLIOGRAPHY

REACTIVE POWER COMPENSATORS

Thyristor-Controlled Reactors and Transformers

1. Hauth, R. L. and Moran, R. J., "Introduction to Static Var Systems for Voltage and Var Control," IEEE Tutorial Text, 78EH0135-4-PWR, 44–85, 1978.
2. Hauth, R. L. and Moran, R. J., "The Performance of Thyristor-Controlled Static Var Systems in HVAC Applications. Part I: Fundamental Relationships," IEEE Tutorial Text, 78EH0135-4-PWR, 1978.
3. Hauth, R. L. and Moran, R. J., "The Performance of Thyristor-Controlled Static Var Systems in HVAC Applications. Part II: Performance Modeling Concepts," IEEE Tutorial Text, 78EH0135-4-PWR, 1978.
4. Ellert, F. J. and Moran, R. J., "HVDC and Static Var Control Applications of Thyristors," IEEE/IAS International Semiconductor Power Converter Conference, Lake Buena Vista, Florida, March 1977.
5. Newall, R. J. et al., "Staged Field Testing of the Victory Hill Static Var Control," *IEEE Trans. Power Appar. Syst.* **99**(2), 426–433 (1980).
6. Hauth, R. L. et al., "Application of a Static Var System to Regulate System Voltage in Western Nebraska," *IEEE Trans. Power Appar. Syst.*, Paper F78-097-8 (1978).
7. McMurray, W., "A Comparative Study of Symmetrical 3-Phase Circuits for Phase-Controlled A.C. Motor Drives," *IEEE Trans., Industry Applications* **IA-10**(3), 403–411 (1974).
8. deMello' F. P., "Thyristor-Controlled Reactors: Analysis of Fundamental Frequency and Harmonic Effects," IEEE Paper A78 106-7, Winter Power Meeting, New York, 1978.
9. L. Gyugyi et al., "Principles and Applications of Static, Thyristor-Controlled Shunt Compensators," *IEEE Trans. Power Appar. Syst.* **97**(5), 1935–1945 (1978).
10. Kaufhold, W. et al., "Static Compensator Performance in Long-Distance Transmission Systems," *Siemens Forsch.-u. Entwickl. Ber., Bd.* **6**(5), 286–292 (1977).
11. Reichert, K. et al., "Reactor Transformer for Controllable Static Compensators," *CIGRE 12-05* (1978).
12. Hammad, A. E. and Mathur, R. M., "A New Generalized Concept for the Design of Thyristor Phase-Controlled Var Compensators. Part I: Steady State Performance," *IEEE Trans. Power Appar. Syst.* **98**(1), 219–226 (1979).

13. Hammad, A. E. and Mathur, R. M., "A New Generalized Concept for the Design of Thyristor Phase-Controlled Var Compensators. Part II: Transient Performance," *IEEE Trans. Power Appar. Syst.* **98**(1), 227–231 (1979).

14. Boidin, M. and Drouin, G., "Performance Dynamiques des Compensateurs Statiques à Thyristors et Principes de Regulation," *Rev. Gen. Electricité* **88**(1), 58–73 (1979).

15. Schweickardt, H. E. et al., "Closed Loop Control of Static Var Sources on EHV Transmission Lines," IEEE Paper A78 135-6, Winter Power Meeting, 1978.

16. Schweickardt, H. E. and Romegialli, G., "The Static Var Source in EHV Transmission Systems and its Control," Brown Boveri Co., Publication No. CH-E 2.0336.OE, 1977.

17. Schweickardt, H. E., "Static Var Sources, Their Alternatives and Use in Transmission Systems," EPRI Seminar, Duluth, Minn., Oct. 24/25, 1978.

18. Reichert, K. et al., "Controllable Reactor Compensator for More Extensive Utilization of High Voltage Transmission Systems," *CIGRE 31-04* (1974).

19. Hochstetter, W., "Properties of Static Compensators for Power Supply Systems," *Siemens Review,* **XLIV**(8) 356–360 (1977).

20. Cronin, J. H. et al., "Transmission Voltage Support with Static Var Generators," IEEE Paper A78 108-3, Winter Meeting, Jan. 1978.

21. Barnes, H. C., "Modeling of Static Shunt Var Systems (SVS) for System Analysis," CIGRE Working Group 31-01, *Electra*, No. 51 (1979).

22. Kapoor, S. C., "Dynamic Stability of Long Transmission Systems with Static Compensators and Synchronous Machines," *IEEE Trans. Power Appar. Syst.* **98**(1), 124–134 (1979).

Thyristor-Switched Capacitors and Hybrids

23. Frank, H. and Petersson, T., "Thyristor-Switched Shunt Capacitors and Their Modeling for Transmission Applications," IEEE Paper A78 105-9, Winter Power Meeting, New York, 1978.

24. Engberg, K. and Ivner, S., "Static Var Systems for Voltage Control During Steady-State and Transient Conditions," EPRI/Hydro-Québec Symposium, Varennes, Sept. 1979.

25. Olwegard, A. et al., "Thyristor-Controlled Shunt Capacitors for Improving System Stability," *CIGRE 32-20* (1976).

26. Gyugyi, L. and Taylor, E., "Characteristics of Static Thyristor-Controlled Compensators for Power Transmission System Applications," *IEEE Trans. Power Appar. Syst.* **99**(5), 1795–1804 (1980).

27. Engberg, K. et al., "Reactors and Capacitors Controlled by Thyristors for Optimum Power System Var Control," EPRI Seminar, Duluth, Oct. 1978.

28. Gyugyi, L., "Static Var Compensator with Continuously Variable Output and Minimum No-Load Losses," EPRI Seminar, Duluth, Oct. 1978.

Synchronous Condensers, Saturated Reactors, and Other Types of Compensator

29. McMurray, W., "Magnetic Frequency Multipliers and Their Rating. Part I: Frequency Triplers," *AIEE Trans.,* 384–390, Sept. 1956.

30. Banks, R. et al., "Applications to EHV Transmission of Static Compensators Using Saturated Reactors," IEEE Canada Conference on Communication and EHV Transmission, Montreal, Nov. 1972.

31. Kusko, A., "Polyphase Closed-Core Reactors," *IEEE Trans.*, 57–64, Jan. 1965.

32. Friedlander, E., "The Development of Saturated Reactors for Network Stabilization as Applicable to Magnet Power Supplies," Second International Conference on Magnet Technology, 1964.

33. Friedlander, E., "Static Network Stabilization – Recent Progress in Reactive Power Control," *GEC J. Science Technol.* **33**(2), 58–65 (1966).

34. George, Y. et al., "Analytical and Numerical Modelling of Saturable Reactors," *ACEC Rev.*, **1-2,** 27–43 (1978).

35. Schirman, A., "Static Reactive Energy Compensators," *ACEC Rev.*, **1,** 3–11 (1972).

36. Becker, H. et al., "Three-Phase Shunt Reactors with Continuously Controlled Reactive Current," *CIGRE 31-33* (1974).

37. George, Y., "Static Compensators of Reactive Energy for Rolling Mill Systems," *ACEC Rev.*, **1-2,** 23–33 (1976).

38. Walshe, K. A. and Cooper, C. B., "Slope Reactance Effects in Single-Phase Saturated Reactors," *Proceedings IEE* **126**(8), 771–776 (1979).

39. Cooper, C. B. and Young, D. J., "Saturated Reactor Compensator for Stabilizing Voltage at Load and Generation Points," *IEE Conference Publication* CP107 (1973).

40. Ainsworth, J. D. et al., "Recent Developments Towards Long Distance AC Transmission Using Saturated Reactors," *IEE Conference Publication* CP107, 242–247 (1973).

41. Friedlander, E., "Transient Reactance Effects in Static Shunt Reactive Compensators for Long AC Lines," *IEEE Trans. Power Appar. Syst.* **98,** 1669–1680 (1976).

42. Friedlander, E. and Jones, K. M., "Saturated Reactors for Long Distance Bulk Power Lines," *Electrical Review*, 940–943, June 1969.

43. Thanawala, H. L., Kelham, W. O., and Williams, W. P., "The Application of Static Shunt Reactive Compensators in Conjunction with Line Series Capacitors to Increase the Transmission Capabilities of Long AC Lines," *CIGRE 31-09* (1976).

44. Ainsworth, J. D. et al., "Long Distance AC Transmission Using Static Voltage Stabilizers and Switched Linear Reactors," *CIGRE 31-01* (1974).

45. Gyugyi, L., "Reactive Power Generation and Control by Thyristor Circuits" *IEEE Trans.*, **IA-15**(5), 521–531 (1980).

46. Laithwaite, E. R. and Kuznetsov, S. B., "The Asynchronous Condenser: a Brushless, Adjustable Power Factor Induction Machine," *IEEE Trans. Power Appar. Syst.*, Paper F80-273-3 (1980).

47. Finlayson, P. T. and Washburn, D. C., "Cycloconverter-Controlled Synchronous Machines for Load Compensation on AC Power Systems," *IEEE Trans. on Industry Applications* **IA-10**(6), 806–813 (1974).

THEORY OF TRANSMISSION SYSTEMS REACTIVE POWER CONTROL

48. Kimbark, E. W., "How to Improve System Stability Without Risking Subsynchronous Resonance," *IEEE Trans. Power Appar. Syst.* **96**(5), 1608–1619 (1977).

49. Barthold, L. O. et al., "Static Shunt Devices for Reactive Power Control," *CIGRE 31-08* (1978).

50. Iliceto, F. and Cinieri, E., "Comparative Analysis of Series and Shunt Compensation Schemes for AC Transmission Systems," *IEEE Trans. Power Appar. Syst.* **96**(6), 1819–1830 (1977).

51. Boyajian, A., "The Physics of Long Transmission Lines," *General Electric Review,* 15–22, July 1949.

52. Baum, F. G., "Voltage Regulation and Insulation for Large Power Long Distance Transmission Systems," *Journal AIEE* **40,** 1017–1032 (1921).

53. Friedlander, E., "Long-Distance Power Transmission by Alternating Current," *Engineering,* Jan. 2, 1942; Jan. 9, 1942; Jan. 16, 1942.

54. Wagner, C. F. and Evans, R. D., "Static Stability Limits and the Intermediate Condenser Station," *AIEE Trans.,* 94–123 (1927).

55. Fortescue, C. L. and Wagner, C. F., "Some Theoretical Considerations of Power Transmission," *Journal AIEE* **43,** 106–113 (1924).

DYNAMIC SHUNT COMPENSATION

56. Grund, C. E., Breuer, G. D., and Petersson, R. P., "AC/DC System Dynamic Performance — Transient Stability Augmentation with Dynamic Reactive Power Compensation," *IEEE Trans. Power Appar. Syst.* **99**(4), 1493–1502 (1980).

57. Elsliger, R. et al., "Transmission of 16000 MW Over a Distance of 1200 km From James Bay to Hydro-Québec's Load Centers," *CIGRE 32-07* (1974).

58. Friedlander, E., "Transient Reactance Effects in Static Shunt Compensators For Long AC Lines," *IEEE Trans. Power Appar. Syst.* **95,** 1669–1680 (1976).

59. Woodford, D. A. and Tarnawecky, M. Z., "Compensation of Long-Distance A.C. Transmission Lines by Shunt Connected Reactance Controllers," *IEEE Trans. Power Appar. Syst.,* **94,** 655–664 (1975).

60. Elsliger, R. et al., "Optimization of Hydro-Québec's 735-kV Dynamic-Shunt-Compensated System Using Static Compensators," IEEE Paper A78 107-5, Winter Power Meeting, New York (1978).

SERIES CAPACITORS

61. Hamann, J. R., Miske, S. A., Johnson, I. B., and Courts, A. L., "A Zinc Oxide Varistor Protective System for Series Capacitors," IEEE Paper 80SM694-0, Summer Power Meeting, Minneapolis, 1980.

62. Breuer, G. D., Rustebakke, H. M., and Gibley, R. A., "The Use of Series Capacitors to Obtain Maximum EHV Transmission Capability," *IEEE Trans. Power Appar. Syst.* **83,** 1090–1102 (1964).

63. Miske, S. A., "A New Technology for Series-Capacitor Protection," *Electric Forum* **5**(1), 18–20 (1979).

64. Jancke, G., Fahlén, N., and Nerf, O., "Series Capacitors in Power Systems," *IEEE Trans. Power Appar. Syst.* **94,** 912–925 (1975).

INDEX